W9-CRB-468

ACS SYMPOSIUM SERIES **589**

Computer-Aided Molecular Design

Applications in Agrochemicals, Materials, and Pharmaceuticals

Charles H. Reynolds, Editor
Rohm and Haas Company

M. Katharine Holloway, Editor
Merck Research Laboratories

Harold K. Cox, Editor
Zeneca Ag Products

Developed from a symposium sponsored
by the Division of Computers in Chemistry
and the Division of Agrochemicals
at the 207th National Meeting
of the American Chemical Society,
San Diego, California,
March 13–17, 1994

American Chemical Society, Washington, DC 1995

Sip/ac
Chim

Library of Congress Cataloging-in-Publication Data

Computer-aided molecular design: applications in agrochemicals, materials, and pharmaceuticals / Charles H. Reynolds, editor; M. Katharine Holloway, editor; Harold K. Cox, editor.

  p.   cm.—(ACS symposium series, ISSN 0097–6156; 589)

"Developed from a symposium sponsored by the Division of Computers in Chemistry and the Division of Agrochemicals at the 207th National Meeting of the American Chemical Society, San Diego, California, March 13–17, 1994."

Includes bibliographical references and indexes.

ISBN 0–8412–3160–5

1. Drugs—Design—Computer simulation—Congresses.
2. Agrochemicals—Design—Computer simulation—Congresses.
3. Organic compounds—Design—Computer simulation—Congresses.
4. Computer-aided design—Congresses. 5. Molecular structure—Computer simulation—Congresses.

   I. Reynolds, Charles H., 1957–   . II. Holloway, M. Katharine, 1957–   . III. Cox, Harold K., 1946–   . IV. American Chemical Society. Division of Computers in Chemistry. V. American Chemical Society. Division of Agrochemicals. VI. Series.

RS420.C67   1995
615′.19—dc20   95–2511
                                                              CIP

This book is printed on acid-free, recycled paper.

Copyright © 1995

American Chemical Society

All Rights Reserved. The appearance of the code at the bottom of the first page of each chapter in this volume indicates the copyright owner's consent that reprographic copies of the chapter may be made for personal or internal use or for the personal or internal use of specific clients. This consent is given on the condition, however, that the copier pay the stated per-copy fee through the Copyright Clearance Center, Inc., 222 Rosewood Drive, Danvers, MA 01923, for copying beyond that permitted by Sections 107 or 108 of the U.S. Copyright Law. This consent does not extend to copying or transmission by any means—graphic or electronic—for any other purpose, such as for general distribution, for advertising or promotional purposes, for creating a new collective work, for resale, or for information storage and retrieval systems. The copying fee for each chapter is indicated in the code at the bottom of the first page of the chapter.

The citation of trade names and/or names of manufacturers in this publication is not to be construed as an endorsement or as approval by ACS of the commercial products or services referenced herein; nor should the mere reference herein to any drawing, specification, chemical process, or other data be regarded as a license or as a conveyance of any right or permission to the holder, reader, or any other person or corporation, to manufacture, reproduce, use, or sell any patented invention or copyrighted work that may in any way be related thereto. Registered names, trademarks, etc., used in this publication, even without specific indication thereof, are not to be considered unprotected by law.

PRINTED IN THE UNITED STATES OF AMERICA

1995 Advisory Board

ACS Symposium Series

M. Joan Comstock, *Series Editor*

Robert J. Alaimo
Procter & Gamble Pharmaceuticals

Mark Arnold
University of Iowa

David Baker
University of Tennessee

Arindam Bose
Pfizer Central Research

Robert F. Brady, Jr.
Naval Research Laboratory

Mary E. Castellion
ChemEdit Company

Margaret A. Cavanaugh
National Science Foundation

Arthur B. Ellis
University of Wisconsin at Madison

Gunda I. Georg
University of Kansas

Madeleine M. Joullie
University of Pennsylvania

Lawrence P. Klemann
Nabisco Foods Group

Douglas R. Lloyd
The University of Texas at Austin

Cynthia A. Maryanoff
R. W. Johnson Pharmaceutical
Research Institute

Roger A. Minear
University of Illinois
at Urbana–Champaign

Omkaram Nalamasu
AT&T Bell Laboratories

Vincent Pecoraro
University of Michigan

George W. Roberts
North Carolina State University

John R. Shapley
University of Illinois
at Urbana–Champaign

Douglas A. Smith
Concurrent Technologies Corporation

L. Somasundaram
DuPont

Michael D. Taylor
Parke-Davis Pharmaceutical Research

William C. Walker
DuPont

Peter Willett
University of Sheffield (England)

RS420
C67
1995
CHEM

Foreword

THE ACS SYMPOSIUM SERIES was first published in 1974 to provide a mechanism for publishing symposia quickly in book form. The purpose of this series is to publish comprehensive books developed from symposia, which are usually "snapshots in time" of the current research being done on a topic, plus some review material on the topic. For this reason, it is necessary that the papers be published as quickly as possible.

Before a symposium-based book is put under contract, the proposed table of contents is reviewed for appropriateness to the topic and for comprehensiveness of the collection. Some papers are excluded at this point, and others are added to round out the scope of the volume. In addition, a draft of each paper is peer-reviewed prior to final acceptance or rejection. This anonymous review process is supervised by the organizer(s) of the symposium, who become the editor(s) of the book. The authors then revise their papers according to the recommendations of both the reviewers and the editors, prepare camera-ready copy, and submit the final papers to the editors, who check that all necessary revisions have been made.

As a rule, only original research papers and original review papers are included in the volumes. Verbatim reproductions of previously published papers are not accepted.

M. Joan Comstock
Series Editor

Contents

AGROCHEMICALS

Preface

COMPUTER-AIDED MOLECULAR DESIGN (CAMD) has become an important tool in many academic and industrial labs. In spite of this, the question of what new products have resulted from CAMD often arises. This question is complicated by the fact that modeling is only part of a multidisciplinary effort. Thus, although modeling is playing an increasing role in product development, it can be difficult to assess its impact in bringing any particular product to market. In many ways this question is analogous to asking what new products have been developed with NMR spectroscopy or X-ray crystallography. It is equally hard to attribute development of specific products to these tools, but most chemists would find working without them unthinkable.

Many experimental chemists ask a related question about CAMD: How can I apply modeling to my research? They have witnessed the explosion of computational papers in the literature and are anxious to bring these techniques to bear on their own research problems. Answers to this question are often buried in seminars and papers that focus primarily on methods and their development. Also, many of the papers published in this area are so laden with specialized terminology that they are inaccessible to the nonspecialist.

Our goal in organizing this symposium and book is to provide answers to these two important questions. Our vehicle for achieving this goal is to highlight case studies that illustrate application of modeling to a variety of chemical problems, with an emphasis on product development. This volume is not meant to be complete, but simply to offer concrete examples that show how CAMD has been used to tackle real-world problems. The contributors have provided numerous examples in which computer simulation has improved mechanistic understanding, provided structural or energetic information, and led the way in rational design of new drugs, agrochemicals, and materials.

The book is divided into three general sections: Pharmaceuticals, Agrochemicals, and Materials. However, some chapters deal with topics that are relevant across sections; for instance, the chapters devoted to organic synthesis and toxicology are pertinent to pharmaceutical as well as agrochemical design.

One strength of this collection is the diversity represented both in terms of applications and computational methods. Despite this diversity, it is possible to discern a few general approaches. The first is

ix

computation as an analytical tool. This approach entails simply computing molecular properties that may be difficult, time consuming, or expensive to measure experimentally. Modern theoretical and computational methods put a wide array of molecular properties within reach. Sometimes the ability to compute energies, structures, or physical properties can be critical to design success.

The second is modeling-based design. This approach can be driven by quantitative structure–activity relationships (QSAR) or more fundamental methods based on structural and energetic calculations. Many examples of modeling-based design are found in this book. Here the connection between modeling and product development is clear, and modeling plays a key role.

Another approach is computer simulation, which can serve as an invaluable tool for probing mechanisms, because it is possible to simulate processes computationally that are difficult or impossible to observe experimentally. As is true for other scientific disciplines, computer simulation is increasingly assuming a role in chemistry comparable and complementary to experiment and theory.

The last approach involves the generation of hypotheses. Even qualitative molecular models can have significant value in generating and evaluating ideas, for example, the de novo design of ligands in pharmaceutical and agrochemical research.

This book was assembled with two audiences in mind: the computational chemist and the experimental chemist, each approaching CAMD from different perspectives, but sharing a common desire to solve important practical problems in the most efficient manner. We hope this collection serves both audiences and provides a fresh perspective on the application of CAMD in chemistry.

We thank our authors for their contributions. We also thank Tripos, Rohm and Haas, and Merck for providing financial support for the symposium on which this book was based.

CHARLES H. REYNOLDS
Rohm and Haas Research
Spring House, PA 19477

M. KATHARINE HOLLOWAY
Merck Research Laboratories
West Point, PA 19486

HAROLD K. COX
Zeneca Ag Products
Richmond, CA 94804

December 19, 1994

Chapter 1

Current Approaches in Computer-Aided Molecular Design

Bruce R. Gelin

MMCC Publishing, P.O. Box 381849, Cambridge, MA 02238

Computer simulation methods in chemistry continue a long history of chemical models and calculations based on them. Great progress in both computer hardware and the software packages has resulted in current emphasis on graphical interfaces for model building, simulation specification and control, and visual analysis of results. A descriptive overview of the most common techniques and their applications is given here. Some approaches which have been repeatedly and successfully applied are characterized as *modeling paradigms*. A brief look at trends and future prospects closes this overview.

Mathematical theories and the corresponding computational evaluations of chemical phenomena are not new. As early as 1929, P.A.M. Dirac wrote that, in view of the contemporary development of quantum mechanics, "the underlying physical laws necessary for the mathematical theory of physics and the whole of chemistry are thus completely known."*(1)* Lacking, of course, were closed-form solutions and the means of carrying out approximate calculations on anything larger than model systems such as the hydrogen atom or H_2^+ ion. For many years, chemists have developed simplifications, approximations, and numerical methods that make it possible to evaluate mathematical theories for realistic molecules. These developments continue to the present day *(2-5)*.

Almost 50 years have passed since the first computation of the steric energy of a molecule in terms of nonbonded interactions. Westheimer and Mayer *(6)* calculated the steric contribution as a means to rationalize the rates of racemization of optically active diphenyl compounds. They used functional forms, and the idea of calculating a molecular energy from them, suggested by Hill *(7)*. It is over 30 years ago that the first use of an electronic computer to calculate and optimize the molecular energy was made by Hendrickson *(8)*. He evaluated simple energy functions for angle bending, torsional twisting, and nonbonded contributions to perform conformational analysis and find minimum-energy geometries for cyclopentane, cyclohexane, and cycloheptane; a footnote indicates that one cycle of energy calculation and structure adjustment for cycloheptane took about 15 seconds on the IBM 709 computer employed in that work.

The immense and continuing progress in computers has certainly played a key role in making chemical computations far more widespread since these early

0097–6156/95/0589–0001$12.00/0
© 1995 American Chemical Society

examples. But progress has not been limited to hardware performance. Growing hardware capabilities, coupled with deeper understanding gained from earlier simulations, permit implementation of improved, or even radically more efficient, software algorithms. An example is the speed increase factor of about 2×10^7 since 1967 in *ab initio* molecular orbital calculations using the GAUSSIAN program and its predecessor, POLYATOM; the advancements are attributed about equally to hardware and software *(9)*.

Scientists have experienced at least three major computing environments, with two almost revolutionary transitions between them. The computer center era of the 1960s and 1970s provided standardized computing, using the FORTRAN language and a batch operating environment. The availability in the late 1970s of time-shared superminicomputers such as the Digital Equipment Corp. VAX allowed researchers to develop programs and obtain results with more rapid, often interactive response, and with significantly less skill required to use the machine. In the late 1980s, integrated graphics workstations began to replace host-terminal architectures. The personal computer has had less impact on chemical modeling and simulation than might have been expected. Apparently the processing speeds and graphics resolution have not been sufficient for modeling applications, and workstations have remained the target platform for most development. Personal computers have been employed for much structure-drawing, information retrieval and presentation, and other chemical utility software, and they can be expected to play a greater role in the future.

The software has followed these computer developments, with many graphical model-building, simulation control, and visualization and analysis programs now available. An important factor in making computational chemistry software available was (and is) the Quantum Chemistry Program Exchange (QCPE) *(10)*, which distributes contributed programs to academic researchers at very low cost. The QCPE is not limited to quantum chemistry codes, but now handles molecular mechanics and many other types of programs as well. Further, a commercial software industry for computational chemistry and molecular modeling has arisen *(11)*, with the first companies founded in the late 1970s and another major group in the mid-1980s; new companies continue to enter the market today. Competition among these companies has increased the variety and improved the quality of programs available. Currently the INTERNET provides a posting and distribution means for programs and other information made available by scientists who wish to encourage use of their work by others *(12)*.

Given this background, it is interesting to consider the present-day range of chemical computations. The largest have required over 1000 hours of central processor time on the fastest computers available; here three examples are cited from the fields of protein dynamics, detergent simulation, and inorganic crystal structure. Mark and van Gunsteren *(13)* modeled the molten globule state of hen egg-white lysozyme by applying molecular dynamics at elevated temperatures to cause gradual denaturation of the protein, starting from its solvated crystal structure; the total number of particles treated was 17,299. Karaborni *et al. (14)* ran molecular dynamics on a system with 31,735 particles, of three types: a non-polar one representing oil, a polar one representing water, and a unit containing one polar and one non-polar particle representing a detergent molecule. A simulation of SiO_2 at various densities in a box 240 Å on a side, using 41,472 particles *(15)*, required over 1200 hours on a massively parallel computer.

At the other end of this range of computations are programs for personal computers (of both the MS-DOS and Macintosh type). Applications range from tasks as simple as drawing chemical structures and placing them in reports, to building models for 3D visualization, to simulations of moderate complexity. In addition to the chemical information utilities mentioned above, a number of molecular mechanics and even quantum mechanics programs are now available for personal computers.

In this overview, we discuss general approaches in the field of computer-aided molecular design, then consider applications in pharmaceuticals, agrochemicals, and materials. Special note is made of some often-used methods, here called *modeling paradigms*, which have been applied repeatedly and successfully. A brief consideration of trends and future prospects closes this overview.

Techniques of Computer-Aided Molecular Design

There is no precise definition of terms such as *molecular modeling, computational chemistry,* or *molecular simulation,* which may be understood variously according to an individual's experience and interests. As a first attempt at classification, those computational methods based on *energetic models* of the system can be distinguished from those dealing primarily with other concepts, such as structure matching, chemical similarity, molecular shape, group contributions, and QSAR.

Among the energetic models, techniques based on electronic structure calculation (molecular orbital, density functional, and semiempirical methods) can be differentiated from the empirical energy function (force field) methods, which do not treat electrons explicitly, but model their effects in terms of analytical functions expressing energy contributions from bond stretching, angle bending, torsions, and nonbonded interactions. There is clearly a trade-off between accuracy of treatment of the physical realities and convenience of evaluation of the model; a single-point energy calculation using energy function methods is many orders of magnitude faster than an *ab initio* electronic structure calculation, but the latter yields much more detailed information.

It is useful to recall the definition of a *model* as a *replica or facsimile of a real object.* Chemists have long used models, from simple paper-and-pencil sketches of structures to the precisely machined CPK models *(16).* But physical models of large molecules are hard to build, expensive, opaque, and difficult to support; while somewhat deformable, they do not have the correct flexibility corresponding to real molecules. For these and other reasons, mathematical models stored in a computer represent a significant advance in the ability of chemists to revise and experiment with molecular models. The complexity and realism of models can span a wide range, but the process of matching a model to the knowledge sought must take into account:

- preserving the essential physical attributes of the system, while removing excessive details
- ease of evaluating the model, interpreting its results, and comparing them to experimental data
- ease of preparing and modifying the model for re-runs representing different assumptions.

The generation of a computer model can be as easy as copying a set of atomic coordinates for a small molecule or a protein. It may be a little harder to build an enzyme-ligand complex, a solvated system, or a periodic solid. An altogether different level of effort is required to construct a dense, amorphous polymer network having chains in a Boltzmann distribution of local conformations, at random orientations, with no long-range order *(17).* Obviously the difficulty of making modifications varies with the complexity of the original model.

Once a model exists, both energetic simulations and non-energy based methods can be applied to it. Among the most common operations performed with energetic models (whether electronic or force-field calculations) are the following:

Energy calculation for a given geometry. This can range in computational effort from a fraction of a second for a force field evaluation of a small molecule (down from the 15 sec in 1961 quoted in Ref. *8*), to tens of seconds for force-field evaluation of a complex multi-molecule system, to many minutes for a highly accurate *ab initio* calculation giving a detailed treatment of the electronic structure and energy of an organic molecule.

Energy minimization. Using gradients of the energy, obtained either by finite differences or from analytical gradient expressions, the atoms are shifted so as to produce a lower energy. After multiple cycles of energy/gradient calculation and coordinate adjustment, the system converges to within pre-set tolerances of the nearest local minimum.

Energy curve or surface mapping. Repeated evaluations of the energy, as one or more coordinates are systematically varied, produce energy barriers, potential energy surfaces, transition paths, etc. If the other coordinates are not permitted to change as the coordinates of interest are varied, the result is a *rigid-geometry map*. If the rest of the system is energy minimized while the coordinates of interest are constrained at each of the series of values, the result is a *flexible-geometry* or *adiabatic map*; this latter process is also called *torsional driving* when the coordinate of interest is a torsion.

Conformational search. The task of finding other local minima or generating an ensemble that properly samples the conformational space of a molecule can range from simple to very difficult. The rotamers of a small molecule with a few single bonds can easily be visualized, but for larger systems, and in the general case, the conformational space is too large or too complex to enumerate, and various combinations of systematic and random searching methods must be employed *(18)*. The fact that an entire recent journal issue was devoted to conformational searching indicates the significance of the problem *(19)*.

Vibrational analysis. The classical normal vibrations problem can be solved if the mass-weighted second-derivative matrix can be constructed. The calculated vibrational frequencies and normal mode descriptions are a valuable link to experimental data. In quantum-mechanical treatments, discrepancies between calculated and experimental frequencies are the basis for deriving scale factors. In force-field calculations, the dependence of the calculated frequencies on the curvatures along coordinates of the force field provides an important means of fitting the empirical energy parameters.

Molecular dynamics. A molecular system can be propagated forward through increments of time according to Newtonian mechanics, given the initial conditions and the forces on each of the atoms during the time increment. Because of the relative rapidity of bond vibrations, very short time steps must be used; thus typical simulation lengths on the order of hundreds of picoseconds (1 ps = 10^{-12} s) for large molecules represent a significant computational investment. Nevertheless a great deal of useful information has been learned about the short-time behavior of biological and non-biological systems, and comparisons to relevant experimental data have generated substantial confidence in the validity of molecular dynamics simulations. A good review of biological applications of dynamics is the volume by Brooks, Karplus, and Pettitt *(20)*; a corresponding volume for the materials sciences is that of Allen and Tildesley *(21)*.

Other types of computations which may use the conformations derived by energy-modeling methods, but do not themselves require or make use of the energy, include the following, to name just a few:

- calculation of *molecular properties* such as electrostatic fields, hydrophobicity, lipophilicity, etc.
- calculation of solvent-accessible or other *molecular surfaces*, and rendition of the molecular properties on these surfaces.
- *comparison of molecules*, using either atom-atom correspondences, other criteria, surfaces, properties, etc.
- *Quantitative Structure-Activity Relationships (QSAR)*, in which the molecular features, or simple properties dependent on them, of a series of related molecules, are correlated with activity or bio-assay data. The ability to consider not only flat structural diagrams but also calculated three-dimensional structures has led to a more advanced version sometimes called 3D QSAR, in which the spatial orientation of groups can be correlated with the activity data.

As such techniques become more affordable from the standpoint of computational requirements, and better validated through experience and many applications, they assume the status of reliable tools that can be used by more and more scientists. Of course, new methods are always under development; there is great current interest, for example, in *de novo* drug design programs which use spatial pharmacophore hypotheses and a library of functional building blocks to construct novel molecules that may improve on the binding or effector properties of known molecules, while also having other more desirable characteristics. Many new methods of molecular simulation and analysis can be anticipated to augment the above schematic outline of available techniques.

Applications

For both biological and non-biological systems, the increasing availability of atomic-level descriptions invites applications of the computational techniques just outlined. The separation of application topics into *life sciences* and *materials science* is convenient, but not necessary, as the two can make use of common techniques and both may be involved in problems such as drug delivery or synthetic biomaterials.

Among the life sciences perhaps the leading application, in terms of intensity of use and investment in equipment, people, and software, has been *rational drug design*. In many cases, modeling activities center on the small medicinal molecules of interest, and the full range of techniques outlined above may be brought to bear on the problem. In those cases where the molecular biology of the system has been characterized, the enzymes and other biological macromolecules involved can be included in the molecular model. In both basic and applied research, modeling techniques have been employed to study structural proteins, immune system proteins (antibodies), carbohydrates, lipids, and nucleic acids. Of interest as well are interacting molecules, as in enzyme-ligand complexes, DNA-drug complexes, DNA-protein recognition, and so on.

Applications in non-biological systems include (among many others) solid catalysts such as zeolites and activated metals; the formation and morphology of crystals; phenomena at surfaces and interfaces; material properties of synthetic polymers; and optical and electronic materials.

A few examples illustrate how techniques developed in one application area have found use in quite different contexts.

Motion of small molecules in media. Investigation of the structure of hemoglobin raised the question of how oxygen molecules could travel from outside the globin to their binding site at the heme group. Simple grid searches of the protein with an O_2 probe showed that available channels were narrow, and a high activation energy would be required for passage of the oxygen through them. Molecular dynamics studies showed that fluctuations significantly enlarge the channels, and helped rationalize the situation (22). Years later, similar dynamics simulation techniques have been employed to model the diffusion of gas molecules in amorphous polymers (23, 24).

Internal motions and relaxations in solids. NMR evidence (25) of aromatic ring flipping in proteins in solution were corroborated by early modeling studies (26) which calculated the potential barrier for such motions using the adiabatic mapping technique described above. Sixteen years later, similar calculations were used to analyze such motions in a glassy polycarbonate (27). In both cases, the ability of the surrounding material matrix to make slight adjustments permitted these large-amplitude motions, which would be impossible in a truly rigid structure.

Dynamical description of structures. General recognition of the existence and significance of microscopic motions, based on many more examples than the above two, has led to attempts to describe structures in terms of more than a single set of fixed coordinates. For biological macromolecules, crystallographers have long been familiar with this issue and are actively studying ensemble descriptions (28). The importance of dynamical fluctuations is also recognized in solid-state contexts (29).

Increasing use of modeling and simulation is fueled in part by the development and validation of new techniques, as mentioned at the end of the previous section. Equally important, however, is the steady growth in structural knowledge at the atomic level, afforded by new applications of diffraction, spectroscopy, microscopy, etc. Both factors will contribute to an increasing breadth of applications in the future.

Modeling Paradigms

Modern computer modeling techniques are certainly attractive: they keep records of the motion of every atom in a system, offer the ability to analyze atomic interactions and energy contributions in complete detail, and afford colorful, full-motion graphical visualization. Also, there are many opportunities for their application, as more chemical systems are characterized at an atomic level. But these two circumstances alone do not guarantee a productive contribution to research. Ten and more years of experience have begun to show useful ways to apply the techniques, in what are here called *modeling paradigms*. These are standardized approaches that have been found to be generally useful, even though the details of each problem are different.

To a large extent, what can be done with modeling depends on what is known about the system of interest. In pharmaceutical research, one modeling approach is appropriate when there is no knowledge of the receptor structure. When the receptor structure is known, additional modeling possibilities are available. Some techniques may be valuable in rationalizing observed assay data, while others may be used to suggest new compounds and predict their activities.

The information sought also dictates the methods that are likely to be useful. In optical and electronic materials problems, it is likely that an electronic description of the system will be necessary; force field calculations may be useful to create a starting geometry, but phenomena that are essentially electronic cannot be addressed with force fields. Similarly, evaluating a molecular sieve as a catalyst may begin with force field based molecular dynamics simulations of the motion of small molecules in the sieve framework, but a detailed description of the catalytic events would require calculations of how the catalyst interacts with the substrate's electronic structure.

Below is a sampling of modeling paradigms that have been useful in various situations. This is by no means a complete listing, but provides some idea of major applications in which simulation and modeling methods have been widely applied, and in which some published experience is available to guide researchers planning their own projects.

Small molecule approach (30). In many pharmaceutical research projects, the only structural knowledge available is of a series of bioactive molecules (inhibitors, agonists, antagonists, etc.). The goal of modeling is to determine a structural explanation for their activities and if possible to elicit the set of structural features, or the pharmacophore, responsible for activity. The modeling paradigm in this case involves building models of the bioactive molecules, examining their conformational possibilities using the conformational search technique, and then by trial-and-error or more advanced means, finding a common alignment that places important functional groups of each molecule in similar spatial orientations. The resulting pharmacophore hypothesis can be used to suggest new candidates as the basis for rational drug design by inspection of the model, by searching a database for (sub-)structural similarity, or by *de novo* design software. As more candidates are tested, and more quantitative information becomes available, structure-activity relations may be developed, using either traditional physico-chemical or 3-dimensional structural descriptors. In part, the rational design cycle is an optimization process intended to produce increasingly specific and active candidates; however, it also encompasses the introduction of completely new and structurally diverse leads which may afford new insights into the design task.

Structure-based design (31). As structure-determination tools and structural knowledge improve, there are more cases in which the receptor structure (alone or complexed with a ligand) is known. In this case, while all the techniques of the small molecule approach are still appropriate, the focus shifts toward analysis of the spatial and energetic aspects of the enzyme-ligand complex and involves modeling of the macromolecular receptor as well as the ligand. New candidates can be developed by the means mentioned above; additionally, it is possible either through co-crystallization or modeling to learn much more about how the candidate interacts with its receptor, and to correlate this information with the assay results. Drug design in this case can be considered as an attempt to manipulate steric bulk, electronic charge, and other factors so as to develop the optimal inhibitor or effector. An inhibitor may only be required to bind irreversibly and block an active site to prevent other molecules from entering it, while an effector may have modified reactivity and physico-chemical characteristics which enable it to initiate or participate in a physiologically important process. Obviously there are additional and more challenging modeling tasks involved in the latter case.

Novel lead generation. The emphasis in the first two paradigms was on *optimization* of existing candidates. Testing diverse new structures may also be valuable in finding truly novel leads, and recent synthetic work to facilitate combinatorial libraries (32, 33) has created increased interest in this approach. The ability to generate large numbers of related structures in a combinatorial family will raise new requirements for rapid screening assays and for new computational tools to record and analyze the large data streams this technique will create.

Integration of databases and modeling. One rich source of leads already in the possession of most companies is their proprietary database of compounds and their properties as learned from prior screening and assays. Publicly available databases also add to the store of available information. Until recently only two-dimensional structures were stored in such databases, and one could search for common substructural elements. Now the three-dimensional implications and possibilities are increasingly taken into account, either by explicit storage of three-dimensional structures in databases or by methods that generate some or all of this information on demand *(34)*. Database searching can be used to find structures conforming to pharmacophoric elements defined by the procedures mentioned thus far, or to search for structures complementary to a hypothesized receptor *(35)*.

Protein homology modeling. Modern molecular biology techniques and the Human Genome Project are producing genetic sequences (and the corresponding protein sequences) at a rapid and increasing rate. The usefulness of protein sequence data is compromised by the lack of a solution to the protein folding problem. A valuable modeling paradigm uses structural knowledge accumulated over years of solving protein structures to build new structures by homology to known ones *(36, 37)*. Well-studied and widely applied techniques are available to evaluate the similarity and degree of homology of a new sequence to those in a sequence database, fold the new sequence onto a known structure, alter sidechains and place them correctly, smooth or anneal regions where deletions have occurred, and add trial structures for insertions. Ref. *(36)* even suggests that homology modeling techniques will become the standard for practical applications, and that protein folding will recede to a problem mainly of theoretical interest.

Polymer structures and properties. While many materials science applications of modeling are more recent than the bio/pharmaceutical studies, standard techniques have emerged in some areas; one is the evaluation of polymer structures and properties (see Ref. *17*). The difficulty of building dense, amorphous polymer networks has been mentioned, but such models have been successfully generated many times. Subsequent molecular mechanics and dynamics studies, either at equilibrium or with applied deformations or strains, provide detailed information about morphology, thermal fluctuations, relaxation processes, and material moduli.

Catalysis simulation. As mentioned at the beginning of this section, molecular sieves have been modeled and studied with molecular dynamics, both to understand the vibrations of the framework and to understand the diffusion of guest molecules. Once binding sites or loci of high occupation have been found, electronic structure calculations can be applied to learn how the host perturbs the guest's electronic structure to induce the catalysis process.

Trends and Prospects

Scientific research has always entailed a healthy interplay between new problems and new techniques. While some researchers concentrate on the development and extensions of methods, others will rely on them only when they have been widely applied and validated. Simulation methods in chemistry are in a state of rapid development, with some of the techniques cited above reduced to routine practice, and others requiring further improvement, evaluation, and testing on model systems. Quantum chemistry methods, despite their long history, are still developing, especially in applications to larger molecules and non-first-row elements. Force fields, with their empirical content, require careful validation. Parameter development activities are a regular feature of force field research, especially as modern force fields attempt to cover a wider range of molecules, in both gas and condensed phases, while accounting simultaneously for vibrations, conformer relative energies, noncovalent interactions, and effects of solvent.

Without going into detail, a number of general directions in computer-aided molecular design techniques can be identified. Underlying them all is the continuing progress in computer speeds and capacities, which shows no sign of slowing for at least the next few years. The desktop computer of the mid-1990s will easily equal the supercomputer of just a decade ago.

Thus compute-intensive methods such as quantum chemistry calculations will be increasingly accessible to all researchers, while specialists can perform calculations of greater accuracy and explore more geometries. Wider experience with different levels of approximate molecular orbital theory should allow more scientists to apply the appropriate method for a particular study with confidence.

Force fields, as alluded to above, are becoming more "universal," both in terms of systems covered and phenomena accounted for. This, combined with competitive improvements in commercial software packages, should increase the range of problems to which force-field methods can be applied by non-experts. It should also be possible, in some cases, for the programs to estimate their own precision, or to warn users when they are about to exceed the range of applicability.

Conformational search is a problem which, because of the combinatorial explosion of possible conformations, cannot be solved in closed form; nevertheless, as with other methods, experience with improved algorithms will give increased assurance of sufficient coverage of conformational space for a particular problem.

In molecular dynamics, the growing body of results obtained over nearly 20 years of experience with biological macromolecules, and compared with appropriate experimental data, generates increased confidence in the essential reliability of the method. A similar accumulation of experience should also improve the reliability of materials simulations. Advances in computer power should be reflected in at least three dimensions: longer simulations of the systems already studied; simulations of new, larger systems; and simulations performed with fewer approximations (e.g., more detailed potentials, better accounting for electrostatics, and more accurate solvent models).

Progress is also driven by the slow but steady increase in structural knowledge, as revealed by advances in diffraction techniques, solution structure determination methods, and other new technologies such as novel forms of microscopy which are capable of near-atomic resolution. A structural model is the prerequisite to simulation and modeling, whether based on energy or not; the trend toward structural explanations of biological and materials phenomena is well established, even if critical results are not always forthcoming as soon as hoped.

Improved visualization techniques and the availability of more computational capability to produce realistic graphics will help researchers to perceive the essential features of problems. The intuition gained will suggest new analytical calculations which will increase the range of tools available for understanding increasingly complex simulations.

The influences of *de novo* design software, integration of database searching with modeling, and combinatorial libraries strongly suggest that rational chemical design will become an increasingly data-intensive process. Mass screening of both natural and combinatorially synthesized samples will create large amounts of data which must be stored and made available in a useful way. Multiplying these factors by the general increase in structural knowledge and the computational means to search the conformational spaces of molecules leads to a future scenario in which the chemical design team will work with much more data than a few structures or structural models of lead compounds and putative receptors. Instead, whole computational libraries will have to be considered in conjunction with detailed three-dimensional representations of receptors. This does not suggest that all members of the chemical design team must become computational chemists; rather, they will need convenient access to high-quality data retrieval and graphics tools to enable them to plan their laboratory work in the increasingly data-intensive context outlined *without* spending undue time learning about computers and theories.

Challenges remain in fundamental applications. Rational design, as for example the design of non-peptide analogs, has proven to be a difficult problem. Optimizing an inhibitor to block an active site has turned out to be more complicated than simply sketching a molecule with enough steric bulk to plug a void; improving the effector properties of a small molecule is an even more complex task, as it may require modifying the reactivity of the molecule as well as its shape and physical properties.

Challenges also remain in making simulations increasingly realistic and relevant. In biological systems this means, in addition to including solvent, considering the characteristics of the biological milieu (such as pH, ionic strength, solubility and partition coefficients, and passage through membranes). It will be necessary to address problems of ADME (absorption, distribution, metabolism, and elimination) which have been heretofore considered too complex to model at an atomistic level. For non-biological systems it means improved large-scale models, longer simulations, an understanding of where electronic effects must be included, and the proper level of theory to apply then.

This overview has presented many reasons for optimism about the promise of computer-aided molecular design at the atomistic level. These must, of course, be weighed against the formidable array of challenges, and the experience that new techniques may not be ready for application as rapidly as sometimes hoped. On balance, it seems that expectations of continued progress are justified and that computer simulation and modeling will play a growing role in pharmaceutical, agrochemical, and materials research.

Acknowledgments

The author wishes to thank individuals too numerous to mention in the computational chemistry software community, including both vendors and users, for many discussions of trends and directions. Gratitude is also due the Medicinal Chemistry Group of the Northeast Section of the ACS for invitations to participate in panel discussions following the two ACS Satellite Television Seminars cited *(30, 31)*.

Literature Cited

1. Dirac, P. A. M. *Proc. Roy. Soc. London* **1929**, *123*, 714.
2. Coulson, C. A. *Valence*; Oxford University Press: London, 1952.
3. Pople, J. A.; Beveridge, D. L. *Approximate Molecular Orbital Theory*; McGraw-Hill: New York, 1970.
4. Clark, T. *A Handbook of Computational Chemistry*; John Wiley & Sons: New York, 1985.
5. *Reviews in Computational Chemistry*; Lipkowitz, K. B.; Boyd, D. B., Eds.; VCH Publishers, Inc.: New York, 1990–1994, Vols. 1–5.
6. Westheimer, F. H.; Mayer, J. E. *J. Chem. Phys.* **1946**, *14*, 733.
7. Hill, T. L. *J. Chem. Phys.* **1946**, *14*, 465.
8. Hendrickson, J. B. *J. Am. Chem. Soc.* **1961**, *83*, 4537.
9. Newsletter of Gaussian, Inc. (Pittsburgh, PA), Summer 1993.
10. Quantum Chemistry Program Exchange, Richard Counts, Indiana University, Bloomington, IN 47405.

11. Gelin, B. R. *Third Annual Industry Report: Computational Chemistry and Molecular Modeling at the Spring 1994 ACS Meeting*; MMCC Publishing: Cambridge, MA, 1994.
12. Varveri, F. S. *J. Chem. Educ.* **1993**, *70*, 204.
13. Mark, A. E; van Gunsteren, W. F. *Biochemistry* **1992**, *31*, 7745.
14. Karaborni, S; van Os, N. M.; Esselink, K.; Hilbers, P. A. J. *Langmuir* **1993**, *9*, 1175.
15. Nakano, A; Bi, L.; Kalia, R. K.; Vashishta, P. *Phys. Rev. Lett.* **1993**, *71*, 85.
16. Koltun, W. L. *Biopolymers* **1965**, *3*, 665.
17. Gelin, B. R. *Molecular Modeling of Polymer Structures and Properties*; Hanser Publishers, Munich, 1994; Chapter 4.
18. Leach, A. R. In *Reviews in Computational Chemistry;* Lipkowitz, K. B. and Boyd, D. B., Eds.; VCH Publishers, Inc.: New York, 1991, Vol. 2; pp 1–13.
19. *J. Mol. Struct. (Theochem)* **1994**, *308*.
20. Brooks, C. L.; Karplus, M.; Pettitt, B. M. *Proteins: A Theoretical Perspective of Dynamics, Structure, and Thermodynamics*; Adv. in Chem. Physics; John Wiley & Sons: New York, 1988; Vol. LXXI.
21. Allen, M.; Tildesley, D. *Computer Simulation of Liquids*; Clarendon Press: Oxford, 1987.
22. Case, D. A.; Karplus, M. *J. Mol. Biol.* **1979**, *132*, 343.
23. Müller-Plathe, F; Rogers, S. C.; van Gunsteren, W. F. *J. Chem. Phys.* **1993**, *98*, 9895.
24. Krishna Pant, P. V.; Boyd, R. H. *Macromolecules* **1993**, *26*, 679.
25. Snyder, G. H.; Rowan, R.; Karplus, S.; Sykes, B. D. *Biochemistry* **1975**, *14*, 3765.
26. Gelin, B. R.; Karplus, M. *Proc. Natl. Acad. Sci. USA* **1975**, *72*, 2002.
27. Hutnik, M.; Argon, A. S.; Suter, U. W. *Macromolecules* **1991**, *24*, 5970.
28. Clarage, J. B.; Phillips, G. N. *Acta Cryst. D.* **1994**, *50*, 24.
29. Catlow, C. R. A. *Comput. Mater. Sci.* **1994**, *2*, 6.
30. The small molecule approach was fully discussed in an ACS Satellite Television Seminar (March 29, 1994) entitled *Molecular Modeling: The Small Molecule Approach* and organized by the ACS Department of Continuing Education. Further details are available in the published work of the four panelists, J. Phillip Bowen, Garland R. Marshall, Yvonne Connolly Martin, and Daniel F. Ortwine.
31. Greer, J.; Erickson, J. W.; Baldwin, J. J.; Varney, M. D. *J. Med. Chem.* **1994**, *37*, 1035. Based on the ACS Satellite Television Seminar *Macromolecular Modeling in the Discovery of New Drugs* (March 6, 1993).
32. Gallop, M. A.; Barrett, R. W.; Dower, W. J.; Fodor, S. P. A.; Gordon, E. M. *J. Med. Chem.* **1994**, *37*, 1233.
33. Gordon, E. M.; Barrett, R. W.; Dower, W. J., Fodor, S. P. A.; Gallop, M. A. *J. Med. Chem.* **1994**, *37*, 1385.
34. Moock, D. E.; Henry, D. R.; Ozkabak, A. G.; Alamgir, M. *J. Chem. Inf. Comput. Sci.* **1994**, *34*, 184; Hurst, T. *ibid.*, 190.
35. Bath, P. A.; Poirrette, A. R.; Willett, P. *J. Chem. Inf. Comput. Sci.* **1994**, *34*, 141; Clark, D. E.; Jones, J.; Willett, P.; Kenny, P. W.; and Glen, R. C.; *ibid.*, 197; and other papers in the same issue of the same journal.
36. Holm, L; Sander, C. *Proteins: Structure, Function, and Genetics* **1994**, *19*, 165.
37. Johnson, M. S.; Srinivasan, N.; Sowdhamini, R.; Blundell, T. L. *Crit. Rev. Biochem. Mol. Biol.* **1994**, *29*, 1.

RECEIVED October 20, 1994

PHARMACEUTICALS

Chapter 2

Molecular Modeling and Quantitative Structure–Activity Relationship Studies in Pursuit of Highly Potent Substituted Octanoamide Angiotensin II Receptor Antagonists

Donald B. Boyd[1], Alan D. Palkowitz, K. Jeff Thrasher,
Kenneth L. Hauser, Celia A. Whitesitt, Jon K. Reel, Richard L. Simon,
William Pfeifer, Sherryl L. Lifer, Kumiko Takeuchi, Vasu Vasudevan,
Aaron D. Kossoy, Jack B. Deeter, Mitchell I. Steinberg,
Karen M. Zimmerman, Sally A. Wiest, and Winston S. Marshall

Lilly Research Laboratories, Eli Lilly and Company,
Indianapolis, IN 46285

Findings from computational studies on novel nonpeptide angiotensin II (AT$_1$) receptor antagonists were confirmed experimentally. To discover novel antihypertensives, several series of substituted 4-phenoxyprolyloctanoamides containing an imidazole ring were derived from substituted 4-amino-N-imidazolyl-2-octanoic acids previously disclosed by our laboratories. The title compounds interact with the AT$_1$ receptor in a highly stereospecific manner and define a subsite of the receptor not accessed by losartan, a well-known nonpeptide AT$_1$ antagonist. Molecular modeling correctly predicted the more active enantiomer of the N-imidazolyl-2-octanoic acids. A quadratic relationship between binding affinity and computed octanol/water partition coefficient for the *para* substituted phenoxy derivatives was found. Optimal *in vivo* pharmacology was achieved with triacids LY301875 (p-CH$_2$COOH, pK$_B$ = 9.6) and LY303336 (p-CH$_2$PO$_3$H$_2$, pK$_B$ = 9.1), both of which are orally bioavailable.

The renin-angiotensin system (RAS) plays an important role in the regulation of blood pressure and fluid balance under normal and various pathophysiological conditions (*1-4*). Angiotensin II (AII) is the octapeptide Asp-Arg-Val-Tyr-Ile-His-Pro-Phe produced as the bioactive end product of the RAS cascade (Figure 1). A powerful vasopressor, AII exerts its effect through membrane-bound receptors coupled to G proteins in smooth muscle and other cells (*5-8*). Interaction of the hormone with guanine nucleotide-binding, regulatory protein-coupled receptors, such as G$_q$ (*9*), stimulates phospholipase C, which in turn leads to production of inositol 1,4,5-trisphosphate, release of Ca^{2+} from the endoplasmic reticulum, and finally contraction

---

[1]Current address: Department of Chemistry, Indiana University–Purdue University at Indianapolis, 402 North Blackford Street, Indianapolis, IN 46202–3274

0097–6156/95/0589–0014$12.50/0
© 1995 American Chemical Society

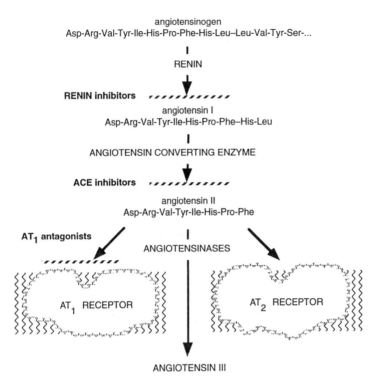

angiotensinogen
Asp-Arg-Val-Tyr-Ile-His-Pro-Phe-His-Leu–Leu–Val-Tyr-Ser-...

RENIN

RENIN inhibitors

angiotensin I
Asp-Arg-Val-Tyr-Ile-His-Pro-Phe–His-Leu

ANGIOTENSIN CONVERTING ENZYME

ACE inhibitors

angiotensin II
Asp-Arg-Val-Tyr-Ile-His-Pro-Phe

AT_1 antagonists   ANGIOTENSINASES

AT_1 RECEPTOR AT_2 RECEPTOR

ANGIOTENSIN III

Figure 1. The renin-angiotensin system (RAS) cascade, which results in the biosynthesis of angiotensin II, can be blocked at the points indicated.

of myofilaments. In addition, AII elicits hypertension indirectly through stimulating release of other hormones, such as aldosterone, vasopressin, and norepinephrine, affecting vasoconstriction and renal sodium absorption.

The antihypertensive pharmaceutical market is huge, amounting to several billion dollars per year. For treating hypertension, the common therapeutic intervention point in the RAS cascade is blockage of angiotensin converting enzyme (ACE). Although ACE inhibitors, such as captopril, enalapril, and lisinopril, have been widely used, the occurrence of minor side effects (*10-12*), such as dry cough, and potentially severe adverse reactions, such as angioedema, has stimulated research interest in other intervention points in the RAS pathway. Moreover, chronic use of ACE inhibitors allows increased renin activity and therefore rebound of AII levels toward normal. Shortcomings of available antihypertensive therapies have been attributed to less than optimal lowering of blood pressure (*13*).

Drug Design Strategies

One strategy for discovering new antihypertensive agents is to inhibit the aspartyl protease renin, but progress by groups using this approach was thwarted by the need to find peptide-based compounds that would be orally bioavailable (*8,14-18*). Another strategy has been to block the binding of AII to the receptor directly responsible for vasoconstriction; a possible advantage is that the intervention would occur regardless of the source of the AII. This latter strategy was originally based on peptidic compounds like saralasin (*19*), which, being peptidic, had poor oral bioavailability and short duration of action. However, recently, dramatic progress has been achieved with nonpeptide structures as described below.

The AT_1 receptor, one of at least two receptor subtypes for AII (*20-23*), mediates blood pressure response and other known physiological actions of AII. Cloning and expression (*24,25*) of the AT_1 receptor have been achieved. [The primary structure and function of another AII receptor, called AT_2, are still under study (*26*).] In principle (*27*), one could rationally design AT_1 inhibitors from three-dimensional structural information about the target receptor protein and/or the natural ligand AII. However, other than the general concept that the AT_1 receptor adopts the familiar tertiary structure with seven transmembrane alpha-helices, there is no detailed structural information on the receptors. Regarding AII itself, this linear octapeptide can adopt a multitude of conformations. There is no consensus on the most probable conformations or on the bioactive conformation (*19,28-32*). Some workers have advanced the opinion that some sort of U-shaped conformation predominates, whereas others have used various linear conformations. Details of the backbone and side chain conformations remain open to further study. Conclusive proof that any proposed conformation is physiologically relevant remains to be established.

In this situation, the more traditional method of screening and following up on leads with structure-activity relationship (SAR) studies is appropriate. This paper reports applications of computational chemistry within the latter context. Much of the literature of computer-aided molecular design (CAMD) deals with method development, or retrospective studies, or simply calculations for the sake of calculations with no immediate or long-term implications for designing molecules. On the other hand, in a large, fast-paced drug discovery effort, computational methodologies must be applied that can obtain answers rapidly and as reliably as possible (*27*). In dynamic, collaborative research, modeling must be able to respond quickly to not only changing priorities, but also to changes in experimental data and interpretation as more replicate experiments are being done. The present work demonstrates that molecular modeling and quantitative structure-activity relationship (QSAR) methodologies can make practical contributions to an ongoing SAR project.

Medicinal Chemistry Background

In the early 1980s, Takeda scientists discovered *N*-benzylimidazole-5-acetic acids (Figure 2) to be weak, but selective, nonpeptide AII antagonists (*33*). Using these leads and a hypothesis based initially on a molecular modeling alignment of the lead structures with a conformation of AII, DuPont scientists executed an elegant SAR program leading to nonpeptide AII receptor antagonists that are not only specific, but also highly potent and orally bioavailable. Compared to the original leads, a 1000-fold improvement in receptor binding affinity was achieved. DuPont's disclosure of EXP7711, DuP 753 (losartan), EXP3174, which is losartan's major metabolite in humans and very bioactive itself (Figure 2), and related compounds (*34-38*) stimulated a stampede of research based mainly on a 2'-tetrazolyl-biphenyl motif (see, e.g., Ref. *39-45*) Losartan is undergoing clinical evaluation as a potential antihypertensive, but it remains to be seen what, if any, advantages it offers relative to existing antihypertensive therapies (*8,46,47*).

Independent of the above well-known work, a large volume screen initiated at Lilly in 1982 revealed LY150921 (pA$_2$ = 4.5, Figure 3) as an initial lead compound in 1984. The pA$_2$ value is the negative of the logarithm of the concentration of an antagonist that gives a twofold shift in the dose-response curve of contraction induced by AII. The pA$_2$ values were determined using rabbit aorta tissue (*47,48*). SAR of about 40 compounds found the octanoic acid tail (LY237117, pA$_2$ = 4.9) to be advantageous. Further synthesis of 150 compounds yielded LY254562 (pA$_2$ = 5.6) with a hydrogen-bonding group on the *ortho* position of the phenyl. Early progress in this series has been disclosed (*49*), and a low energy conformation of AII was constructed in hopes of rationalizing the data (*50*). An acidic group directly on the *ortho* position, i.e., LY266099, which was obtained after an SAR of another 140 compounds, further improved potency to $pA_2 = 6.2$. Continued modifications on the aromatic ring showed that acidic substituents were helpful to potency, particularly at the *ortho* position, and, in fact, introduction of the *ortho*-sulfonic acid of LY280134 ($pA_2 = 7.1$) resulted in a significant improvement. Further elaboration of the SAR is given elsewhere (*51-59*).

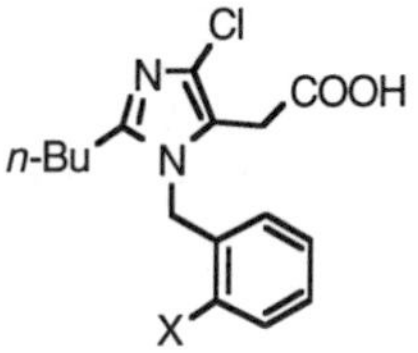

cpd	X	IC$_{50}$ (μM)
S-8307	Cl	40
S-8308	NO$_2$	15

cpd	R$_5$	R$_2$'	IC$_{50}$ (μM)
EXP7711	CH$_2$OH	COOH	0.30
losartan	CH$_2$OH	CN$_4$H	0.019
EXP3174	COOH	CN$_4$H	0.0013

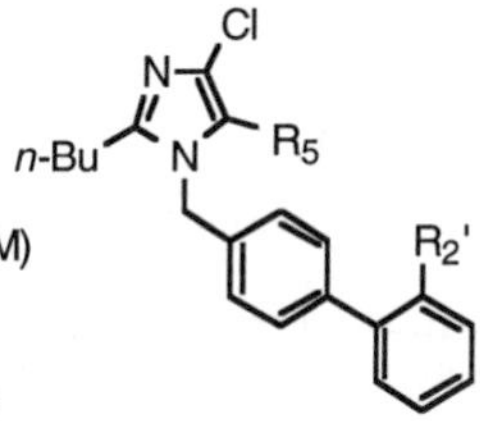

Figure 2. Chemical structure of Takeda leads from patent literature and the DuPont Merck compounds.

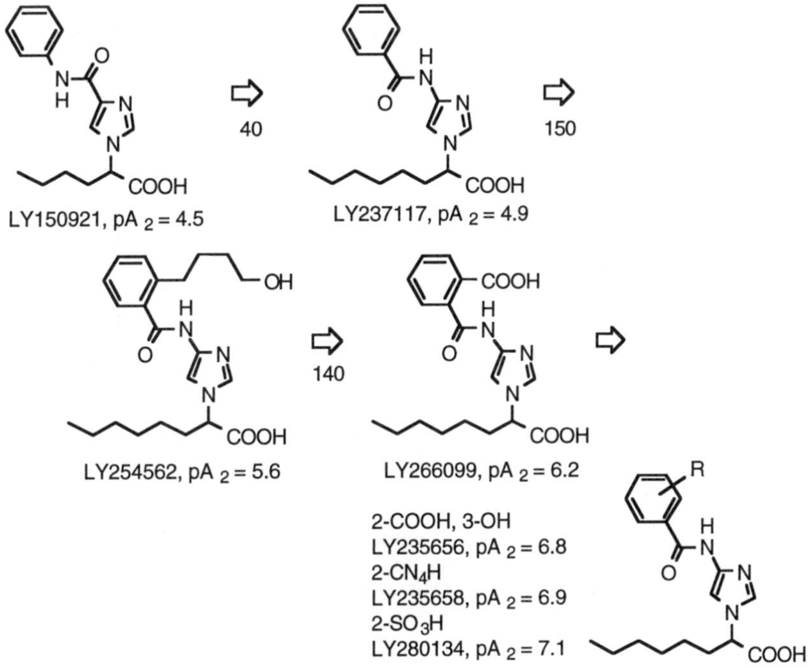

LY150921, pA$_2$ = 4.5   40

LY237117, pA$_2$ = 4.9   150

LY254562, pA$_2$ = 5.6   140

LY266099, pA$_2$ = 6.2

2-COOH, 3-OH
LY235656, pA$_2$ = 6.8
2-CN$_4$H
LY235658, pA$_2$ = 6.9
2-SO$_3$H
LY280134, pA$_2$ = 7.1

Figure 3. Early history of the AII project at Lilly.

Computational Methodology

The techniques used here have been described in the book series "Reviews in Computational Chemistry" (60). Molecular modeling was done with the SYBYL software system, Version 5.5 (61-65). Octanol/water partition coefficients, in the form $logP_{O/W}$, were computed from molecular topology using the CLOGP program (65-68). Data was analyzed with the statistical program JMP 2.0 (69).

Elucidation of Molecular Shape and Pharmacophore

In 1989 when molecular modeling was initiated on the N-imidazole-2-octanoic acid series (Figure 3), it had been thought that these compounds were completely different structurally and otherwise from DuPont's 2'-tetrazolyl-biphenyl series, which was beginning to be disclosed (Figure 2). Molecular modeling first addressed the basic question of what conformational similarities and differences existed between the two sets of compounds.

A systematic conformational search was done on losartan and a representative member of the N-imidazole-2-octanoic acid series (Figure 4). Each structure, built by sketching it into the computer, was energy minimized with the TRIPOS force field using defaults, except as noted. Torsional angle adjustments were made to the models to assure they were in reasonable (i.e., extended) starting conformations, and the minimizations were repeated until the structures were satisfactory. With SYBYL, it is necessary to run a minimization 300, 600, or more iterations before phenyl rings are flat and other obvious distortions are relieved. The calculations were done on the unionized solute molecule without solvent molecules because of the known shortcomings of the TRIPOS force field in dealing with charge interactions (70).

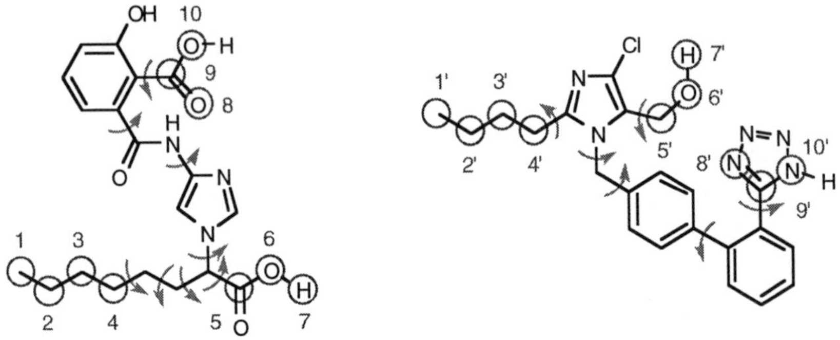

Figure 4. Chemical structures of LY235656 (left) and losartan (right) shown with rotatable bonds used in systematic conformational searching and paired atoms used in flexible fitting.

Energy minimization gives one conformation; the next task was to explore conformational space, the objective being to find the global energy conformation of each structure. The two three-dimensionalized structures from the energy minimizations were subjected to systematic conformational searches. Doing the systematic search on more than 10 rotatable bonds would take prohibitively long on a VAX superminicomputer, so only the important torsional angles (Figure 4) were varied, the other torsional angles and all bond lengths and bond angles being fixed at their previously optimized values. With an understanding of the conformational adaptability of alkyl chains and an indeterminacy of their receptor-bound conformation, torsional angles between the terminal four carbons of the chains were not varied. An angle increment of 30° was used for each rotatable bond through 360°. In trial runs, no energy cutoff (i.e., no conformations were excluded), a high energy cutoff (10 kcal/mole), and a low energy cutoff (4 kcal/mole) were tried, but use of a cutoff was, of course, immaterial to the main objective. The set of torsional angles giving the lowest energy for rigid rotations will be referred to as the global energy minimum conformation; this conformation could be reminimized to fine-tune the bond lengths and angles, but, as we will see in a moment, the conformation will be subjected to further adjustments when it is fit to the other molecule.

Overall, the SYBYL-predicted global energy minimum conformation of losartan appeared reasonable. The two rings of the biphenyl moiety of losartan were predicted to prefer a twisted conformation. Such a conformation is consistent with a subsequently reported crystallographic structure (*41*) of EXP7711, which is the *ortho*-carboxyl acid analog of losartan.

The global energy minimum structures of losartan and LY235656 were flexibly fitted together. In contrast to rigid least-squares fitting, flexible fitting allows each of the structures being aligned to relax bond lengths, bond angles, and torsional angles, subject only to the constraints of the molecular mechanics force field and additional spring constants between atoms designated to overlap (Figure 4). Four carbon atoms of the respective alkyl chains, three atoms of the respective *ortho* acidic groups, and three atoms of the respective hydrogen bonding functionalities (C-O-H) were used to align the molecules, thereby giving roughly equal importance to each potential pharmacophoric region. A force constant of 5 kcal/mole-Å^2 between designated atoms was used, which is low enough not to introduce unreasonable structural distortions, but large enough to bring the paired atoms in proximity if conformationally feasible (*71,72*).

The resulting alignment shows a strong and surprising degree of similarity in size and shape of losartan and the *N*-imidazole-2-octanoic acid series (Figure 5). The discovery was a major revelation. Although the modeling result was unanticipated, it was consistent with two experimental observations: both the *N*-imidazole-2-octanoic acid series and the losartan series bind to the AT$_1$ receptor, and correlations between binding affinity and certain molecular properties (*vide infra*) are shared by the two series.

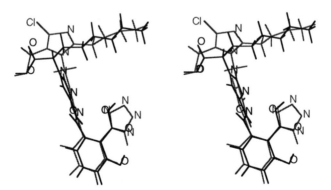

Figure 5. Stereo molecular graphics of the structures from flexibly fitting the global energy minimum conformations of LY235656 (black) and losartan (gray) with the TRIPOS force field.

The conformations depicted in Figure 5 deserve further comment. They are not necessarily the only conformations of the two molecules that may align, and the overlapping structures are not necessarily relevant to the receptor-bound conformation of the antagonists. It is quite probable that in reality the two molecules could have the alkyl chains in any number of other overlapping conformations. The chains are flexible enough and close enough in space that they could fit in the same hydrophobic groove of the AII receptor.

According to the TRIPOS force field, losartan's predicted conformation has the *ortho*-tetrazole group folded back toward the *n*-butyl chain, and the molecule adopts a V-shape. The same V-shape was predicted for the *N*-imidazole-2-octanoic acids and most of the other molecules treated. This aspect of the shape is thought to be an artifact of the force field overestimating nonbonded attractive forces in the absence of solvent. Nevertheless, the SYBYL calculations should suffice for purposes of illustrating similarities and differences in molecular shape. In other words, the goal was to obtain a comparison of conformations, so as to answer the question of whether it was *feasible* for the two series of compounds to bind to the same three-dimensional receptor. It is possible that in the receptor the phenyl rings bearing the *ortho*-acid could be flipped *ca.* 180° from the way they are depicted in Figure 5. Such a conformational change would maintain overlap of the acidic groups but would increase the distance between the hydrophilic, anionic group and the nonpolar, hydrophobic alkyl chain. Otherwise the overall molecular shape depicted in Figure 5 is quite feasible as a bioactive one.

The molecular modeling results allow the pharmacophore shown in Figure 6 to be inferred. The butyl and hexyl chains, the carboxylate and hydroxymethyl, and carboxylate and tetrazole moieties match well. On the other hand, the imidazole rings of the respective molecules do not overlap, and, in fact, are not even in the same plane. The implications of this are that there is no pharmacophoric atom in the imidazole (cf. all the imidazole surrogates that have been used in losartan analogs) or that the receptor site residues can change conformation to accommodate either position of the imidazole. The modeling suggests that it would be possible to direct SAR at modification of the central part of the *N*-imidazole-2-octanoic acid series without disturbing the pharmacophore defined by the peripheral groups.

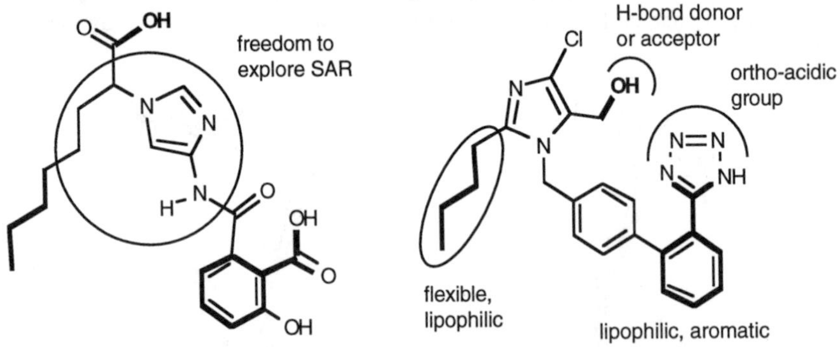

Figure 6. Pharmacophore for angiotensin II receptor antagonism as initially determined by similarity in shape and functionality of the Lilly and DuPont series.

Ligand Design Based on Pharmacophore

The discovery of the similarity of the molecules shown in Figure 5 immediately suggested that it should be possible to evaluate molecular modifications in the *N*-imidazole-2-octanoic acid series using as a criterion whether the new structures could achieve the common pharmacophoric arrangement that we found. New structures that were under consideration for synthesis could be evaluated by molecular modeling, and the results could be factored into the prioritization of which structures should be prepared and tested.

For each such structure tested in the "computational screen," the following steps were employed: (1) construction of molecular model on the computer, (2) thorough energy minimization, (3) systematic conformational energy search using the energy minimized structure, (4) flexible alignment of the resulting global energy minimum conformation with that of losartan. Details of the molecular modeling were the same as described above.

For the flexible fitting step, the SYBYL program produces molecular mechanics energies of the two structures before and after the fit. Because the starting geometry in each case is at or near the global energy minimum, the final energy of the fit structures represents the *increase* in energy necessary for the new structure to adopt a pharmacophore like that in losartan. The TRIPOS force field used by SYBYL is adequate for many purposes, but it was felt that it was not accurate enough to discriminate between small energy differences in the structures under consideration. Consequently, the following decision tree was used. If the energy increase was less than about 5 kcal/mole, a structure was deemed capable of adopting the target pharmacophore easily. On the other hand, if the energy increase was greater than 10 kcal/mole, then that was considered to be prohibitively expensive in terms of energy. If the energy increase was between 5 and 10 kcal/mole, this was a gray area where the molecular modeling could not make a definitive statement.

Computer resources allowed only about one molecule to be run through the protocol per week (with step 3 above being rate limiting), while the large team of synthetic chemists could generate new compounds at a faster rate than was possible to evaluate computationally. There is insufficient room in this paper to describe each of the structures that were evaluated during the course of research. Generally, however, it was found that the protocol, despite its simplicity, correctly predicted whether or not a modification was advantageous if the modification involved conformational and shape issues. For modifications affecting other determinants, such as lipophilicity or charge distribution, then the molecular modeling was unrelated to the activity-limiting property. The next two sections present two examples of how our molecular modeling protocol proved useful.

Prediction of Bioactive Enantiomer

Early in the project, it was not known if one enantiomer of the *N*-imidazole-2-octanoic acids would have more activity than the other. Molecular modeling was applied to address this question. Using the protocol described above, the (*R*) enantiomer, which was the one used in the original modeling, fit the losartan template well as already reported. However, the (*S*) enantiomer, if forced into the model pharmacophoric conformation, was computed to have an energy increase in the prohibitive range. On this basis, it was predicted that there would be a difference in biological activity of the enantiomers, and (*R*) would be more active than (*S*).

As the chemistry advanced, new members of the series were synthesized, separated chromatographically, and tested separately. One isomer consistently had an approximately tenfold higher affinity constant in the rabbit aorta assay than the other. However, it was impossible from NMR or other available analytical means to assign absolute stereochemistry, so they were simply labeled A and B.

About two years after the prediction of the more active enantiomer, an intermediate LY298162 (Figure 7) was finally found that formed crystals suitable for small molecule x-ray diffraction. The compound was an intermediate that could be related unambiguously to the *less* active enantiomer. The crystal structure (Figure 7) proved the compound to be (*S*). Thus, the molecular modeling prediction was correct, i.e., the compound better able to meet the pharmacophoric requirements had the (*R*) configuration at the 2 position of the octanoic acid.

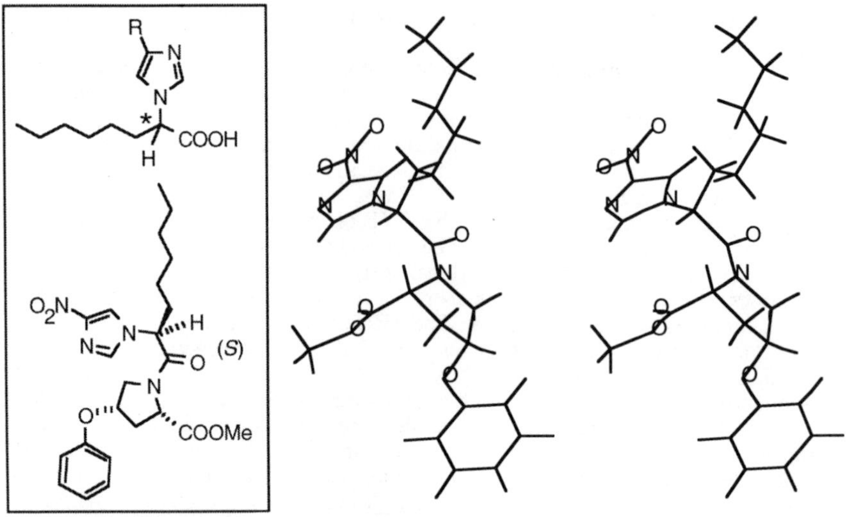

Figure 7. (left) Chiral center in the octanoic acid series and the chemical structure of LY298162 which was determined by x-ray crystallography. (right) Stereo molecular graphics of the crystalline state structure of LY298162. The chiral center in the octanoamide is (*S*). The two chiral centers in the proline ring are also (*S*) as in the natural form of the amino acid. Note that the hexyl chain is not all anti; there is one gauche interaction, probably due to crystal packing forces.

Modification of the Central Region

Molecular modeling suggested that there would be ways to rigidify or otherwise modify the central region of the *N*-imidazole-2-octanoic acids (Figure 6). Replacing the amide group with appropriate aromatic rings offered the possibility to keep the pharmacophoric groups in the desired spatial position, but on new scaffolds (Figure 8). This line of reasoning stimulated the design and synthesis of new series with the imidazole replaced by indole, benzimidazole, and indazole rings. Pleasingly, some of these modifications led to improvement in the pA_2 values (Figure 9).

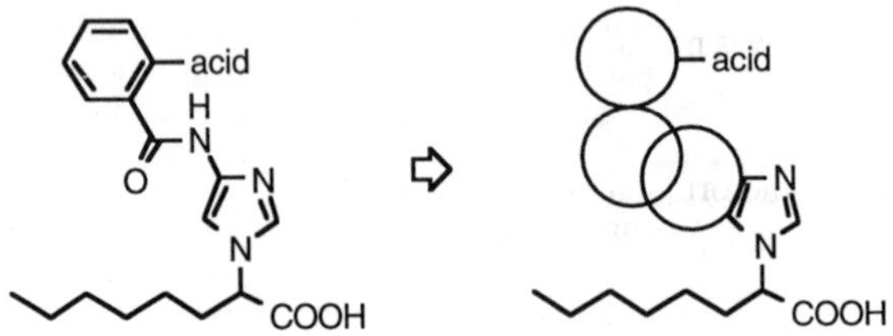

Figure 8. Concepts for introducing aromatic rings fused to one of the existing rings in the *N*-imidazole-2-octanoic acids series so as to constrain the conformational possibilities and to change polarity in the region between the imidazole and the *ortho*-substituted terminal phenyl ring.

Molecular modeling again showed a dramatic similarity in shape that the *N*-heterocycle-2-octanoic acids and the 2'-tetrazolyl-biphenyl series are able to reach. An indole analog in our series and DuP 532, which is a more potent, second generation analog of losartan (*73*) (Figure 10), were used in the computational protocol. The alignment of Figure 11 shows the imidazole rings of the two molecules do not overlap at all and, in fact, are nearly orthogonal. However, the lipophilic alkyl chain, the *ortho* acidic group, and the hydrogen bonding functionality coincide in space extremely well. It is also worth noting that the bicyclics render the imidazole series more biphenyl-like because an aromatic ring replaces the amide linkage.

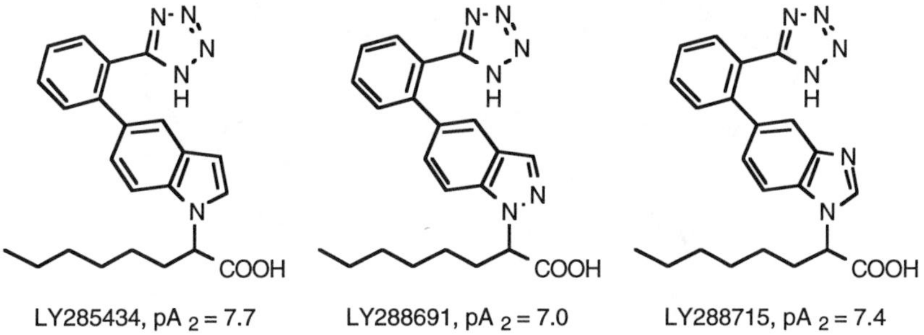

LY285434, pA$_2$ = 7.7   LY288691, pA$_2$ = 7.0   LY288715, pA$_2$ = 7.4

Figure 9. Angiotensin II receptor antagonism of benzoheterocyclic derivatives of the *N*-imidazole-2-octanoic acids. The indole showed higher binding affinity than the indazole, the benzimidazole, and the corresponding *ortho*-tetrazole-substituted imidazole LY235658 (Figure 3).

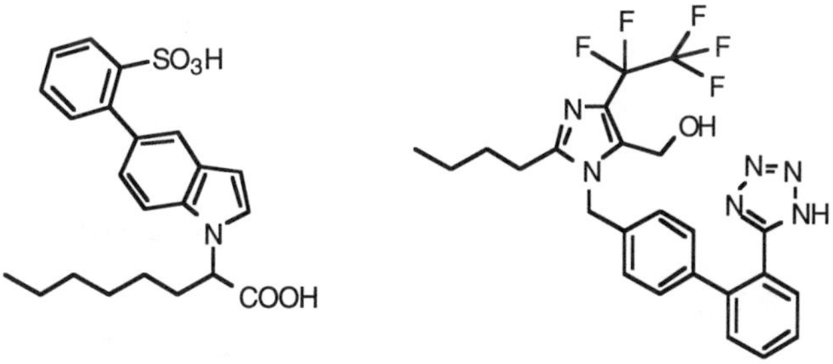

Figure 10. Chemical structures of an early member of the indole series (left) and DuP 532 (right), which has 4-pentafluoroethyl instead of the 4-chloro substituent of losartan. These structures were used in molecular modeling.

Substituents Effects

QSAR analysis is useful for showing trends between a molecular property and biological activity. These trends can help guide the SAR away from dead-ends and toward promising molecular modifications. After the *ortho*-sulfonic acid derivative (LY280134, Figure 3) was synthesized and tested, a relationship between *in vitro* activity and acidity of the *ortho* substituents was quantitated. The pK$_a$ of this group and the compound's pA2 value could be fit with a linear regression. The initial relationship was demonstrated using approximate pK$_a$ values from reference compounds (*74-76*). This relationship was later confirmed experimentally by determination of the pK$_a$'s of the derivatives shown in Figure 12. It was thus possible to use this information to redirect the SAR from further substitutions on the phenyl ring, which already seemed optimized in terms of bioactivity, toward other regions of the molecule.

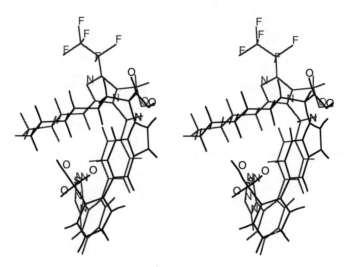

Figure 11. Stereo molecular graphics of the overlap of an indole compound (black) and DuP 532 (gray). The other benzoheterocycles fit well like the indole. The V-shape of the molecules has already been discussed in the text.

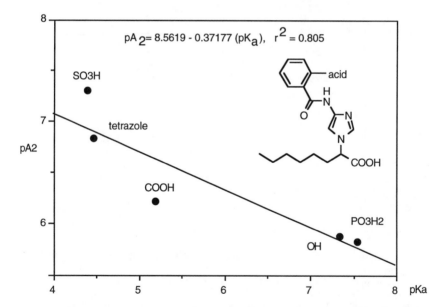

Figure 12. Angiotensin II antagonism in the *N*-imidazole-2-octanoic acid series increases as the *ortho* group on the terminal phenyl ring becomes more acidic. The regression equation and line are shown.

Subsequent to our finding the inverse pA_2-pK_a relationship for our series, workers at DuPont reported that they had independently found a similar relationship for their *N*-imidazolylmethylbiphenyl series (*37*). The relationship they found, using estimated pK_a's, is shown in Figure 13. Potency increases roughly with more acidic *ortho* substituents.

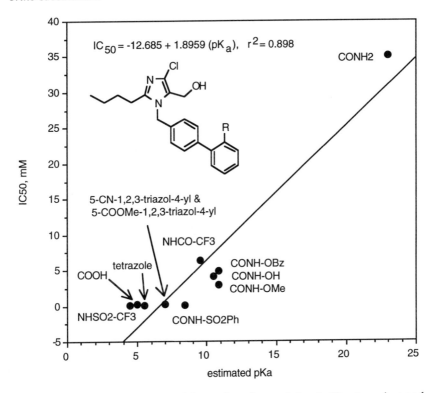

Figure 13. Regression relating IC_{50} values for angiotensin II antagonism and acidity of the *ortho* group on the terminal phenyl ring in the *N*-imidazolylmethylbiphenyl series.

Another QSAR that was found involved the lipophilicity of the *N*-imidazole-2-octanoic acid series with various substituents on the acid-bearing terminal phenyl ring (Figure 14). Compounds that were more hydrophilic (lower $logP_{O/W}$ values) were generally less active. This relationship, albeit weak, is consistent with work on the losartan series which concluded that the terminal phenyl ring fits into a hydrophobic pocket (*34*). Making this end of the ligand nonplanar, i.e., replacing the phenyl ring with various cycloalkenyls, also reduces activity (*47,48*).

As the SAR evolved in the octanoic acid series, the octanoic acid itself was derivatized in various ways including linkage to a proline through an amide bond. This modification was introduced with the hope of reducing serum albumin binding and increasing oral bioavailability. Because the chemistry for the imidazole series was easier than for the benzoheterocyclic series, and because the bioactivity profile of proline-containing imidazole compounds proved to be good, SAR work shifted back to the imidazole series. Chemistry on the proline ring itself showed that introduction of a 4-phenoxy substituent on the proline further enhanced potency.

QSAR analysis of the substituent variation at the *para* position of the phenoxy ring suggested that bioactivity was related parabolically with $logP_{O/W}$ values (Figure

15). Making this region of the ligand too hydrophilic or too hydrophobic was deleterious. Continued SAR on the phenoxy ring (Figure 15) confirmed the existence of a parabolic relationship between pA2 values and logP$_{o/w}$ values. LY301875 (p-CH$_2$COOH, pK$_B$ = 9.6) was among the compounds made.

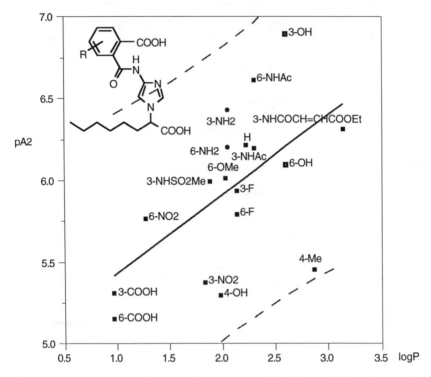

Figure 14. Angiotensin II antagonism in the 2-COOH set of N-imidazole-2-octanoic acids generally increases as the substituents on the terminal phenyl ring make the compounds more lipophilic. logP$_{o/w}$ values were computed. The regression equation is pA2 = (4.95±0.37) + (0.49±0.17)logP, r^2 = 0.33, p = 0.0130, n = 18; 95% confidence limits are shown.

Subsite of AT$_1$ Receptor Reached by the Phenoxyproline Octanoamides

As will be reported elsewhere in detail (59), the *para*-carboxymethyl congener (LY301875) and other compounds in the phenoxyproline octanoamide series were evaluated by extensive pharmacology. Not only must a ligand display high potency, but also other chemical and biological characteristics, such as oral bioavailability, to be worthy of interest. It was deemed on scientific grounds that the compound had the potency and other properties to be appropriate for clinical evaluation.

LY301875 is interesting structurally in that it is a triacid and has three chiral centers. Interaction of the structure with the AT$_1$ receptor is highly stereospecific as determined by preparing all eight diastereomers of LY301875 (51). The pA2 values of the compounds are compared in Figure 16. Activity drops drastically if either chiral center on proline is (R), whereas the chiral center on the octanoamide chain is less influential. As depicted in Figure 17, it is thought that the long flexible alkyl chain can bend to fit into an apparently narrow, lipophilic pocket of the receptor regardless of whether the octanoamide chiral center is (*R*) or (*S*).

Molecular modeling (Figure 18) compares the structures of LY301875 and losartan. The phenoxyproline side chain occupies a receptor subsite beyond the reach of losartan and its analogs. The activity data for our series demonstrate that ligand interaction with this subsite contributes significantly to binding. The subsite is called P* ("P-star") in recognition of its discovery with the *para*-substituted phenoxyproline side chain.

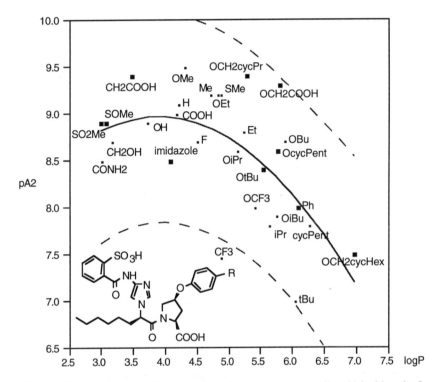

Figure 15. Angiotensin II antagonism in the phenoxyproline *N*-imidazole-2-octanoic acid series shows a quadratic dependence on lipophilicity as it is modified by substituents on the *para* position of the phenoxy ring. The relationship between pA$_2$ values and lipophilicity of the *para* substituted phenoxyproline *N*-imidazole-2-octanoic acids was initially discovered with the compounds designated by small points (n = 19, r^2 = 0.49, p = 0.0046). Compounds made thereafter are shown as large squares. The parabolic regression line (n = 29, r^2 = 0.39, p = 0.0016) and the 95% confidence limits are shown. The optimum logP$_{o/w}$ value is close to that of CH$_2$COOH, which is the substituent of LY301875. The logP$_{o/w}$ values were computed for the unionized forms of the structures. Although these values are not reflective of physiologic pH, the relationship found for the initial set of compounds was predictive. The possibility exists, however remote, that the acidic groups are not all ionized in the micro environment of the receptor site.

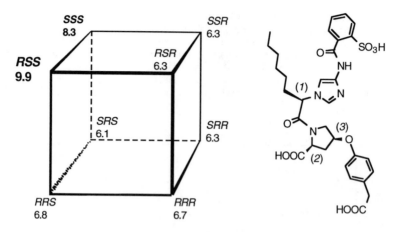

Figure 16. Binding affinities (expressed as pA2 values) to the AT1 receptor for the 8 diastereomers, each of which is represented at one apex of the cube. An epimerization of a chiral center corresponds mathematically to moving from one apex to another. The three chiral centers, in order of the labels at each apex, are marked in the chemical structure. The most active isomer is *RSS* (LY301875).

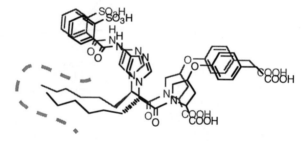

Figure 17. The alkyl chain of the (*RSS*) and (*SSS*) diastereomers bending to fit in the same narrow, lipophilic pocket of the receptor.

Comparison to the Three-Dimensional Structure of Angiotensin II

We conclude with one additional molecular modeling study. Numerous previous attempts to deduce the conformations of the octapeptide AII involved applying physical chemical techniques and synthesizing peptides with residues replaced, deleted, or conformationally constrained (see, e.g., *19,29,77*). These attempts unfortunately did not lead to a consensus on the bioactive conformation. Recently, the first x-ray crystallographic study of AII was reported (*78*). Different from any of the previously hypothesized conformations, the x-ray structure is twisted into somewhat of a corkscrew when complexed to an anti-anti-idiotypic antibody reportedly mimicking an AII receptor. The crystalline state conformation may not be the true bioactive one either, but it is of interest to see what, if any, correspondence may exist between it and our antagonist.

Numerous possibilities can be envisioned for how LY301875 and AII might align. The possibility shown in Figure 19 has the lipophilic alkyl chain in the region of the Ile side chain, the *ortho*-sulfonic acid near the C-terminal carboxylic acid, and the phenoxy group arbitrarily proximal to the Tyr side chain. This alignment is somewhat consistent with those other authors have assumed (*31,32,38*), except that the imidazole ring of the antagonist was not forced to be overlap with the histidine side chain. The precise interactions of these ligands with the AT_1 receptor need to be explored by site directed mutagenesis and other techniques.

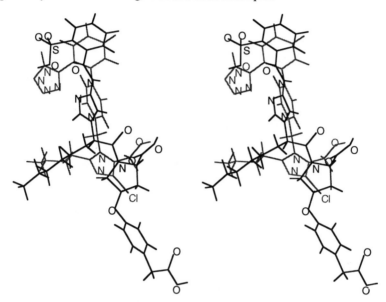

Figure 18. Stereo view of the overlap of the phenoxyproline octanoamide LY301875 (black) and losartan (gray) obtained by flexibly fitting the *ortho* acidic groups (sulfonic acid and tetrazole), the alkyl chains (hexyl and butyl), and the hydrogen bonding groups (proline carboxyl and hydroxymethyl). The shape and dimensions of the two compounds are similar except for the 4-phenoxyproline side chain which extends into the P^* subsite.

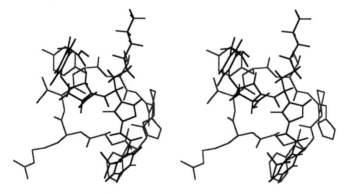

Figure 19. Stereo molecular graphics of an alignment of the crystallographic conformation of angiotensin II and LY301875. The residue sequence in the octapeptide is Asp-Arg-Val-Tyr-Ile-His-Pro-Phe. The Asp_1 residue is at the lower center "rear" of the figure, whereas Phe_8 is in front of it.

Conclusions

The computational chemistry work involved with this project illustrates the realities of their application in drug discovery. The most sophisticated academic techniques are not always suited for contributing to a project in a pharmaceutical research environment (79). Molecular modeling and QSAR proved they could provide insight and demonstrated predictive abilities.

The structure-activity relationship has evolved along the branching tree structure shown in highly abbreviated form in Figure 20. A novel series of chiral nonpeptide AT_1 antagonists was identified that interacts with the receptor in a very stereospecific manner. These ligands, exemplified by LY301875 (p-CH$_2$COOH, pK_B = 9.6), LY303336 (p-CH$_2$PO$_3$H$_2$, pK_B = 9.1), and related compounds, define a subsite of the AT_1 receptor. Structure-activity information derived with the aid of molecular modeling and QSAR is highlighted in Figure 21. The recent literature shows that the majority of the medicinal chemistry stemming from the disclosure of losartan has involved keeping the 2'-tetrazolyl-biphenyl template and replacing the imidazole ring by surrogates (45). In contrast, the series we have reported here is distinct and may offer unique pharmacological or clinical characteristics.

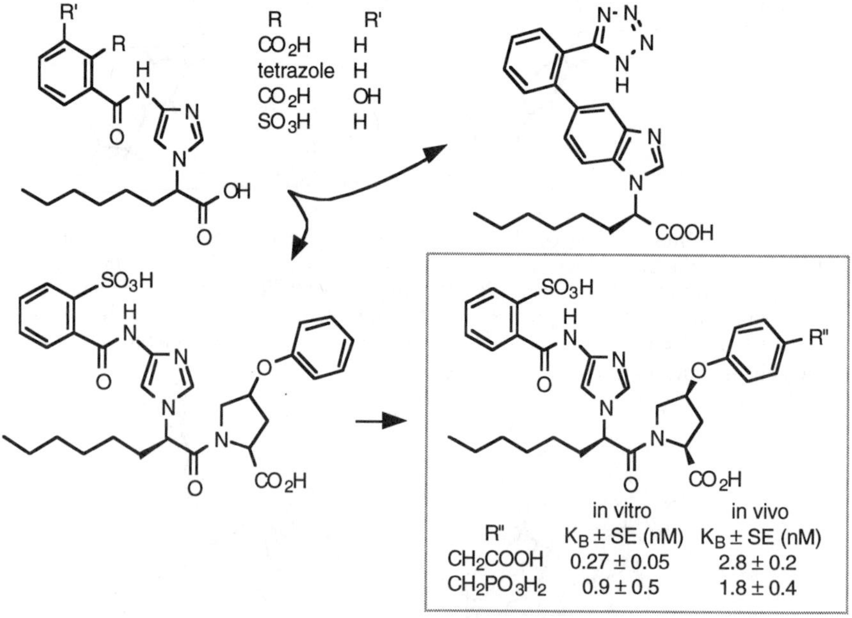

R"	in vitro $K_B \pm SE$ (nM)	in vivo $K_B \pm SE$ (nM)
CH$_2$COOH	0.27 ± 0.05	2.8 ± 0.2
CH$_2$PO$_3$H$_2$	0.9 ± 0.5	1.8 ± 0.4

Figure 20. Evolution of the SAR culminating in the phenoxyproline octanoamides exemplified by two highly potent AT_1 receptor antagonists with oral bioavailability and no agonist effect. The *in vitro* binding constants are measured with isolated rabbit thoracic aorta tissue; the *in vivo* data are from pithed rats.

Acknowledgments

We wish to thank W. A. Spitzer, D. W. Robertson, M. T. Reamer, C. J. Paget, Jr., F. Mohamadi, M. M. Spees, A. Hunt, S. J. Dominiani, J. E. Fritz, G. J. Cullinan, R. F. Brown, J. J. Nunes, M. E. Quimby, J. W. Paschal, M. A. Heathman, S.

Burris, C. J. Barnett, T. M. Wilson, and K. G. Bemis for contributing to some of the groundwork. Drs. M. Perelman, R. G. Harrison, G. V. Kaiser, W. B. Lacefield, R. W. Burchfield, M. Haslanger, and R. W. Souter helped make this work possible.

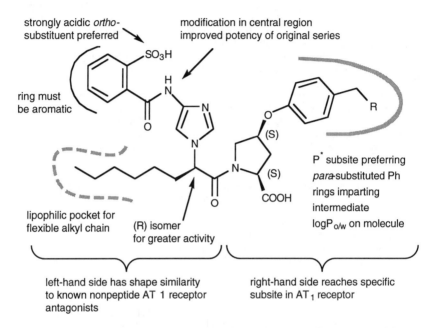

Figure 21. Summary of SAR information on the phenoxyproline octanoamides.

Literature Cited

1. Peach, M. J. *Physiol. Rev.* **1977**, *57*, 313-370.
2. Vallotton, M. B. *Trends Pharmacol. Sci.* **1987**, *8*, 69-74.
3. Foote, E. F.; Halstenson, C. E. *Ann. Pharmacother.* **1993**, *27*, 1495-1503.
4. Steinberg, M. I.; Wiest, S. A.; Palkowitz, A. D. *Cardiovasc. Drug Rev.* **1993**, *11*, 312-358.
5. Weber, V.; Monnot, C.; Bihoreau, C.; Corvol, P.; Clauser, E. *Horm. Res.* **1990**, *34*, 101-104.
6. Vallotton, M. B.; Capponi, A. M.; Johnson, E. I. M.; Lang, U. *Horm. Res.* **1990**, *34*, 105-110.
7. Siemens, I. R.; Swanson, G. N.; Fluharty, S. J.; Harding, J. W. *J. Neurochem.* **1991**, *57*, 690-700.
8. Timmermans, P. B. M. W. M.; Wong, P. C.; Chiu, A. T.; Herblin, W. F.; Benfield, P.; Carini, D. J.; Lee, R. J.; Wexler, R. R.; Saye, J. A. M.; Smith, R. D. *Pharmacol. Rev.* **1993**, *45*, 205-251.
9. Catt K.; Abbott, A. *Trends Pharmacol. Sci.* **1991**, *12*, 279-281.
10. Chin, H. L.; Buchan, D. A. *Ann. Intern. Med.* **1990**, *112*, 312-313.
11. McEwan, J. R.; Fuller, R. W. *J. Cardiovasc. Pharmacol.* **1989**, *13* (Suppl. 3), S67-S69.
12. Skidgel, R. A.; Engelbrecht, S.; Johnson, A. R.; Erdos, E. G. *Peptides* **1984**, *5*, 769-776.

13. Hansson, L. *Am. J. Hypertension* **1991**, *4*, 84S-87S.
14. Kleinert, H. D.; Rosenberg, S. H.; Baker, W. R.; Stein, H. H.; Klinghofer, V.; Barlow, J.; Spina, K.; Polakowski, J.; Kovar, P.; Cohen, J.; Denissen, J. *Science* **1990**, *257*, 1940-1943.
15. Kleinert, H. D.; Baker, W. R.; Rosenberg, S. H.; Stein, H. H. *J. Cellular Biochem.* **1993**, Suppl. 17C, 210.
16. Gupta, S. K.; Granneman, G. R.; Boger, R. S.; Hollenberg, N. K.; Luther, R. R. *Drug Metab. Dispos.* **1992**, *20*, 821-825.
17. Camezind, E.; Nussberger, J.; Juillerat, L; Munafo, A.; Fischli, W.; Coassolo, P.; van Brummelen, P.; Kleinbloesem, C. H.; Waeber, B.; Brunner, R. R. *J. Cardiovasc. Pharmacol.* **1991**, *18*, 299-307.
18. Smith, A. B. III, Hirschmann, R.; Pasternak, A.; Akaishi, R.; Guzman, M. C.; Jones, D. R.; Keenan, T. P.; Sprengeler, P. A.; Darke, P. L.; Emini, E. A.; Holloway, M. K.; Schleif, W. A. *J. Med. Chem.* **1994**, *37*, 215-218.
19. Moore, G. J. In *Comprehensive Medicinal Chemistry*, Hansch, C.; Sammes, P. G.; Taylor, J. B.; Emmett, J. C., Eds.; Pergamon Press: Oxford, UK, 1990; Vol. 3; pp 961-980.
20. Chiu, A. T.; Herblin, W. F.; McCall, D. E.; Ardecky, R. J.; Carini, D. J.; Duncia, J. V.; Pease, L. J.; Wong, P. C.; Wexler, R. R.; Johnson, A. L.; Timmermans, P. B. M. W. M. *Biochem. Biophys. Res. Commun.* **1989**, *165*, 196-203.
21. Whitebread, S.; Mele, M.; Kamber, B.; de Gasparo, M. *Biochem. Biophys. Res. Commun.* **1989**, *163*, 284-291.
22. Wong, P. C.; Hart, S. D.; Zaspel, A. M.; Chiu, A. T.; Ardecky, R. J.; Smith, R. D.; Timmermans, P. B. M. W. M. *J. Pharmacol. Exp. Ther.* **1990**, *255*, 584-592.
23. Wiest, S. A.; Rampersaud, A.; Zimmerman, K.; Steinberg, M. I. *J. Cardiovasc. Pharmacol.* **1991**, *17*, 177-184.
24. Sasaki, K.; Yamano, Y.; Bardhan, S.; Iwai, N.; Murray, J. J.; Hasegawa, M.; Matsuda, Y.; Inagami, T. *Nature* **1991**, *351*, 230-233.
25. Murphy, T. J.; Alexander, R. W.; Griendling, K. K.; Runge, M. S.; Bernstein, K. E. *Nature* **1991**, *351*, 233-236.
26. Mantlo, N. B.; Kim, D.; Ondeyka, D.; Chang, R. S. L.; Kivlighn, S. D.; Siegl, P. K. S.; Greenlee, W. J. *Bioorg. Med. Chem. Lett.* **1994**, *4*, 17-22.
27. Balbes, L. M.; Mascarella, S. W.; Boyd, D. B. In *Reviews in Computational Chemistry*, Lipkowitz, K. B.; Boyd, D. B., Eds.; VCH Publishers: New York, NY, 1994; Vol. 5, pp 337-379.
28. Smeby, R. R.; Fermandjian, S. In *Chemistry and Biochemistry of Amino Acids, Peptides, and Proteins: A Survey of Recent Developments*, Weinstein, B., Ed., Marcel Dekker: New York, NY, 1978; Vol. 5, pp 117-162.
29. Howard, R. B.; Smeby, R. R. In *Handbook of Hypertension, Pathophysiology of Hypertension - Regulatory Mechanisms*, Zanchetti, A.; Tarazi, R. C., Eds.; Elsevier: Amsterdam, The Netherlands, 1986; Vol. 8, pp 389-397.
30. Matsoukas, J. M.; Bigam, G.; Zhou, N.; Moore, G. J. *Peptides* **1990**, *11*, 359-366.
31. Weinstock, J.; Keenan, R. M.; Samanen, J.; Hempel, J.; Finkelstein, J. A.; Franz, R. G.; Gaitanopoulos, D. E.; Girard, G. R.; Gleason, J. G.; Hill, D. T.; Morgan, T. M.; Peishoff, C. E.; Aiyar, N.; Brooks, D. P.; Fredrickson, T. A.; Ohlstein, E. H.; Ruffolo, R. R., Jr.; Stack, E. J.; Sulpizio, A. C.; Weidley, E. F.; Edwards, R. M. *J. Med. Chem.* **1991**, *34*, 1514-1517.
32. Pierson, M. E.; Freer, R. J. *Peptide Res.* **1992**, *5*, 102-105.
33. Furukawa, Y.; Kishimoto, S.; Nishikawa, K. U.S. Patents 4,340,598 and 4,355,040, issued to Takeda Chemical Industries, Japan, 1982.

34. Duncia, J. V.; Chiu, A. T.; Carini, D. J.; Gregory, G. B.; Johnson, A. L.; Price, W. A.; Wells, G. J.; Wong, P. C.; Calabrese, J. C.; Timmermans, P. B. M. W. M. *J. Med. Chem.* **1990**, *33*, 1312-1329.
35. Carini, D. J.; Duncia, J. V.; Johnson, A. L.; Chiu, A. T.; Price, W. A.; Wong, P. C.; Timmermans, P. B. M. W. M. *J. Med. Chem.* **1990**, *33*, 1330-1336.
36. Timmermans, P. B. M. W. M.; Wong, P. C.; Chiu, A. T.; Herblin, W. F. *Trends Pharmacol. Sci.* **1991**, *12*, 55-62.
37. Carini, D. J.; Duncia, J. V.; Aldrich, P. E.; Chiu, A. T.; Johnson, A. L.; Pierce, M. E.; Price, W. A.; Santella, J. B. III; Wells, G. J.; Wexler, R. R.; Wong, P. C.; Yoo, S.-E.; Timmermans, P. B. M. W. M. *J. Med. Chem.* **1991**, *34*, 2525-2547.
38. Duncia, J. V.; Carini, D. J.; Chiu, A. T.; Johnson, A. L.; Price, W. A.; Wong, P. C.; Wexler, R. R.; Timmermans, P. B. M. W. M. *Med. Res. Rev.* **1992**, *12*, 149-191.
39. Bovy, P. R.; Collins, J. T.; Olins, G. M.; McMahon, E. G.; Hutton, W. C. *J. Med. Chem.* **1991**, *34*, 2410.
40. Bühlmayer, P.; Criscione, L.; Fuhrer, W.; Furet, P.; de Gasparo, M.; Stutz, S.; Whitebread, S. *J. Med. Chem.* **1991**, *34*, 3105-3114.
41. Bradbury, R. H.; Allott, C. P.; Dennis, M.; Fisher, E.; Major, J. S.; Masek, B. B.; Oldham, A. A.; Pearce, R. J.; Rankine, N.; Revill, J. M.; Roberts, D. A.; Russell, S. T. *J. Med. Chem.* **1992**, *35*, 4027-4038.
42. Atwal, K. S.; Ahmed, S. Z.; Bird, J. E.; Delaney, C. L.; Dickinson, K. E. J.; Ferrara, F. N.; Hedberg, A.; Miller, A. V.; Moreland, S.; O'Brien, B. C.; Schaeffer, T. R.; Waldron, T. L.; Weller, H. N. *J. Med. Chem.* **1992**, *35*, 4751-4763.
43. Murray, W. V.; Lalan, P.; Gill, A.; Addo, M. F.; Lewis, J. M.; Lee, D. K. H.; Rampulla, R.; Wachter, M. P.; Hsi, J. D.; Underwood, D. C. *Bioorg. Med. Chem. Lett.* **1992**, *2*, 1775-1779.
44. Bovy, P. R.; Reitz, D. B.; Collins, J. T.; Chamberlain, T. S.; Olins, G. M.; Corpus, V. M.; McMahon, E. G.; Palomo, M. A.; Koepke, J. P.; Smits, G. J.; McGraw, D. E.; Gaw, J. F. *J. Med. Chem.* **1993**, *36*, 101-110.
45. Greenlee, W. J.; Mantlo, N. B., Eds. *Bioorg. Med. Chem. Lett.* **1994**, *4* (1), 1-222.
46. Ohlstein, E. H.; Gellai, M.; Brooks, D. P.; Vickery, L.; Jugus, J.; Sulpizio, A.; Ruffolo, R. R., Jr.; Weinstock, J.; Edwards, R. M. *J. Pharmacol. Exp. Ther.* **1992**, *262*, 595-601.
47. Lin, H.-S.; Rampersaud, A. A.; Zimmerman, K.; Steinberg, M. I.; Boyd, D. B. *J. Med. Chem.* **1992**, *35*, 2658-2667.
48. Lin, H.-S.; Rampersaud, A. A.; Zimmerman, K.; Steinberg, M. I.; Boyd, D. B. *J. Chin. Chem. Soc.* (Taipei) **1993**, *40*, 273-282.
49. Lifer, S. L.; Marshall, W. S.; Mohamadi, F.; Reel, J. K.; Simon, R. L.; Steinberg, M. I.; Whitesitt, C. A. U.S. Patent 5073566, 1991, European Patent Application 438869, 1991; Canadian Patent Application 2030961, 1991.
50. Mohamadi, F.; Gesellschen, P. D.; Steinberg, M. I., unpublished work.
51. Steinberg, M. I.; Palkowitz, A. D.; Thrasher, K. J.; Reel, J. K.; Zimmerman, K. M.; Whitesitt, C. A.; Simon, R. L.; Hauser, K. L.; Lifer, S. L.; Pfeifer, W.; Takeuchi, K.; Wiest, S. A.; Vasudevan, V.; Bemis, K. G.; Deeter, J. B.; Marshall, W. S.; Boyd, D. B. *Bioorg. Med. Chem. Lett.* **1994**, *4*, 51-56.
52. Thrasher, K. J.; Boyd, D. B.; Lifer, S.; Marshall, W. S.; Palkowitz, A. D.; Pfeifer, W.; Reel, J.; Simon, R.; Steinberg, M. I.; Whitesitt, C. A. European Patent Application 93304264.0, June 2, 1993.
53. Thrasher, K. J.; Boyd, D. B.; Lifer, S.; Marshall, W. S.; Palkowitz, A. D.; Pfeifer, W.; Reel, J.; Simon, R.; Steinberg, M. I.; Vasudevan, V.; Whitesitt, C. A. European Patent Application 93304266.5, June 2, 1993.

54. Reel, J. K.; Simon, R. L.; Pfeifer, W.; Whitesitt, C. A.; Lifer, S. L.; Steinberg, M. I.; Palkowitz, A. D.; Thrasher, K. J.; Hauser, K. L.; Zimmerman, K. M.; Takeuchi, K.; Wiest, S. A.; Boyd, D. B.; Marshall, W. S. Abstracts of Papers, 207th American Chemical Society National Meeting, March 13-18, 1994, San Diego, CA, MEDI 206.

55. Thrasher, K. J.; Palkowitz, A. D.; Reel, J. K.; Hauser, K. L.; Whitesitt, C. A.; Steinberg, M. I.; Zimmerman, K. M.; Simon, R. L.; Pfeifer, W.; Boyd, D. B.; Lifer, S. L.; Wiest, S. A.; Marshall, W. S. Abstracts of Papers, 207th American Chemical Society National Meeting, March 13-18, 1994, San Diego, CA, MEDI 207.

56. Pfeifer, W.; Whitesitt, C. A.; Simon, R. L.; Hauser, K. L.; Lifer, S. L.; Wiest, S. A.; Vasudevan, V.; Reel, J. K.; Palkowitz, A. D.; Thrasher, K. J.; Zimmerman, K. M.; Steinberg, M. I.; Boyd, D. B.; Marshall, W. S. Abstracts of Papers, 207th American Chemical Society National Meeting, March 13-18, 1994, San Diego, CA, MEDI 204.

57. Simon, R. L.; Whitesitt, C. A.; Zimmerman, K. M.; Steinberg, M. I.; Vasudevan, V.; Wiest, S. A.; Lifer, S. L.; Pfeifer, W.; Reel, J. K.; Palkowitz, A. D.; Thrasher, K. J.; Hauser, K. L.; Boyd, D. B.; Marshall, W. S. Abstracts of Papers, 207th American Chemical Society National Meeting, March 13-18, 1994, San Diego, CA, MEDI 203.

58. Zimmerman, K. M.; Steinberg, M. I.; Wiest, S. A.; Palkowitz, A. D.; Thrasher, K. J.; Reel, J. K.; Whitesitt, C. A.; Simon, R. L.; Pfeifer, W.; Boyd, D. B.; Hauser, K. L.; Lifer, S. L.; Marshall, W. S. Abstracts of Papers, 207th American Chemical Society National Meeting, March 13-18, 1994, San Diego, CA, MEDI 209.

59. Palkowitz, A. D.; Steinberg, M. I.; Thrasher, K. J.; Reel, J. K.; Hauser, K. L.; Zimmerman, K. M.; Wiest, S. A.; Whitesitt, C. A.; Simon, R. L.; Pfeifer, W.; Lifer, S. L.; Boyd, D. B.; Barnett, C. J.; Wilson, T. M.; Deeter, J. B.; Takeuchi, K. J.; Riley, R. E.; Miller, W. D.; Marshall, W. S. *J. Med. Chem.* **1994**, *37*, in press.

60. Lipkowitz, K. B.; Boyd, D. B., Eds., *Reviews in Computational Chemistry*, VCH Publishers, New York, NY; Volume I, 1990; Volume II, 1991; Volume III, 1992; Volume IV, 1993; Volume V, 1994; Volume 6, 1995.

61. Motoc, I.; Dammkoehler, R. A.; Mayer, D.; Labanowski, J. *Quant. Struct.-Act. Relat.* **1986**, *5*, 99-105.

62. Labanowski, J.; Motoc, I.; Naylor, C. B.; Mayer, D.; Dammkoehler, R. A. *Quant. Struct.-Act. Relat.* **1986**, *5*, 138-152.

63. Van Opdenbosch, N.; Cramer, R. III; Giarrusso, F. F. *J. Mol. Graphics* **1985**, *3*, 110-111.

64. Manual to SYBYL Molecular Modeling Software, Tripos Associates: 1699 Hanley Road, St. Louis, MO, 1992.

65. Boyd, D. B. In *Reviews in Computational Chemistry*, Lipkowitz, K. B.; Boyd, D. B., Eds.; VCH Publishers: New York, NY, 1994; Vol. 5, pp 381-428.

66. Hansch, C.; Leo, A. *Substituent Constants for Correlation Analysis in Chemistry and Biology*, Wiley-Interscience: New York, NY, 1979.

67. Leo, A. J. In *Comprehensive Medicinal Chemistry*; Hansch, C.; Sammes, P. G.; Taylor, J. B.; Ramsden, C. A, Eds.; Pergamon Press: Oxford, UK, 1990; Vol. 4, pp 295-319.

68. DayMenus Software Manual, Version 3.63, Daylight Chemical Information Systems, Inc.: 18500 Von Karman Avenue, Irvine, CA, 1991.

69. JMP User's Guide; SAS Institute, Inc.: SAS Campus Drive, Cary, NC, 1989.

70. Clark, M.; Cramer, R. D. III; Van Opdenbosch, N. *J. Comput. Chem.* **1989**, *10*, 982-1012.

71. Jungheim, L. J.; Boyd, D. B.; Indelicato, J. M.; Pasini, C. E.; Preston, D. A.; Alborn, W. E., Jr. *J. Med. Chem.* **1991**, *34*, 1732-1739.

72. Robertson, D. W.; Boyd, D. B. In *Advances in Second Messengers and Phosphoprotein Research*, Strada, S. J.; Hidaka, H., Eds.; Raven Press: New York, NY, 1991; Vol. 25, pp 321-340.

73. Chiu, A. T.; Carini, D. J.; Duncia, J. V.; Leung, K. H.; McCall, D. E.; Price, W. A., Jr.; Wong, P. C.; Smith, R. D.; Wexler, R. R.; Timmermans, P. B. M. W. M. *Biochem. Biophys. Res. Commun.* **1991**, *177*, 209-217.

74. Perrin, D. D. *Dissociation Constants of Organic Bases in Aqueous Solution*, Butterworths: London, UK, 1965.

75. Albert, A.; Serjeant, E. P. *Determination of Ionization Constants: Laboratory Manual*, 3rd Ed., Chapman and Hall: London, UK, 1984.

76. Weast, R. C., Ed. *CRC Handbook of Chemistry and Physics*, CRC Press: Boca Raton, FL, 1980.

77. Samanen, J.; Cash, T.; Narindray, D.; Brandeis, E.; Adams, W., Jr.; Weideman, H.; Yellin, T.; Regoli, D. *J. Med. Chem.* **1991**, *34*, 3036-3043.

78. Garcia, K. C.; Ronco, P. M.; Verroust, P. J.; Brünger, A. T.; Amzel, L. M. *Science* **1992**, *257*, 502-507.

79. Boyd, D. B. In *Encyclopedia of Computer Science and Technology*, Kent, A.; Williams, J. G., Eds.; Marcel Dekker: New York, NY, 1995; Vol. 33, in press.

RECEIVED October 11, 1994

Chapter 3

Structure-Based Design of Human Immunodeficiency Virus-1 Protease Inhibitors

Correlating Calculated Energy with Activity

M. Katharine Holloway and Jenny M. Wai

Molecular Systems Department, Merck Research Laboratories, West Point, PA 19486

We have found that a simple calculated energy value, E_{inter}, correlates well with the observed *in vitro* enzyme activity of a series of HIV-1 protease inhibitors. This correlation was derived employing a test dataset of 33 inhibitors with modifications at the P_1' and P_2' sites. It has proved valuable in the structure-based design of subsequent HIV-1 protease inhibitors which exhibit significant structural variation. In particular, it has been successful in a truly predictive sense, *i.e.* predictions of activity were made prior to synthesis. Several examples of this are illustrated, including a precursor (**41**) to a current clinical candidate, L-735,524 (**42**).

Of critical importance to the application of computer-aided molecular design is the accurate assessment of the relative energy of interaction between two or more molecules of interest. In the design of pharmaceuticals, these molecules may correspond to an enzyme and an inhibitor, a receptor and an antagonist, an antibody and an antigen, or a DNA helix and an intercalator. Ideally, one would like the energetic evaluation to be as rapid as possible, while maintaining a high level of accuracy such that useful predictions of activity may be made in advance of biological testing. In the absence of accurate energetic evaluation, only qualitative modeling is possible, which, while useful, cannot reliably address the question of binding affinity.

Historically, several approaches have been employed to calculate and/or predict binding affinity. The free energy of binding can be calculated directly via Free Energy Peturbation (FEP) calculations (*1*). This approach has been reported to yield free energies accurate to ± 1 kcal/mol with respect to experiment. However, due to the amount of computer time required, these calculations are impractical for routine assessment of the binding affinity of proposed compounds.

Other simpler approaches have included Comparative Molecular Field Analysis (CoMFA) (2) and the Hypothetical Active Site Lattice (HASL) (*3*) method. While these approaches are rapid relative to FEP, they involve correlation with multiple calculated properties, fields, or sites, and their accuracy can be limited by several factors, *e.g.* the size and diversity of the training set of structures and the choice of alignment for these structures.

0097–6156/95/0589–0036$12.00/0
© 1995 American Chemical Society

We report herein a single simple predictor of relative binding affinity for HIV-1 protease inhibitors which has been highly effective in a structure-based design program.

HIV-1 Protease

The HIV-1 protease is an aspartyl protease which is responsible for processing the polyproteins coded for by the *gag* and *pol* genes of the HIV-1 virus, thus leading to production of the HIV-1 structural proteins (*e.g.* p17, p24, p7, and p6) and enzymes (reverse transcriptase, integrase, and the protease itself) (*4*). Inactivation of HIV-1 protease has been demonstrated to result in the production of noninfectious virions (*5*). Thus, HIV-1 protease inhibitors have been an attractive drug design target for treatment of the Acquired ImmunoDeficiency Syndrome (AIDS) (*6*).

Structurally, the HIV-1 protease is a symmetrical homodimer, which contains 99 residues and one characteristic Asp-Thr-Gly sequence per monomer. Figure 1 shows two views of an X-ray structure of the native form of the HIV-1 protease (*7*). In the

first view, one can clearly see two β hairpin loops at the top of the structure which are commonly referred to as the flaps due to their dynamic nature. These flaps are presumably open initially to allow diffusion of the substrate or inhibitor into the active site and subsequently close to form both hydrogen-bonding and hydrophobic contacts with the small molecule. The second view is looking through the flaps at the active site, with the binding cleft roughly from top to bottom. The two catalytic aspartates are located directly below the flaps on the floor of the active site.

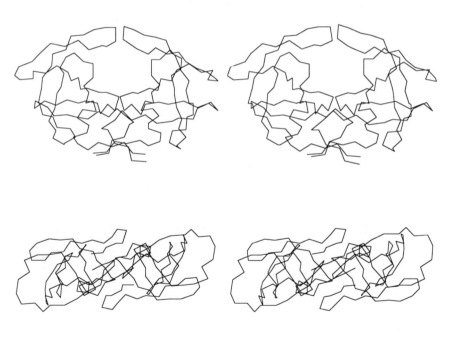

Figure 1. Two views of the α-carbon trace of the native HIV-1 protease.

Qualitative Inhibitor Modeling

Initially, we began our modeling of HIV-1 protease inhibitors in a qualitative sense. They were docked in the active site of the native enzyme shown in Figure 1 using conformations which were based on X-ray structures of renin inhibitors bound in the active sites of fungal aspartyl proteases such as endothiapepsin (8) and *Rhizopus* pepsin (9). The first compound which was modeled was L-685,434 (1), an early lead, which was very potent *in vitro* (IC_{50} = 0.25 nM) but lacked cell potency (CIC_{95} = 400 nM) (10,11).

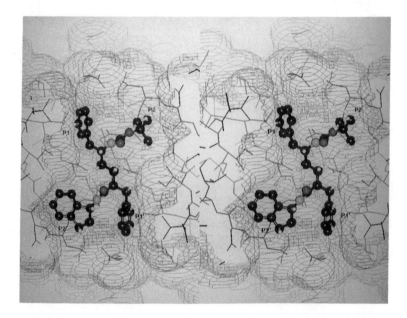

1, R = H

2, R = $OCH_2CH_2N\overset{\frown}{\underset{\smile}{}}O$

Based on the model of **1** in the active site, shown in Figure 2, it became clear that solubilizing groups which might affect the physicochemical and thus pharmacokinetic properties could be attached at the *para* position of the P_1' substituent. This led to an

Figure 2. L-685,434 (**1**) as modeled in the native enzyme active site. A molecular surface is included to illustrate the fit of the inhibitor in the active site cavity.

inhibitor, L-689,502 (**2**), of comparable *in vitro* potency (IC_{50} = 0.45 nM) but enhanced cell potency (CIC_{95} = 12 nM) which was the first safety assessment candidate in this program (*11*). Unfortunately, it failed in safety assessment due to hepatotoxicity.

When a subsequent X-ray structure was solved for the complex of **2** with HIV-1 protease (*11*), it was clear that we had accurately modeled the bound conformation for this inhibitor as shown in Figure 3. In the X-ray structure, the para substituent on the P_1' aromatic group does indeed point out away from the active site towards solvent. Its exposure to solvent is further supported by the fact that the morpholine group is disordered and thus is absent from this plot.

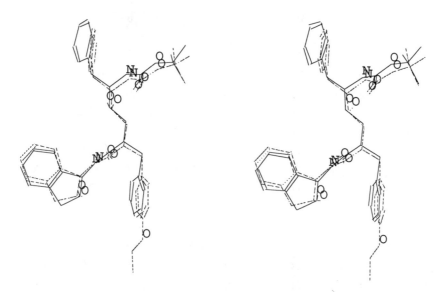

Figure 3. A comparison of the model of **1** (solid) and the X-ray structure of **2** (dashed).

Semi-Quantitative Inhibitor Modeling

Our success with qualitative predictions led us to try a more quantitative approach. We looked for a correlation not only between the modeled and X-ray structures, but also between the calculated energy and the observed *in vitro* activity.

This was accomplished via the following protocol. Protonation states for the titratable enzyme residues were selected based on the fact that the enzyme assay is carried out at a pH of 5.5. Thus we chose to make all titratable residues charged, with the exception of Tyr_{59} and Asp_{A25}, one of the pair of catalytic aspartates (*12*). A representative set of inhibitors was then selected to form a test dataset. This included 16 inhibitors with modifications in P_1' and 16 inhibitors with modifications in P_2', which are shown in Tables I and II.

Each group contained as wide a spread in activity and structure as possible. A model of **1** was constructed as described above and all subsequent models were derived from it. The inhibitor models were minimized in the enzyme active site using the MM2X force field (*13*), a variant of MM2 (*14*). In all calculations the inhibitor was completely flexible and the enzyme was completely rigid. Dielectric constants of 1.5 for intramolecular interactions and 1.0 for intermolecular interactions were employed.

Table I. Experimental IC_{50} Values and Calculated Enzyme•Inhibitor Intermolecular Energies for the P_1' Training Set of HIV-1 Protease Inhibitors.

No	R_1	R_2	R_3	IC_{50} nM	pIC_{50}[a]	E_{Nat}[b]	E_{AcPep}[c]	E_{502}[d]
1	CH_2Ph	H	H	0.25	9.6021	-108.3	-134.8	-145.1
3	CH_2Ph	CH_3	H	7.7	8.1135	-106.3	-131.1	-140.4
4	$CH_2CH_2CH_2Ph$	H	OH	0.19	9.7212	-113.0	-139.2	-143.6
5	CH_2-4-CF_3Ph	H	H	0.26	9.5850	-114.0	-141.3	-149.6
6	(E)-$CH_2CH=CHPh$	H	H	0.23	9.6383	-113.4	-138.3	-147.1
7	$CH_2C_6F_5$	H	H	0.6	9.2218	-115.3	-139.1	-149.4
8	CH_2-4-CH_3Ph	H	H	0.29	9.5376	-109.1	-135.4	-146.5
9	CH_2-4-NH_2Ph	H	H	0.31	9.5086	-110.4	-137.1	-146.1
10	CH_2-4-NO_2Ph	H	H	0.27	9.5686	-118.4	-147.7	-151.4
11	H	H	H	2934	5.5325	-98.6	-125.0	-129.2
12	CH_2-4-OHPh	H	H	0.16	9.7959	-110.7	-136.6	-149.7
13	$CH_2CH=CH_2$	H	H	27.5	7.5607	-102.9	-131.5	-137.8
14	CH_2-4-IPh	H	H	0.72	9.1427	-113.4	-140.4	-148.4
15	$CH_2C(O)Ph$	H	H	5.42	8.2660	-114.2	-141.3	-150.3
16	CH_2-4-pyridyl	H	H	0.53	9.2757	-111.6	-134.4	-144.9
17	CH_2SPh	H	H	0.25	9.6021	-112.3	-138.9	-146.0
18	CH_2-4-t-butylPh	H	H	0.17	9.7696	-113.3	-137.0	-150.9

[a] $pIC_{50} = -\log(IC_{50})$
[b] Intermolecular energy (kcal/mol) calculated in the native HIV-1 protease active site.
[c] Intermolecular energy (kcal/mol) calculated in the acetylpepstatin inhibited HIV-1 protease active site.
[d] Intermolecular energy (kcal/mol) calculated in the L-689,502 inhibited active site.

We then looked for a correlation between the calculated intermolecular component of the energy (E_{inter}) and the observed IC_{50} (15), assuming that E_{inter} might be proportional to the enthalpy of binding (ΔH_{bind}) and that the entropy of binding (ΔS_{bind}) might be small or constant, thus giving us something which might be proportional to the free energy of binding (ΔG_{bind}), the value we were actually interested in.

$$\Delta G_{bind} = \Delta H_{bind} - T\Delta S_{bind}$$

Table II. Experimental IC_{50} Values and Calculated Enzyme•Inhibitor Intermolecular Energies for the P_2' Training Set of HIV-1 Protease Inhibitors.

No	R	IC_{50} nM	pIC_{50}[a]	E_{Nat}[b]	E_{AcPep}[c]	E_{502}[d]
19		114	6.9431	-94.3	-122.3	-131.5
20		9.53	8.0209	-98.2	-128.1	-132.9
21		34.25	7.4653	-97.4	-124.7	-135.5
22		690	6.1612	-103.1	-135.0	-134.1
23		161	6.7932	-96.6	-127.7	-130.8
24		66.3	7.1785	-106.1	-134.6	-139.3
25		212.42	6.6728	-91.7	-133.6	-144.0
26		121.8	6.9144	-99.2	-130.6	-134.2
27		0.7	9.1549	-109.9	-135.6	-146.7
28		0.18	9.7447	-109.2	-136.4	-145.5
29		40.5	7.3925	-94.5	-135.5	-134.5

Continued on next page

Table II. (Continued)

No	R	IC_{50} nM	pIC_{50}[a]	E_{Nat}[b]	E_{AcPep}[c]	E_{502}[d]
30		30000	4.5229	-88.4	-120.1	-124.1
31		130	6.8861	-94.4	-122.3	-129.8
32		146	6.8356	-96.5	-123.2	-134.1
33		0.1	10.000	-111.7	-139.3	-149.1
34		38.6	7.4134	-108.1	-133.2	-138.4

[a] pIC_{50} = - log (IC_{50})
[b] Intermolecular energy (kcal/mol) calculated in the native HIV-1 protease active site.
[c] Intermolecular energy (kcal/mol) calculated in the acetylpepstatin inhibited HIV-1 protease active site.
[d] Intermolecular energy (kcal/mol) calculated in the L-689,502 inhibited active site.

Tables I and II give the E_{inter} values for the compounds in the test dataset as minimized in three different enzyme active sites, the native enzyme (Nat), the acetylpepstatin inhibited enzyme (AcPep) (*16*), and the L-689,502 inhibited enzyme (502) (*11*). We observed a very good correlation between E_{inter} and pIC_{50}, *i.e.* -log(IC_{50}), for all three as shown in Figure 4. Surprisingly, a good correlation was observed even with the native enzyme active site where the flaps are open, rather than interacting with the inhibitor, as illustrated in Figure 1.

Both the R^2 and the cross-validated R^2 values are listed below for each correlation. In all three cases these are very comparable, indicating that no one datapoint is overly influential to the correlation; thus, the correlation should be predictive.

Native:
$$pIC_{50} = -0.15435(E_{inter}) - 8.069 \tag{1}$$
$R=0.8524, R^2=0.7265,$ crossvalidated $R^2=0.6910$

Acetylpepstatin inhibited:
$$pIC_{50} = -0.17302(E_{inter}) - 14.901 \tag{2}$$
$R=0.7623, R^2=0.5811,$ crossvalidated $R^2=0.5244$

L-689,502 inhibited:
$$pIC_{50} = -0.16946(E_{inter}) - 15.707 \tag{3}$$
$R=0.8852, R^2=0.7835,$ crossvalidated $R^2=0.7551$

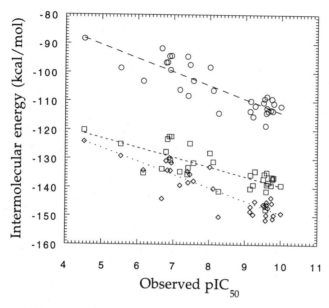

Figure 4. Plot of calculated enzyme•inhibitor intermolecular energy vs. experimental enzyme inhibition (pIC$_{50}$) for the training set of inhibitors, **1**, **3-34** (circles = native enzyme active site, squares = acetylpepstatin inhibited enzyme active site, diamonds = L-689,502 inhibited enzyme active site).

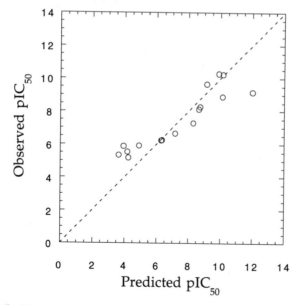

Figure 5. Plot of predicted pIC$_{50}$ vs. observed pIC$_{50}$ values for the predicted set of inhibitors. The line is one of unit slope, *i.e.* predicted pIC$_{50}$ = observed pIC$_{50}$.

In fact, using these correlation equations, we were able to make predictions of activity for inhibitors prior to synthesis, *i.e.* true predictions and not *post hoc* explanations of activity. Because this was done over a period of time some of the predictions were made with the earliest correlation equation, that of the native enzyme, some with the acetylpepstatin inhibited enzyme, and some with the L-689,502 inhibited enzyme. The accuracy of these predictions is illustrated in Figure 5. Here, the line is one of unit slope, not a correlation line. There is only one significant outlier which will be discussed in more detail subsequently. It must be emphasized that we made many more predictions than are shown in Figure 5. Frequently, when our prediction for a compound was unfavorable, it was not synthesized; or when a prediction was favorable, the exact compound which was modeled was not synthesized, but rather an analog.

Examples. In order to illustrate the power and the limitations of this simple approach, some specific examples of predictions are listed in Table III.

Like many others (*17*), we hypothesized that a symmetrical inhibitor might bind more tightly to the symmetrical active site of the HIV-1 protease. Thus, **35** was designed as a symmetrical version of **1**. The activity of **35** was predicted using the earliest correlation equation and it is the outlier seen in Figure 5. We were concerned that our prediction was poor due to differences in binding between our model and experiment. However, when the X-ray structure was solved (*18*), this was obviously not the case, as shown in Figure 6. There are several other possible explanations for the overprediction of activity: (1) the use of the native enzyme model in Equation 1, rather than one of the inhibited enzyme models in Equations 2 or 3; (2) the presence of an additional hydrogen bond to the active site which would be overemphasized in a gas-phase molecular mechanics calculation; (3) the addition of significant favorable van der Waals interactions via the second aminoindanol moiety at the N-terminus; or (4) the existence of a higher barrier to obtaining the bioactive conformation necessary for binding. Of the four possibilities, the first may be eliminated since the activity predicted using the other two models is also exaggerated. However, factor (4) may play a key role, since **35** experienced a large decrease in its intramolecular enthalpy when minimized outside of the active site, an indication that the bound conformation may be significantly higher in energy than the global minimum.

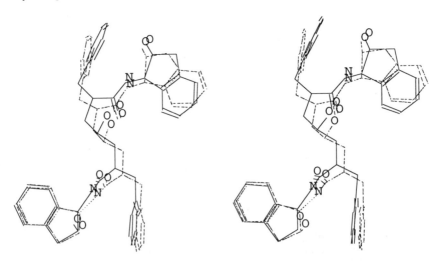

Figure 6. A comparison of the modeled and X-ray structures of **35**.

TABLE III. Calculated Enzyme•Inhibitor Intermolecular Energies and Predicted and Observed pIC$_{50}$ Values for the Predicted Set of HIV-1 Protease Inhibitors.

No	Structure	E_{inter}[a] kcal/mol	Corr. Eq.[b]	Predicted pIC$_{50}$[c]	Observed pIC$_{50}$
35		-123.4	(1)	12.012	9.1612
36		-125.9	(2)	4.9099	5.8965
37		-143.6	(2)	8.7383	8.2676
38		-148.9	(2)	9.8847	10.2676
40		-141.6	(2)	8.3058[d]	7.2774
41		-143.2	(2)	8.6518	8.1163

[a]Calculated intermolecular energy in the indicated active site.

[b]Correlation equation which was used for prediction based upon minimization in the corresponding enzyme active site.

[c]pIC$_{50}$ = -log (IC$_{50}$).

[d]The prediction of activity was originally made for a closely related compound; the "predicted" activity reported here was calculated at a later date for the compound which was actually synthesized.

We have illustrated the worst prediction first. However, one can see in Table III that we were able to accurately predict that the 6-membered lactam ring in **36** would fit poorly in the active site although the analogous 5-membered lactam was a nanomolar inhibitor (IC_{50} = 37 nM) (*19*). In the best case we were able to distinguish small differences in activity between two diastereomers, **37** and **38**. These latter compounds were based on the Roche inhibitor Ro-31-8959, **39**, but incorporate a novel amino acid residue at P_2 in place of asparagine (*20*).

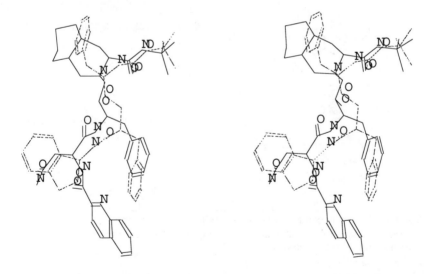

39   **42,**

The last two examples involve an interesting feature of HIV-1 protease complexes with inhibitors. Due to the symmetrical nature of the enzyme, many inhibitors are observed to bind in the active site in two directions, both in an N->C and a C->N orientation with respect to the flaps, which in their closed H-bonded form introduce the asymmetry which is the direction marker. Comparing models of **39**, oriented in the N->C and C->N fashions, in the active site led to the design of an active "reversed" Roche analog, **40**, in which all the binding elements remain the same, but the amide bond directionality is reversed. Similarly, comparing models of **1** and **39**, oriented in the N->C and C->N fashions, respectively, as shown in Figure 7,

Figure 7. A comparison of the model of **1** oriented in an N -> C fashion and the model of **39** oriented in a C -> N fashion.

led to the hypothesis that novel hybrid inhibitors such as **41** could be developed which incorporated the C-terminal halves of each. As predicted, **41** was an active HIV-1 protease inhibitor. It was also the parent structure of the Phase II clinical candidate, L-735,524 (**42**) (*21*). Figure 8 compares the models of **41** and **42** as minimized in the HIV-1 protease active site.

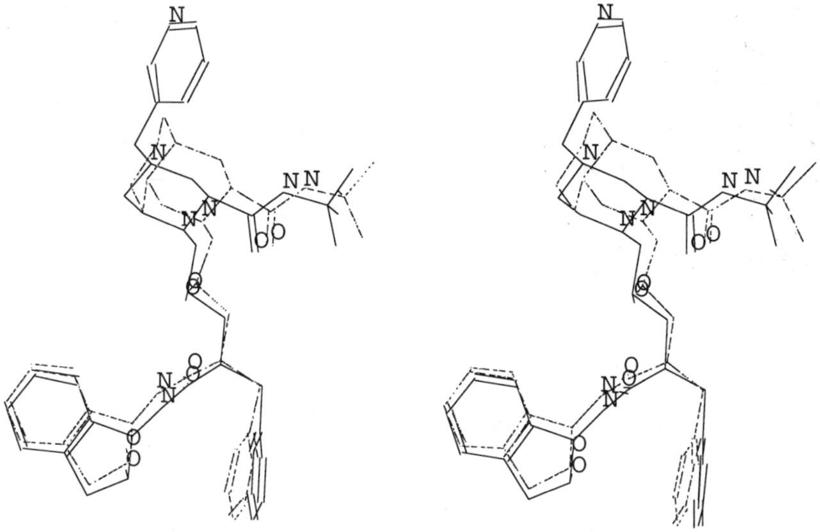

Figure 8. A comparison of the models of **41** and **42** as bound in the enzyme active site.

Improving the Model. Although we experienced remarkable predictive success employing a simple approach, we obviously neglected several key factors in binding, *e.g.* the flexibility of the enzyme active site, the difference in energy between the solution and bound conformations of the inhibitor, and the solvation/desolvation of the inhibitor and the enzyme. Thus, we hoped to improve the correlation, and thus our ability to make accurate predictions, by incorporating some of these effects into our computed energies. Table IV lists the results of some of these calculations.

In order to assess the effect of the flexibility of the enzyme active site, we had to employ a different molecular mechanics program and thus a different force field because the program OPTIMOL, within which the MM2X force field resides, does not generally allow for flexibility of the protein context molecule. Thus, we repeated the initial calculations with the rigid active site using the CHARMm force field (*22,23*) prior to performing minimization of the enzyme•inhibitor complex. This afforded an interesting comparison of the results derived from two different force fields, indicating that the MM2X force field is superior to CHARMm for this specific application, *i.e.* the correlation coefficients for the acetylpepstatin and L-689,502 inhibited enzymes were 0.4780 (0.7623) and 0.7213 (0.8852) for CHARMm (MM2X), respectively.

Further CHARMm calculations were performed to evaluate the importance of relaxation of the enzyme active site. Several residues (8, 22-23, 25-32, 45-51, 76, 78-82, and 87 in each monomer) which line the active site cavity were allowed to minimize concurrent with the inhibitor. The resulting correlation coefficient for the energies obtained in this manner was 0.6432 for the acetylpepstatin inhibited enzyme,

Table IV. Correlation coefficients obtained when employing a variety of computational approaches.

Computational Protocol	Correlation Coefficient	
	AcetylPepstatin	L-689,502
E_{inter}, MM2X	0.7623	0.8852
E_{inter}, CHARMm	0.4780	0.7213
E_{inter}, MM2X charges, CHARMm		0.8262
E_{inter}, flexible enzyme, CHARMm	0.6432	
E_{inter} - Inh_{flex} ($\varepsilon = 1$), MM2X	0.7679	
E_{inter} - Inh_{flex} ($\varepsilon = 50$), MM2X	0.7656	
$Solv_{inh}$, BMIN		0.2814
$Solv_{total}$, BMIN		0.3439
E_{inter} (MM2X) & $Solv_{inh}$ (BMIN)		0.8863
E_{inter} (MM2X) & $Solv_{total}$ (BMIN)		0.8880
Inhibitor Surface Area	0.5504	
Inhibitor Volume	0.5647	

significantly better than that obtained with the rigid enzyme active site using the CHARMm force field. However, this was still slightly worse than that obtained using the MM2X force field with the rigid active site.

We postulated that the better correlation obtained with the MM2X force field might be due to the consistent charging scheme employed. In order to explore this hypothesis we performed CHARMm calculations using the MM2X charges. As expected, this led to an improvement in the correlation coefficient, from 0.7213 (CHARMm/CHARMm charges) to 0.8262 (CHARMm/MM2X charges) for the L-689,502 inhibited enzyme active site. One might naturally speculate that coupling the use of MM2X charges and allowing the enzyme active site to relax might increase the overall correlation enough to be of interest. We plan to repeat these calculations employing the Merck Molecular Force Field (MMFF), an improved version of MM2X, which has been incorporated into the CHARMm program (*23*).

We also attempted to correct for the energetic cost of the inhibitor attaining the bound conformation by subtracting from E_{inter} the difference in energy between the free and bound inhibitor conformations. The energy of the uncomplexed inhibitor was assessed at two different dielectric constants, 1 and 50. Although each improved the correlation slightly ($R = 0.7679$ and 0.7656 for $\varepsilon = 1$ and $\varepsilon = 50$, respectively, versus 0.7623 for the acetylpepstatin inhibited enzyme), neither showed a significant enough effect to justify the inclusion of another term.

Incorporation of solvation effects via the BATCHMIN (*24*) GB/SA continuum solvation method (*25*) using the MM2 force field also was ineffective at improving the correlation more than marginally, independent of whether the solvation of the inhibitor only ($R = 0.8863$) or the total solvation of binding ($R = 0.8880$) was considered. This was consistent with a low correlation between the computed solvation energy and the observed activity ($R = 0.2814$ and 0.3439 for the inhibitor and total solvation, respectively). Although this was disappointing, it may simply indicate that we have not properly approached the computation of solvation energy for these inhibitors.

Thus, unfortunately we have yet to improve significantly upon our initial simple predictor of activity, E_{inter}, despite the fact that other energetic factors are clearly key contributors to binding affinity.

Conclusions

In summary, we have illustrated that the qualitative approach to modeling of HIV-1 protease inhibitors was successful in facilitating the design of the first safety assessment candidate in this program, **2**. However, the semi-quantitative approach to modeling of HIV-1 protease inhibitors, via the derivation of a correlation between E_{inter} and IC_{50}, was successful in predicting the activity of a variety of proposed inhibitors prior to synthesis. This facilitated the design of the current Phase II clinical candidate, L-735,524 (**42**).

Literature Cited

1. For a recent review of FEP calculations, see Kollman, P. *Chem. Rev.* **1993**, *93*, 2395-2417.
2. (a) Waller, C. L.; Oprea, T. I.; Giolitti, A.; Marshall, G. R. *J. Med. Chem.* **1993**, *36*, 4152; (b) Oprea, T. I.; Waller, C. L.; Marshall, G. R. *J. Med. Chem.* **1994**, *37*, 2206.
3. Doweyko, A. M. *J. Med. Chem.* **1994**, *37*, 1769-1778.
4. Farmerie, W. G.; Loeb, D. D.; Casavant, N. C.; Hutchinson, C. A., III; Edgel, M. H.; Swanstrom, R. *Science* **1987**, *236*, 305-308.
5. Kohl, N. E.; Emini, E. A.; Schleif, W. A.; Davis, L. J.; Heimbach, J. C.; Dixon, R. A. F.; Scolnick, E. M.; Sigal, I. *Proc. Nat. Acad. Sci. U.S.A.* **1988**, *85*, 4686.
6. Huff, J. R. *J. Med. Chem.* **1991**, *34*, 2305-2314.
7. Navia, M. A.; Fitzgerald, P. M. D.; McKeever, B. M.; Leu, C.-T.; Heimbach, J. C.; Herber, W. K.; Sigal, I. S.; Darke, P. L.; Springer, J. P. *Nature* **1989**, *337*, 615-620.
8. Foundling, S. I.; Cooper, J.; Watson, F. E.; Cleasby, A.; Pearl, L. H.; Sibanda, B. L.; Hemmings, A.; Wood, S. P.; Blundell, T. L.; Valler, M. J.; Norey, C. G.; Kay, J.; Boger, J.; Dunn, B. M.; Leckie, B. J.; Jones, D. M.; Atrash, B.; Hallet, A.; Szelke, M. *Nature* **1987**, *327*, 349-352.
9. Suguna, K.; Padlan, E. A.; Smith, C. W.; Carlson, W. D.; Davies, D. R. *Proc. Natl. Acad. Sci. USA* **1987**, *84*, 7009-7013.
10. Lyle, T. A.; Wiscount, C. M.; Guare, J. P.; Thompson, W. J.; Anderson, P. S.; Darke, P. L.; Zugay, J. A.; Emini, E. A.; Schleif, W. A.; Quintero, J. C.; Dixon, R. A. F.; Sigal, I. S.; Huff, J. R. *J. Med. Chem.* **1991**, *34*, 1228-1230.
11. Thompson, W. J.; Fitzgerald, P. M. D.; Holloway, M. K.; Emini, E. A.; Darke, P. L.; McKeever, B. M.; Schleif, W. A.; Quintero, J. C.; Zugay, J. A.; Tucker, T. J.; Schwering, J. E.; Homnick, C. F.; Nunberg, J.; Springer, J. P.; Huff, J. R. *J. Med. Chem.* **1992**, *35*, 1685-1701.
12. The latter protonation state was chosen based on pH rate profiles which suggest that the catalytic aspartates of the fungal aspartyl proteases Penicillopepsin and *Rhizopus* Pepsin and the HIV-1 protease share one negative charge. See (a) Hofmann, T.; Hodges, R. S.; James, M. N. G. *Biochemistry* **1984**, *23*, 635-643; (b) Hyland, L. J.; Tomaszek, T. A., Jr.; Meek, T. D. *Biochemistry* **1991**, *30*, 8454-8463.
13. (a) Halgren, T. A. *J. Am. Chem. Soc.* **1992**, *114*, 7827-7843; (b) Holloway, M. K.; Wai, J. M.; Halgren, T. A.; Fitzgerald, P. M. D.; Vacca, J. P.; Dorsey, B. D.;

Levin, R. B.; Thompson, W. J.; Chen, L. J.; deSolms, S. J.; Gaffin, N.; Ghosh, A. K,; Giuliani, E. A.; Graham, S. L.; Guare, J. P.; Hungate, R. W.; Lyle, T. A.; Sanders, W. M.; Tucker, T. J.; Wiggins, M.; Wiscount, C. M.; Woltersdorf, O. W.; Young, S. D.; Darke, P. L.; Zugay, J. A. *J. Med. Chem.*. in press.
14. (a)Allinger, N. L. *J. Am. Chem. Soc.* **1977**, *99*, 8127. (b) Burkert, U.; Allinger, N. L. *Molecular Mechanics*; American Chemical Society: Washington, DC, 1982.
15. IC_{50} values reported in this manuscript were determined following the protocol described in Heimbach, J. C.; Garsky, V. M.; Michelson, S. R.; Dixon, R. A. F.; Sigal, I. S.; Darke, P. L. *Biochem. Biophys. Res. Commun.* **1989**, *164*, 955-960.
16. Fitzgerald, P. M. D.; McKeever, B. M.; VanMiddlesworth, J. F.; Springer, J. P.; Heimbach, J. C.; Leu, C.-T.; Herber, W. K.; Dixon, R. A. F.; Darke, P. L. *J. Biol. Chem.* **1990**, *265*, 14209-14219.
17. (a) Erickson, J.; Neidhart, D. J.; VanDrie, J.; Kempf, D, J.; Wang, X. C.; Norbeck, D. W.; Plattner, J. J.; Rittenhouse, J. W.; Turon, M.; Widburg, N.; Kohlbrenner, W. E.; Simmer, R.; Helfrich, R.; Paul, D. A.; Knigge, M. *Science* **1990**, *249*, 527-533; (b) Humber, D. C.; Cammack, N.; Coates, J. A. V.; Cobley, K. N.; Orr, D. C.; Storer, R.; Weingarten, G. G.; Weir, M. P. *J. Med. Chem.* **1992**, *35*, 3080-3081; (c) Jadhav, P. K.; Woerner, F. J. *Bioorg. Med. Chem. Letters* **1992**, *2*, 353-356; (d) Spaltenstein, A.; Leban, J. J.; Furfine, E. S. *Tet. Lett.* **1993**, *34*, 1457-1460; (e) Dreyer, G. B.; Boehm, J. C.; Chenera, B.; DesJarlais, R. L.; Hassell, A. M.; Meek, T. D.; Tomaszek, T. A., Jr. *Biochem.* **1993**, *32*, 937-947.
18. Bone, R. F.; Vacca, J. P.; Anderson, P. S.; Holloway, M. K. *J. Am. Chem. Soc.* **1991**, *113*, 9382-9384.
19. Vacca, J. P.; Fitzgerald, P. M. D.; Holloway, M. K.; Hungate, R. W.; Starbuck, K. E.; Chen, L. J.; Darke, P. L.; Anderson, P. S.; Huff, J. R. *Biorg. Med. Chem. Lett.* **1994**, *4*, 499-504.
20. Thompson, W. J.; Ghosh, A. K.; Holloway, M. K.; Lee, H. Y.; Munson, P. M.; Schwering, J. E.; Wai, J.; Darke, P. L.; Zugay, J.; Emini, E. A.; Schleif, W. A.; Huff, J. R.; Anderson, P. S. *J. Am. Chem. Soc.* **1993**, *115*, 801-803.
21. Vacca, J. P.; Dorsey, B. D.; Schleif, W. A.; Levin, R. B.; McDaniel, S. L.; Darke, P. L.; Zugay, J.; Quintero, J. C.; Blahy, O. M.; Roth, E.; Sardana, V. V.; Schlabach, A. J.; Graham, P. I.; Condra, J. H.; Gotlib, L.; Holloway, M. K.; Lin, J.; Chen, I.-W.; Vastag, K.; Ostovic, D.; Anderson, P. S.; Emini, E. A.; Huff, J. R. *Proc. Natl. Acad. Sci. USA* **1994**, *91*, 4096-4100.
22. CHARMm Version 21.1.7b
23. Available from Molecular Simulations Inc., 16 New England Executive Park, Burlington, MA 01803-5297
24. Available from W. Clark Still, Department of Chemistry, Columbia University, New York, NY 10027.
25. Still, W. C.; Tempczyk, A.; Hawley, R. C.; Hendrickson, T. *J. Am. Chem. Soc.* **1990**, *112*, 6127-6129.

RECEIVED October 11, 1994

Chapter 4

From Maps to Models

A Concerted Computational Approach to Analysis of the Structure–Activity Relationships of Amiloride Analogues

Carol A. Venanzi[1], Ronald A. Buono[1], William J. Skawinski[1],
Thomas J. Busanic[1], Thomas J. Venanzi[2], Randy J. Zauhar[3,4],
and Victor B. Luzhkov[5]

[1]Department of Chemical Engineering, Chemistry, and Environmental Science, New Jersey Institute of Technology, Newark, NJ 07102
[2]Department of Chemistry, College of New Rochelle, New Rochelle, NY 10805
[3]Biotechnology Institute and Department of Molecular and Cell Biology, 519 Wartik Laboratory, Pennsylvania State University, University Park, PA 16802

In order to interpret structure-activity data on the binding of amiloride analogues to the sodium channel in the frog skin, work in this laboratory has been directed towards a concerted computational study of amiloride and its analogues using quantum mechanics, molecular dynamics, static solvation model studies, and innovative model-building with stereolithography. Based on molecular electrostatic potentials of analogues with pyrazine ring modifications, a pharmacophore has been identified for analogues which form a stable blocking complex with the channel. The validity of this pharmacophore was tested and found to hold for two analogues with sidechain elongations. Quantum mechanical conformational analysis and molecular dynamics and static solvation studies of amiloride were carried out to determine the degree of planarity of amiloride as well as the relative energy of various conformers in solution. The A1 free base conformer (OCCN=180°) was found to be more stable than the A4 (OCCN=0°) in solution. This sheds light on earlier NMR studies which were unable to distinguish between the two conformers in solution. The protonated species of amiloride was found on the average to be planar in solution. This has important implications for its mode of binding to proteins and nucleic acids.

[4]Current address: Tripos Associates, Inc., St. Louis, MO 63144
[5]Current address: Institute of Chemical Physics, Chernogolovka, Moscow Region, Russia 142432

0097–6156/95/0589–0051$12.00/0
© 1995 American Chemical Society

Introduction: Structure-Activity Studies of Amiloride Analogues.

Amiloride, 1, a novel acylguanidinium diuretic, has been shown to be a potent inhibitor of sodium transport in a variety of cellular and epithelial transport systems *(1-4)*. In particular, amiloride has been used to probe the mechanism of taste transduction in lingual epithelia *(5-19)*. Since the molecular structure of the epithelial sodium channel is not known, information on the amiloride binding site on the ion channel protein must be intuited from structure-activity studies of amiloride analogues. For example, Cuthbert has pointed out the importance of the guanidinium group for blocking epithelial sodium channels *(20)*. Li et al. *(21,22)* have used noise analysis of the stationary sodium current transversing frog skin to study the efficacy of amiloride analogues with pyrazine ring modifications and/or guanidinium sidechain modifications to block sodium transport. Some of their data is summarized in Table I. The numbers in parentheses are the pK_a-corrected values which are based on only the concentration of the protonated species. The data correlate changes in the substituent at the 5- or 6-position of the pyrazine ring or changes in the sidechain to differences in the microscopic association constant, k_{on}, and the dissociation constant, k_{off}. The table shows that substitution of other halogens or hydrogen for chlorine has little effect on k_{on}, but a considerable effect on k_{off}. A relatively low value for k_{off}, and a consequently long block time, indicates that the analogue interacts with the channel binding site for a relatively long time, forming a stable complex with the channel that prevents sodium transport through the channel (i.e. a blocking complex). Analogue 5, with a hydrogen at position 6, forms the least stable blocking complex with the channel protein. Substitution at position 5 with either a hydrogen (analogue 6) or chlorine, 7, affects both k_{on} and k_{off}, with 7 forming the poorer blocking complex with the channel. Of the two molecules with elongated sidechains, analogue 18 has a pK_a-corrected k_{on} similar to amiloride, while that of 19 is much smaller. This suggests *(22)* that they have a different type of initial molecular interaction with the channel. The k_{off} values indicate that 19 is a slightly better blocker than amiloride, while 18 is worse. Li, et al. *(21,22)*, building on the plug-type model of Cuthbert *(23)*, have suggested a two-step model for the analogue-channel interaction: (1) The guanidinium sidechain invades the channel entrance and interacts with an anionic site to form an encounter complex; (2) Then, either no block occurs and the molecule is released, or the substituent at the 6-position of the pyrazine ring binds to an electropositive site on the channel, forming a stable blocking complex.

Molecular Electrostatic Potential (MEP) Maps. In order to interpret this data at the molecular level, we have initiated a step-wise computational approach to the problem *(24-26)*. For analogues 1-7 in Table I, we carried out the following procedure using ab initio quantum mechanics: (1) geometry optimization of the free base and protonated conformers of amiloride in order to determine which conformer in each species was the most stable and should, therefore, be used for the subsequent MEP study. This data also provided the basic structural template for the other pyrazine ring analogues; (2) calculation of the molecular electrostatic potential maps of the A1 free base form (OCCN=180°) of 1 in order to determine the site of protonation (the other pyrazine ring analogues were assumed to be protonated in the same position); (3) calculation of the molecular electrostatic potential maps of the protonated forms of 1-7 in order to determine features important to the formation of the encounter complex; (4) construction of a model encounter complex (formate anion in a chelate-type orientation with the guanidinium group of the analogue) and calculation of the molecular electrostatic potential maps of this complex in order to determine the steric and electrostatic features of the analogues (i.e. the pharmacophore) important to the formation of a stable blocking complex with the ion channel. Details of the calculations are given in the original papers *(24,25)*. An overview of the work, the assumptions implicit in this approach, and the computational methods employed is

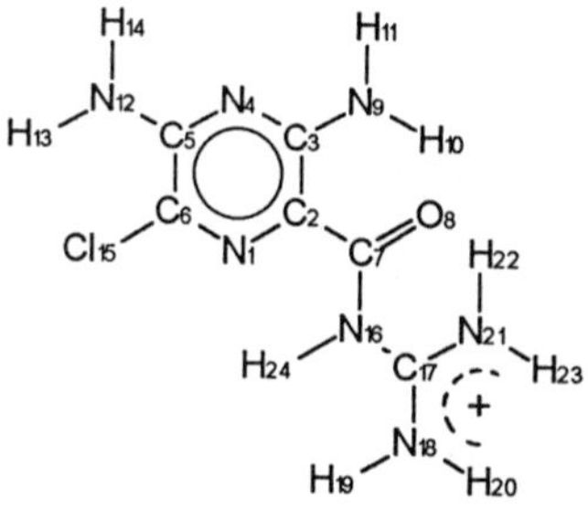

1

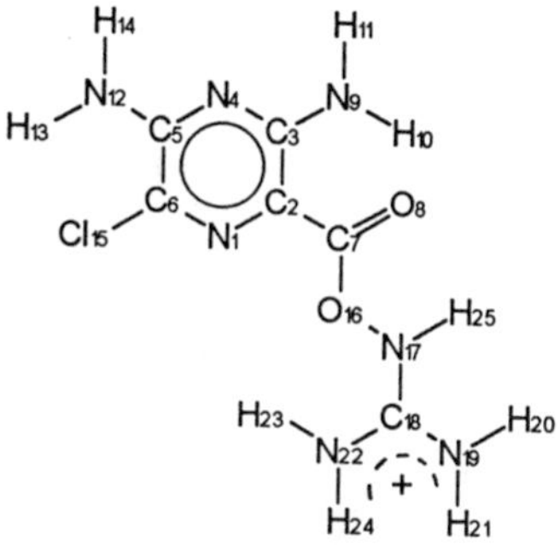

18

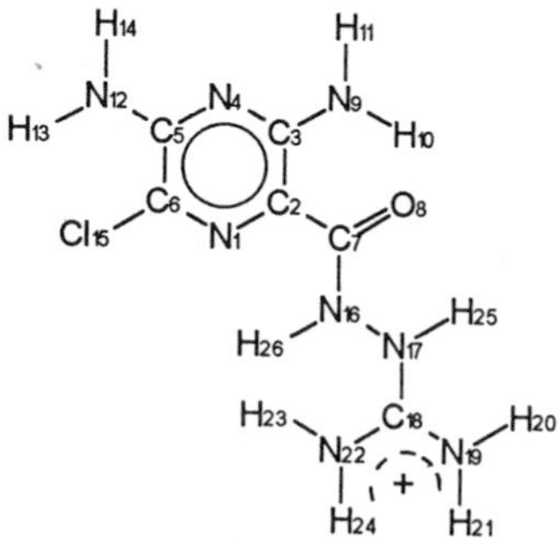

19

Table I. Structure-Activity Data for Binding to Sodium Channel in Frog Skin

Analogue	Substitution at Position		k_{on}[a]	k_{off}[b]	Block time[c]
	5	6			
1	-NH$_2$	-Cl	13.17±0.25	3.93±0.19	255
2	-NH$_2$	-Br	14.19±1.09	5.58±0.92	179
3	-NH$_2$	-I	11.43±0.90	17.41±0.40	57
4	-NH$_2$	-F	13.54±0.65	32.20±1.57	31
5	-NH$_2$	-H	14.47±0.68	176.25±17.73	6
6	-H	-Cl	3.32±0.44 (3.42)	10.89±1.35	92
7	-Cl	-Cl	5.16±0.46	151.10±16.48	7

Analogue	Elongation with	k_{on}[a]	k_{off}[b]	Block time[c]
18	-O	1.22±0.07 (13.4)	20.67±3.72	48
19	-NH-	2.16±0.11	3.41±0.55	293

SOURCE: adapted from refs. 21 and 22.
[a]In units s^{-1}μM^{-1}.
[b]In units s^{-1}.
[c]In units ms.

given in a recent review *(26)*. For completeness, the assumptions are summarized here:

(1) The receptor provides a binding site which is complementary in both molecular shape and molecular electrostatic potential to that of the analogue.

(2) The most potent analogue provides the steric and electrostatic template for the pharmacophore.

(3) Electrostatic forces are dominant in directing the analogue into the receptor site.

(4) Low energy conformers of the analogue, not just the global energy minimum conformation calculated in the gas phase, should be considered as templates for the pharmacophore.

Although this approach has some features in common with the Active Analogue Approach of Marshall*(27-30)*, it is, in contrast, not a traditional QSAR analysis.

Analysis of the MEP maps of the protonated species indicated that those analogues that have k_{on} values similar to amiloride have strong, distinguishing minima in the MEP pattern off the carbonyl oxygen, off N_4, and off the amino groups at positions 3 and 5 of the pyrazine ring. Analogues which have k_{on} values which differ from amiloride lack two or more of these features and exhibit a much more positive pattern over the pyrazine ring. This can be seen, for example, by comparison of the MEP maps of protonated amiloride, **1** (Figure 1a), **5** (Figure 1b), and **6** (Figure 1c).

Analysis of the MEP maps of the model encounter complexes indicated that:

(1) A stable blocking complex is formed with analogues that have a deep, localized minimum off the 6-position of the pyrazine ring;

(2) The stability of the blocking complex is directly related to the depth of the minimum;

(3) Substitution at position 5 affects not only the depth but also the location and size of the minimum off position 6;

(4) Steric factors may influence the optimal binding of the 6-position ligand to the ion channel.

Points (1)-(3) can be seen, for example, by comparison of the MEP maps of the model encounter complexes of **1** (Figure 2a), **5** (Figure 2b), and **6** (Figure 2c). Point (4) can be seen by comparison of the MEP maps of the encounter complexes of **2** (Figure 2d) and **3** (Figure 2e). The latter two maps are very similar and yet the values of k_{off} (and the block time) are different. Since iodine has a much larger van der Waals radius than bromine, this suggests that steric factors may also influence the binding to the ion channel.

Although Li and coworkers identified the guanidinium group and the substituent at the 6-position of the pyrazine ring as being implicated in binding, the unique feature of our work is that it provides additional information at the molecular level by relating the stability of the blocking complex to the size, depth, and location of the minimum in the molecular electrostatic potential off the 6-position. Taking amiloride as the template for a molecule which forms a stable blocking complex with the channel, it seems that the positions of the proton donors H_{20} and H_{23}, along with the position of the minimum off chlorine, may identify the relative spatial location of complementary sites on the channel. The above features can be incorporated into a pharmacophore to identify a molecule that could form a stable blocking complex with the sodium channel.

The Pharmacophore Hypothesis. There are several features to the amiloride pharmacophore:

(1) Strong, distinguishing minima in the MEP map off the carbonyl oxygen, and off positions 3, 4, and 5 of the pyrazine ring;

(2) A broad, positive MEP maximum localized over the sidechain;

(3) A deep, localized MEP minimum off the 6-position of the pyrazine ring;

(4) A fixed distance between the proton donors of the chelating guanidinium group and the minimum off position 6 of the pyrazine ring.

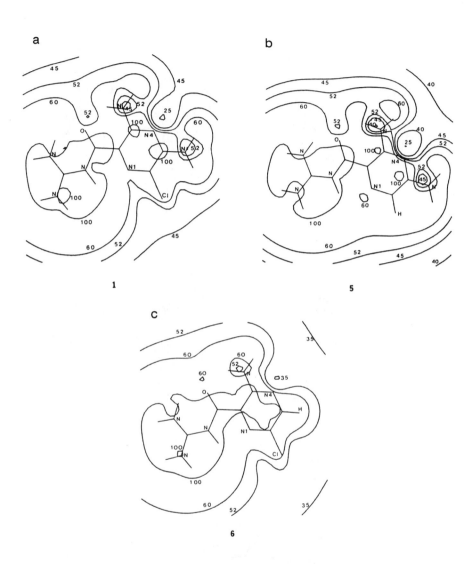

Figure 1. Molecular electrostatic potential maps of protonated analogues. 3-21G* basis set.   Contours in units of kcal/mol. a. **1**. b. **5**. c. **6**.   (Reproduced with permission from ref. *(25)*. Copyright 1992 American Chemical Society.)

Figure 2. Molecular electrostatic potential maps of encounter complexes. STO-3G basis set. Contours in units of kcal/mol. a. **1**. b. **5**. c. **6**. d. **2**. e. **3**. (Reproduced with permission from ref. *(25)*. Copyright 1992 American Chemical Society.)

Perhaps the most interesting question is: how can analogues **18** and **19**, which have elongated sidechains, fulfill condition (4) of the pharmacophore? If **18** and **19** are assumed to be planar, overlap of the guanidinium group of either **18** or **19** with that of amiloride shows that the chlorine atoms of **18** and **19** are located in a very different region of space than that of amiloride. Similarly, if the chlorines are overlapped, the chelating hydrogens of **18** and **19** are offset from those of amiloride. This seems to indicate that it might be difficult for **18** and **19** to orient in such a fashion as to simultaneously bind to the anionic site as well as to the electropositive site on the channel off chlorine. Our latest studies in this direction are described in the section on Testing the Pharmacophore Hypothesis. But first, solvation studies of amiloride were carried out in order to determine whether the pharmacophore hypothesis needed to be refined by consideration of the solvent effect on the structure and relative energy of the amiloride conformers.

Refinement of the Pharmacophore Hypothesis.

The amiloride pharmacophore was developed on the basis of gas phase conformational analysis and molecular electrostatic potential maps. The results of the conformational analysis using the 3-21G* basis set showed that, for both the free base and protonated species, the global energy minimum conformations (A1 and F1, respectively, each with OCCN=180°) were planar with hydrogen bonding between $O_8...H_{10}$ and $O_8...H_{22}$ *(24)*. For the free base species, A1 was shown to be 2.50 kcal/mol more stable than A4 (OCCN=0°) with a barrier to rotation around the $-C_2-C_7-$ bond of 19 kcal/mol *(24)*. The assumption was made that it was unlikely that substitution at the 5- or 6-position of the pyrazine ring would significantly affect the height of the barrier or the relative energy of the A1-like and A4-like conformers and, therefore, that the free base forms of the analogues would probably be more stable in the A1-like than the A4-like conformer. Since Dreiding models showed steric repulsion between the pyrazine ring amino hydrogen, H_{10}, and the sidechain proton, H_{24}, in the F4 conformer (OCCN=0°), all the protonated analogues were assumed to bind in an F1-like conformation. The molecular electrostatic potential analysis was carried out on the analogues in the planar A1-like and F1-like conformers and the resulting maps seemed useful in interpreting the structure-activity data. However, two questions arose which needed to be answered in order to decide whether the pharmacophore should be refined:

(1) Could solvent disrupt the intramolecular hydrogen bonding pattern noted in these conformers, thereby stabilizing nonplanar conformers which might be involved in binding to the channel? Since the analogue-channel interaction itself cannot be modeled due to lack of data on the molecular structure of the channel protein, the water-analogue interaction may serve as an indicator of the degree to which nonplanar conformers of the analogue can be stabilized by interaction with the surrounding environment.

(2) Could solvent affect the relative ordering of the energies calculated in the gas phase? In particular, would solvent stabilization result in A4, the conformer with the larger dipole moment, being more stable than A1? The NMR studies of Smith et al. *(31)* were unable to distinguish between these two conformers in solution.

In order to investigate these questions, we carried out molecular dynamics and static solvation studies of the free base and protonated conformers of amiloride. The details of the calculations are given in a recent publication *(32)* and are summarized in another ACS Symposium volume *(33)*. Only the most relevant points will be given below.

Torsional Studies.
As with the free base conformers, gas phase 3-21G* conformational analysis was carried out on protonated amiloride, **1**, in order to determine the barrier to rotation around the $-C_2-C_7-$ bond and to provide data for parameterization of the relevant torsional potential function in the GROMOS *(34)* molecular dynamics package. The study showed that the energy of the protonated

species rises slowly from 0 to 4 kcal/mol as the OCCN torsional angle changes from $180°$ to $160°$. Beyond this region, the energy increases steeply and monotonically, reaching a maximum of 33 kcal/mol at the F4 (OCCN=$0°$) conformer due to steric repulsion between H_{10} and H_{24}. This indicates that, although F1 (shown as structure **1**) is the conformer of lowest energy, there are other nonplanar conformers within $\pm20°$ of F1 which may be accessed with little expenditure of energy and could be significantly stabilized by interaction with solvent.

Solvation Studies. Molecular dynamics simulations and static solvation studies were carried out on the free base and protonated conformers of amiloride. The static solvation models included the Langevin Dipole (LD) technique *(35-38)*, which treats water in an inner region as polarizable point dipoles and as a continuum in an outer region, as well as the Induced Polarization Charge Boundary Element (IPCBE) Method *(39-43)* and the Self-Consistent Reaction Field (SCRF) Method *(44-50)*, both of which treat water as a continuum. Although all the techniques allow for geometry optimization of the solute in the presence of the solvent, in order to avoid problems with differences in the basis sets or force fields, the LD, IPCBE, and SCRF methods were applied to the fixed, 3-21G*-optimized gas phase geometries.

    Molecular Dynamics Simulations. Molecular dynamics simulations of 30 ps in length were carried out for the A1, A4, and F1 conformers. Comparison of the internal energy for A1 and A4 shows that A1 is more stable than A4 by 3.4 kcal/mol, in agreement with the gas phase results of 2.5 kcal/mol. On the other hand, the A4-water interaction energy term is 1.4 kcal/mol more stable than the A1-water interaction term. However, summation of these two terms predicts A1 to be more stable in solution than A4 by 2.0 kcal/mol. Analysis of the OCCN torsional angle for the A1, A4, and F1 trajectories shows that it tends to vary by less than $\pm10°$ from planarity in each case. So the molecular dynamics results support the use of the planar A1-like and F1-like conformers for calculation of the electrostatic potential maps and indicate that no refinement of the pharmacophore is required from the standpoint of solvent stabilization of nonplanar conformers.

    Static Solvation Models. The results of the three static solvation models support the findings of the MD simulations that A1 is more stable in water than A4. The LD and IPCBE methods were used to study all the conformers from the torsional barrier study for both the free base and protonated species. For both techniques, three sets of atomic point charges were used: 3-21G* charges adjusted to the GROMOS charge group concept *(34)*, 3-21G* Mulliken charges, and 3-21G* potential-derived charges. The electrostatic contribution to the hydration free energy was calculated to be very similar in both techniques for each conformer. When the gas phase relative energy difference was added to the difference in hydration free energy for A1 and A4, both methods predicted the A1 conformer to be more stable than A4 by about 0.4-1.8 kcal/mol, depending on the atomic charge set. (The only exception was the LD method with the 3-21G* Mulliken charge set which predicted A4 to be more stable by 0.1 kcal/mol.) In a separate calculation using the AM1 Hamiltonian incorporated into the LD method, the results showed A1 to be more stable than A4 by 2.1 kcal/mol. The SCRF method, using the 3-21G* basis set, also predicted the A1 conformer to be more stable than A4 by 1.1 kcal/mol. In the absence of definitive experimental data, the agreement of four disparate solvent models, ranging from discrete water molecules to a continuum, seems to indicate that solvent has little effect on the planarity and relative energy ordering of the conformers and that the assumptions of the gas phase study are valid.

Testing the Pharmacophore Hypothesis on Amiloride Analogues with Sidechain Elongations.

As the high barrier for rotation around the primary torsional angle involving the $-C_2-C_7-$ bond in both the free base and protonated species of amiloride shows, there is considerable conjugation between the pyrazine ring and acylguanidinium sidechain. In analogues **18** and **19**, the insertion of -O- or -NH-, respectively, between the carbonyl and guanidinium groups may affect the degree of conjugation and allow the molecules to adopt conformations in which the guanidinium group is nonplanar with respect to the pyrazine ring. Since it is clear that planar **18** and **19** cannot fit the amiloride pharmacophore, investigation of alternative conformers may provide an indication of why **19** in particular forms a more stable blocking complex with the channel than amiloride.

Conformational Analysis. The 3-21G* and 6-31G* basis sets were used with the GAUSSIAN92 *(51)* program to study rotation around the secondary torsional angle involving the $-C_7-X_{16}-$ bond and the tertiary torsional angle involving the $-X_{16}-N_{17}-$ bond, where X=O for **18** and X=N for **19** (Skawinski, W.J. and Venanzi, C.A., manuscript in preparation). Although a thorough analysis of the conformational potential energy surface has not yet been completed, preliminary results seem to indicate that, for the 6-31G* basis set, the global energy minimum for **18** occurs at

$N_{17}-O_{16}-C_7-O_8 = 3.5°$
$C_{18}-N_{17}-O_{16}-C_7 = 142.9°$
$H_{25}-N_{17}-O_{16}-C_7 = -5.2°$

and for **19** at

$N_{17}-N_{16}-C_7-O_8 = -17.3°$
$C_{18}-N_{17}-N_{16}-C_7 = 95.3°$
$H_{25}-N_{17}-N_{16}-C_7 = -52.3°$.

Figure 3 shows that the global energy minimum conformers of both **18** and **19** are nonplanar structures, with **19** being significantly more out-of-plane. Superposition of **18** or **19** onto amiloride by fitting the chlorine atom and the chelating hydrogens (H_{21} and H_{24} of **18** and **19**) to the chlorine atom and H_{23} and H_{20} of amiloride shows that the bent, nonplanar structures of **18** and **19** can indeed fulfill the spatial requirements of condition (4) of the pharmacophore, although as a result, neither the guanidinium groups nor the pyrazine rings of the fitted partners are coplanar.

Molecular Electrostatic Potential Maps. In order to investigate whether **18** and **19** could reproduce the other features of the pharmacophore, particularly condition (3), molecular electrostatic potential maps of the analogue-formate anion complex were calculated in the STO-3G basis set using the SPARTAN 3.0 program *(52)* and compared to the STO-3G maps calculated for the pyrazine ring analogues *(25)*. The results showed a similarity in all the features of the maps of **18**, **19**, and amiloride, especially in the location of a deep, localized MEP minimum off the 6-position of the pyrazine ring (Skawinski, W.J., Busanic, T., and Venanzi, C.A., manuscript in preparation). This begins to explain why **19**, in particular, is as good a blocker as amiloride. But further study of the maps and the molecular structures is needed in order to interpret the difference in the blocking capabilities of **18** and **19**.

New Directions in Modeling: Stereolithography.

Since the molecular structure of the amiloride binding site on the channel protein is not known, one way to intuit information about the site is to assume that it is complementary in molecular shape and molecular electrostatic potential pattern to that of the most potent analogue(s). Amiloride and **19** form the most stable blocking complexes with the channel and we have shown above how **19** can fit important features of the amiloride pharmacophore. However, **19** is nonplanar whereas

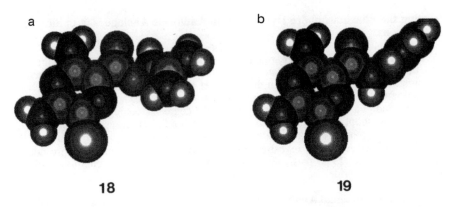

Figure 3. Gas phase global energy minimum conformers, 6-31G* basis set. a. **18**. b. **19**.

amiloride is planar. What may this tell us about the binding site? Is it an open cavity where all that really matters is that a potent analogue be able to fit the spatial requirements of the three-point binding involving the chelating hydrogens and the chlorine? Since amiloride appears to bind in a planar orientation to DNA *(53)*, binding in a sequence-selective fashion to sites rich in adenine and thymine residues, does this give us some clues about the nature of amiloride binding to the ion channel? Is it important that the guanidinium group and or the pyrazine ring form some type of stacking complex with the amino acid residues of the channel protein? On the other hand, it is known from analysis of protein crystal structures *(54-58)* that, although the chelate-type orientation between the carboxylate moiety of Asp and the guanidinium group of Arg is the most common type of molecular interaction between the two functional groups, other orientations are possible. Perhaps **19** interacts with the anionic site on the channel using a pair of hydrogens other than H_{21} and H_{24}?

In order to investigate these issues further, in the absence of explicit knowledge of the molecular structure of the ion channel, we are using the technique of stereolithography *(59)* to build plastic models of the global energy minimum conformers of amiloride analogues and of their complementary molecular shapes. We have also applied the technique to other molecules of biological interest, such as cyclodextrins and novel amino acid analogues *(60,61)*. The first step in the stereolithography procedure is to process the molecular coordinates so that the solid volume is "sliced" into a series of thin layers. This information is passed on to the stereolithography apparatus, which consists of a cube-shaped vessel filled with a liquid polymer. Within this vessel is a movable platform whose height is computer-controlled. At the start of the process, the platform is at a maximum height within the vessel so that a thin layer of liquid polymer lies above its surface. This liquid polymer is cured by interaction with ultraviolet light to form a solid material. A computer-controlled ultraviolet laser draws the image of the first slice of the model onto the thin liquid film, producing a solid model of the slice. Then the platform moves down one step, resulting in a thin layer of liquid being positioned above the first slice produced. The laser then draws the next layer of the model which becomes bonded to the first and the process is repeated until the entire model has been fabricated layer-by-layer. The model is partially cured in the apparatus and then moved to an ultraviolet oven for final curing. The advantage of this technique is that it can not only produce a model of a *calculated* molecular structure, but it can also produce a model of the shape that is complementary to that molecular structure, i.e. the "footprint" left behind by the molecular shape in the plastic. For example, Figure 4 shows how the model of **19** in

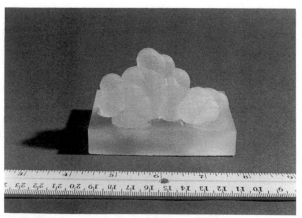

Figure 4. Stereolithography model of the 6-31G* gas phase global energy minimum conformer of **19** in the amiloride complementary shape. (Photo credit: Bill Wittkop.)

its nonplanar, gas phase   global energy minimum conformation can fit into the amiloride complementary shape with its chelating hydrogens and chlorine atom aligned at the locations where those of amiloride would be found.   Although the pyrazine ring of **19** is clearly not coplanar with the complementary shape of the amiloride pyrazine ring, the models show that **19** can fit the steric requirements of the amiloride pharmacophore. The rigid plastic model of the amiloride complementary molecular shape does not take into account the dynamical motion of a protein binding site and is not meant to be a realistic model of the receptor site. However, it gives a first approximation to the spatial requirements that define the amiloride pharmacophore and, in combination with the MEP and superposition studies described above, leads to some understanding of how **19**, which has a global energy minimum conformer so different from that of amiloride, could be just as effective as a sodium channel blocker. Ultimately the models of the analogues will be encoded with information on their molecular electrostatic potentials and will be used to intuit the complementary MEP of the site.

Future Studies of Amiloride Analogues.

Completion of the conformational analysis of **18** and **19** may reveal that for each analogue there is a wide range of torsional angles which define conformers with relatively low energy. In that case, it is possible that the analogues may bind in one of these conformations, rather than the global energy minimum structure. Therefore, the MEP, superposition, and stereolithography analysis described above will be applied to these conformers, as well, in order to attempt to determine the difference in the efficacy of **18** and **19** as sodium channel blockers.

Li et al. *(21)* have studied many other analogues with sidechain elongations. We plan to investigate these as well, using the concerted computational approach described here, in order to obtain a better understanding at the molecular level of the structure-activity relationships of amiloride analogues.

Acknowledgments. This work was funded by grants to C.A.V. from the National Science Foundation, the New Jersey Commission on Science and Technology, the Campbell Institute for Research and Technology, the National Academy of Sciences Cooperation in Applied Science and Technology program, and by generous grants of computer time from New Jersey Institute of Technology and the Pittsburgh Supercomputing Center.

Literature Cited.

(1)   Garty, H.; Benos, D. J. *Physiological Rev.* **1988**, *68*, 309.
(2)   Benos, D. J. In *Na+/H+ Exchange*; S. Grinstein, Ed.; CRC: Boca Raton, 1988; pp 121.
(3)   *Cation Transport Probes: The Amiloride Series*; Kleyman, T. R.; Cragoe, E. J., Jr., Ed.; Academic: New York, 1990; Vol. 191, pp 739.
(4)   Kleyman, T. R.; Cragoe, E. J., Jr. *J. Membrane Biol.* **1988**, *105*, 1.
(5)   DeSimone, J. A.; Heck, G. L.; DeSimone, S. K. *Science* **1981**, *214*, 1039.
(6)   Schiffman, S. S.; Lockhead, E.; Maes, F. W. *Proc. Natl. Acad. Sci. USA* **1983**, *80*, 6136.
(7)   DeSimone, J. A.; Heck, G. L.; Mierson, S.; DeSimone, S. K. *J. Gen. Physiol.* **1984**, *83*, 633.
(8)   Heck, G. L.; Mierson, S.; DeSimone, J. A. *Science* **1984**, *223*, 403.
(9)   DeSimone, J. A.; Ferrell, F. *Physiol. Soc.* **1985**, R52.
(10)   Schiffman, S. S.; Simon, S. A.; Gill, J. M.; Beeker, T. G. *Physiol. Behav.* **1986**, *36*, 1129.
(11)   Hellekant, G.; DuBois, G. E.; Roberts, T. W.; Wel, H. v. d. *Chemical Senses* **1988**, *13*, 89.
(12)   Heck, G. L.; Persaud, K. C.; DeSimone, J. *Biophys. J.* **1989**, *55*, 843.
(13)   DeSimone, J. A.; Heck, G. L.; Persuad, K. C.; Mierson, S. *Chemical Senses* **1989**, *1*, 13.
(14)   Avenet, P.; Lindemann, B. *J. Membrane Biol.* **1988**, *105*, 245.
(15)   Desor, J. A.; Finn, J. *Chemical Senses* **1989**, *14*, 793.
(16)   Schiffman, S. S.; Frey, A. E.; Suggs, M. S.; Cragoe, E. J., Jr.; Erickson, R. P. *Physiol. Behav.* **1990**, *47*, 435.
(17)   Schiffman, S. S.; Suggs, M. S.; Cragoe, E. J., Jr.; Erickson, R. P. *Physiol. Behav.* **1990**, *47*, 455.
(18)   Ye, Q.; Heck, G. L.; DeSimone, J. *Science* **1991**, *254*, 724.
(19)   Tennissen, A. M. *Physiol. Behav.* **1992**, *51*, 1061.
(20)   Cuthbert, A. W. *Molecular Pharmacol.* **1976**, *12*, 945.
(21)   Li, J. H.-Y.; Cragoe, E. J., Jr.; Lindemann, B. *J. Membrane Biol.* **1987**, *95*, 171.
(22)   Li, J. H.-Y.; Cragoe, E. J., Jr.; Lindemann, B. *J. Membrane Biol.* **1985**, *83*, 45.
(23)   Cuthbert, A. W. *Experientia* **1976**, *32*, 1321.
(24)   Venanzi, C. A.; Plant, C.; Venanzi, T. J. *J. Comput. Chem.* **1991**, *12*, 850.
(25)   Venanzi, C. A.; C.Plant; Venanzi, T. J. *J. Med. Chem.* **1992**, *35*, 1643.
(26)   Venanzi, C. A.; Venanzi, T. J. In *Mechanisms of Taste Transduction*; S. A. Simon and S. D. Roper, Ed.; CRC: Boca Raton, 1993; pp 428.
(27)   Waller, C. L.; Marshall, G. R. *J. Med. Chem.* **1993**, *36*, 2390.
(28)   DePriest, S. A.; Mayer, D.; Naylor, C. B.; Marshall, G. R. *J. Am. Chem. Soc.* **1993**, *115*, 5372.
(29)   Marshall, G. R. In *Proceedings of the 6th International Symposium on Medicinal Chemistry*; M. A. Simkins, Ed.; Cotswold: Oxford, England, 1979; pp 225.

(30)   Marshall, G. R.; Barry, C. D.; Bosshard, H. E.; Dammkoehler, R. A.; Dunn, D. A. In *ACS Symposium Series, Computer-Aided Drug Design*American Chemical Society: Washington, DC, 1979; Vol. 112; pp 205.

(31)   Smith, R. L.; Cochran, D. W.; Gund, P.; Cragoe, E. J., Jr. *J. Am. Chem. Soc.* **1979**, *101*, 191.

(32)   Buono, R. A.; Venanzi, T. J.; Zauhar, R. J.; Luzhkov, V. B.; C.A.Venanzi *J. Am. Chem. Soc.* **1994**, *116*, 1502.

(33)   Venanzi, C. A.; Buono, R. A.; Luzhkov, V. B.; Zauhar, R. J.; Venanzi, T. J. In *Structure and Reactivity in Aqueous Solution: Characterization of Chemical and Biological Systems*; C. J. Cramer and D. G. Truhlar, Ed.; American Chemical Society: Washington, D.C., 1994; pp 260.

(34)   *GROMOS (1987)*, van Gunsteren, W. F.; Berendesen, H. J. C., Biomos, B.V.: Groningen, The Netherlands.

(35)   Warshel, A.; Russel, S. T. *Quart. Rev. Biophys.* **1984**, *17*, 283.

(36)   Russel, S. T.; Warshel, A. *J. Mol. Biol.* **1985**, *185*, 389.

(37)   Luzhkov, V.; Warshel, A. *J. Comput. Chem.* **1992**, *13*, 199.

(38)   Lee, F. S.; Chu, Z. T.; Warshel, A. *J. Comput. Chem.* **1993**, *14*, 161.

(39)   Fox, T.; Roesch, N.; Zauhar, R. J. *J. Comput. Chem.* **1993**, *14*, 253.

(40)   Zauhar, R. J.; Morgan, R. S. *J. Mol. Biol.* **1985**, *186*, 815.

(41)   Zauhar, R. J.; Morgan, R. S. *J. Comput. Chem.* **1988**, *9*, 171.

(42)   Zauhar, R. J.; Morgan, R. S. *J. Comput. Chem.* **1990**, *11*, 603.

(43)   Zauhar, R. J. *J. Comput. Chem.* **1991**, *12*, 575.

(44)   Wong, M. W.; Frisch, M. J.; Wiberg, K. B. *J. Am. Chem. Soc.* **1991**, *113*, 4776.

(45)   Wong, M. W.; Wiberg, K. B. *J. Chem. Phys.* **1991**, *95*, 8991.

(46)   Wong, M. W.; Wiberg, K. B.; Frisch, M. J. *J. Am. Chem. Soc.* **1992**, *114*, 523.

(47)   Wong, M. W.; Wiberg, K. B.; Frisch, M. J. *J. Am. Chem. Soc.* **1992**, *114*, 1645.

(48)   Wong, M. W.; Wiberg, K. B.; Frisch, M. J. *J. Am. Chem. Soc.* **1993**, *115*, 1078.

(49)   Cieplak, A. S.; Wiberg, K. B. *J. Am. Chem. Soc.* **1992**, *114*, 9226.

(50)   Wiberg, K. B.; Wong, M. W. *J. Am. Chem. Soc.* **1993**, *115*, 1078.

(51)   *Gaussian92, Release A*, Frisch, M. J.; Trucks, G. W.; Head-Gordon, M.; Gill, P. M. W.; Wong, M. W.; Foresman, J. B.; Johnson, B. G.; Schlegel, H. B.; Robb, M. A.; Replogle, E. S.; Gompers, R.; Andres, J. L.; Raghavachari, K.; Binkley, J. S.; Gonzalez, C.; Martin, R. L.; Fox, D. J.; Defrees, D. J.; Baker, J.; Stewart, J. J. P.; Pople, J. A., Gaussian, Inc.: Pittsburgh, PA.

(52)   *Spartan 3.0*, Hehre, W., Wavefunction, Inc.: Irvine, CA.

(53)   Bailly, C.; Cuthbert, A. W.; Gentle, D.; Knowles, M. R.; Waring, M. J. *Biochemistry* **1993**, *32*, 2514.

(54)   Salunke, D. M.; Vijayan, M. *Int. J. Peptide Protein Res.* **1981**, *18*, 348.

(55)   Tintelnot, M.; Andrews, P. *J. Comp.-Aided Mol. Des.* **1989**, *3*, 67.

(56)   Singh, J.; Thornton, J. M.; Snarey, M.; Campbell, S. F. *FEBS Letters* **1987**, *224*, 161.

(57)   Eggleston, D. S.; Hodgson, D. J. *Int. J. Peptide Protein Res.* **1985**, *25*, 242.

(58)   Milburn, M. W.; Privé, G. C.; Milligan, D. L.; Scott, W. G.; Yeh, J.; Jancarik, J.; Koshland, D. E., Jr.; Kim, S.-H. *Science* **1991**, *254*, 1342.

(59)   Kochan, D. *Computers in Industry* **1992**, *20*, 133.

(60)   Skawinski, W. J.; Busanic, T. J.; Ofsievich, A. D.; Venanzi, T. J.; Luzhkov, V. B.; Venanzi, C. A. *Information Technologies and Disabilities* **1994**, *1*, article 6.

(61)   Skawinski, W. J.; Busanic, T. J.; Ofsievich, A. D.; Venanzi, T. J.; Luzhkov, V. B.; Venanzi, C. A. *J. Mol. Graphics,* in press.

RECEIVED October 11, 1994

Chapter 5

De Novo Design
Ligand Construction and Prediction of Affinity

Tudor I. Oprea, Chris M. W. Ho, and Garland R. Marshall

Center for Molecular Design, Washington University, Lopata 510, One Brookings Drive, St. Louis, MO 63130

The availability of a therapeutic target of known three-dimensional structure challenges the computational chemist to design and predict the affinity of novel ligands. A set of computational tools (CAVITY, FOUNDATION, SPLICE, DBMAKER, FLOOD) have been developed to discover sets of molecular fragments which optimize binding in subdomains of the active site. These are then spliced together combinatorially and optimized in size to fit within the active site cavity. A crucial aspect is the method of predicting the affinity so that compounds can be prioritized for retrosynthetic analysis and synthesis. We have used CoMFA, a 3D QSAR paradigm, to develop a predictive model for HIV protease. The robustness of this model was confirmed by a different 3D-QSAR method (HASL), and has good predictive and explanatory power. Novel ligands for HIV protease have been designed subject to chemical constraints which minimize their synthetic difficulty. Their relative affinities have been predicted using the CoMFA model and the series is undergoing synthesis and binding assays.

Receptors are macromolecules specialized in recognizing a specific molecular pattern from the large number of molecular species with which it could interact. Pharmacological receptors are generally transmembrane molecules that can be activated by specific signal molecules (e.g., agonists); a process followed by a specific biochemical or biological response from the cell (organ) associated with the receptor. Enzymatic binding sites have substrate (ligand) specificity, and biochemical reactions are triggered upon substrate binding. In immunology, antibodies can be considered as receptors that specifically bind antigens - and immunoglobulins activate cellular responses in the presence of antigens (e.g., mast cell degranulation). Therapeutic manipulation of these macromolecular targets (collectively termed receptors) using specific ligands (drugs) is the objective of the pharmaceutical industry. Efforts aim at orally-available, potent, stereospecific drugs, that should act as agonists, antagonists or modulators on the desired molecular (and cellular) target. With the advent of powerful computers, an increasing majority of these ligands are sought using a rational approach, computer-aided drug design.

0097–6156/95/0589–0064$12.00/0
© 1995 American Chemical Society

The development of software for computer-aided drug design has focused on several aspects:

• molecular modeling - the ability to produce accurate models for small molecules in vacuo or in solvent (based on quantum mechanical or empirical methods), using geometry optimization, conformational analysis, molecular dynamics and related tools;

• tertiary structure prediction - the ability to successfully predict and/or calculate macromolecular structures (largely based on increasingly available X-ray and NMR determinations);

• molecular graphics methods - the representation of molecular properties using various techniques, so that essential information is visualized on the screen;

• docking and prediction of affinity - the identification of the binding site(s) and the computation of intermolecular interactions, with the use of molecular recognition and docking techniques; such methods may use flexible or rigid docking, free-energy perturbation methods, or structure-activity relationship models to predict the binding affinity;

• chemical databases - storage and manipulation of chemical information, with the use of advanced computational methods (e.g., clustering packages for multiple conformer and similarity searches); chemical database searches can suggest new lead compounds and may help evaluate synthetic feasibility; the spinoff of chemical database searching is a method known as *de novo* ligand design (when new molecular structures are software-generated, then evaluated by computer methods for binding affinity and by medicinal chemists for synthetic feasibility). Such methods are collectively termed rational design programs (RDPs).

The use of computers in drug design needs to be an economically viable solution in the face of other methods of drug discovery, such as screening and serendipity. Direct drug design techniques start from the known 3D-structure of the target macromolecule, in most cases co-crystallized with a lead compound. Efforts are then directed towards improving the binding affinity of the ligand. In most cases, novel structures are required to ensure patent protection. Indirect drug design techniques start from experimental (e.g., biophysical, pharmacological) evidence for the studied macromolecular target of unknown 3D structure, and mainly utilize biological data for series of ligands, searching for common patterns in the structural and property space (in order to define the pharmacophore). Both methods are aimed at obtaining more potent ligands by differentiating features that increase binding affinity from those that contribute to agonism, partial agonism and/or antagonism, and to outline key features for receptor (stereo)selectivity. When such features are manipulated by synthetic efforts, then tested on the desired target, a crucial step in selecting candidates for synthesis is the (accurate) estimate of binding affinity.

The majority of ligands bind to receptors in a reversible, non-covalent manner. With the increasing availability of 3D structures for macromolecules of therapeutic interest, direct molecular modeling techniques have become more important. Ligand specificity is obtained by matching the molecular architecture of the receptor binding site (Fischer's lock and key theory). The 3D structure of the ligand in the receptor binding site reveals the orientation of key functional groups (pharmacophore) that a ligand must present to ensure receptor recognition and selectivity. These pharmacophoric features are matched by corresponding receptophore features (key residues of the receptor that are essential for agonist and/or antagonist binding). Receptophore elements are, for most cases, identified using site-directed mutagenesis.

The challenge of the medicinal chemist is to develop molecular structures that match pharmacophore and receptophore features, while limiting side effects or catabolic susceptibility. However, even if pharmacophore and receptophore data are available, alterations in the chemical structure that would lead to improved potency are not obvious. In fact, due to competitive proprietary pressures, a requirement for non-obvious alterations to allow patent protection is implicitly imposed.

Numerous computer-aided design tools have been developed to aid medicinal chemists. Based on the underlying strategy, these RDPs usually belong to one of three main categories: 1. Scanners, 2. Builders, or 3. Hybrids.

Scanners are database searching programs and represent the majority of the RDPs available. A query specifies the 3D pharmacophore pattern for the desired ligands, then a chemical database is searched. Recovered compounds should place the query elements in the correct orientation. Depending upon the extent of the database, novel compounds possessing unique geometry can be recovered, providing insight as well as a foundation for further refinement. Numerous three-dimensional database search and retrieval systems have been developed(1-12). Notable programs include ALADDIN(5), 3DSEARCH(6), MACCS-3D(7), CHEM-X(8), and SYBYL/3DB(9). Recent reviews by Martin(13,14) and Borman(15) thoroughly discuss this technique and its application.

Builders are RDPs that essentially spawn and evolve ligands from a seed point within the receptor cavity. Generated ligands are the result of an iterative process of growth, followed by geometry optimization and refinement. These include the programs: GROW(16), LEGEND(17), GenStr/GroupBuild(18), and LEAPFROG(9). GROW and LEGEND add randomly-selected atom types or predefined substructures (e.g., amino acid residues) to the growing structure at sites which are sterically clear of either the receptor or the ligand itself. Upon completion, minimizations are performed to relax the structure within the active site. GenStr/GroupBuild and LEAPFROG use pre-selected functional groups and iterative force-field optimizations to ensure a reasonable ligand geometry upon completion.

The third category of RDPs combine techniques from both scanners and builders, thus, they utilize a hybrid strategy. There are several advantages to this approach. Most receptors have complex geometric and potential energy features(19), including numerous receptophore elements, e.g., hydrogen-bonding loci or hydrophobic subpockets. In performing a 3D search, probability alone dictates that the odds of retrieving an Erhlich's "magic bullet" that matches all pharmacophoric loci are extremely small. The odds are further diminished as the majority of database-retrieval systems maintain only a single, static conformer per structure.

Hybrid RDPs partition the active site of the target receptor into receptophoric-element containing subsites. Chemical fragments (building blocks) complementary to each subsite are then designed or retrieved from databases. These fragments are then linked to form ligands. The advantage of this approach is that ligand diversity is augmented through the combinatorial assembly of numerous sub-components. Programs using this approach are LUDI(20,21), BUILDER(22-24), the "linked fragment approach" of Verlinde(25), the MCSS method(26), and the use of transferable fragment properties(27-29), among others.

All these approaches have inherent strengths and weaknesses. Scanners have been popular because they evolved from pre-existing 2D structure searching methods. Their advantage is that answers are quickly obtained, providing ideas for compounds which might never have been considered as candidates for a specific target. When data on synthetic methods associated with similar compounds is included in the database, chemists can benefit from previous synthetic work. The disadvantage of scanners is that a pre-existing chemical database is required, and that the solution space is biased by the contents of that database (e.g., chemical diversity). Pharmaceutical companies have access to in-house databases (and synthetic means); however, these are limited by previous work. Commercially available databases are used by smaller companies and academic researchers, and can be combined with in-house databases to enhance chemical diversity.

To address the above limitations, the builder strategy was developed. A naive builder is influenced only by the active site environment. No chemical database is needed, for the benefit of researchers with limited access to such databases. Any class of structures could be potentially generated; however, the developer must provide the heuristics (e.g., when to perform a specific operation, what atom types to use, where to attach the next functional group, etc.), logistics (e.g., attach atom, break bond, modify torsion angle, fuse ring, etc.), and a molecular mechanics framework to maintain self consistency. Thus, builders are biased by the creativity of the program developer. Scanner RDPs suggest structures for which synthetic procedures are known. This is not necessarily true for builder programs, although true novelty of the suggested structure is more likely.

Although chemical databases are required for hybrid RDPs, ligand diversity is assured through the combinatorial assortment of building blocks. The sampled solution space of the database can be user-specified. Active site regions are individually complemented, hence, the proposed ligands are more likely to match the entire pharmacophore. The principal weakness of hybrid RDPs arises from the volume of data one must process. Consider a scenario where one hundred potential ligand fragments are recovered. Given that all are chemically valid, each structure must first be screened to ensure steric and electrostatic complementarity with the active site. Structures that conflict sterically with the receptor are deleted, although a fair percentage can be recovered by pruning appropriate atoms. What remains is the difficult task of scrutinizing fragments to find combinations that produce suitable ligands. For 100 components, there are nearly 5000 unique pairs of structures, along with triplets, quadruplets, etc., to consider. Segments of several different fragments may be necessary to piece together an optimized, legitimate ligand.

One of the common problems with all *de novo* design strategies is the prediction (forecast) of binding affinity, or the use of a scoring function to estimate the free energy of binding. A large number of solutions is generally available (especially when structures that incompletely match the pharmacophoric pattern are examined), and the evaluation of the binding affinity (or free energy) becomes the only measure to prioritize compounds for retrosynthetic analysis and synthesis. The use of free-energy perturbation methods to calculate the ΔG of binding requires costly molecular dynamics calculations and seems practically limited to only minor modifications of a compound with known activity, whereas a simple molecular-mechanics estimate is mainly enthalpic in nature. When available, internally consistent and predictive QSAR models can be used to evaluate software-generated compounds.

We summarize here our ongoing research in this area: a ligand design system based on four programs, CAVITY(30), FOUNDATION(31), SPLICE(32), and DBMAKER(33), used in conjunction with a 3D QSAR model for HIV-1 protease inhibitors to predict binding affinity.

Overview of ligand design

Our hybrid ligand design process(34) is summarized in Figure 1. Given the crystal structure of a ligand-receptor system, an active-site cavity shell is first generated with the program CAVITY. The cavity shell simplifies the visual display, revealing regions where current ligands can be modified, and allowing the evaluation of final products (e.g., steric contacts). The initial step is to define the binding-site cavity with the 3D

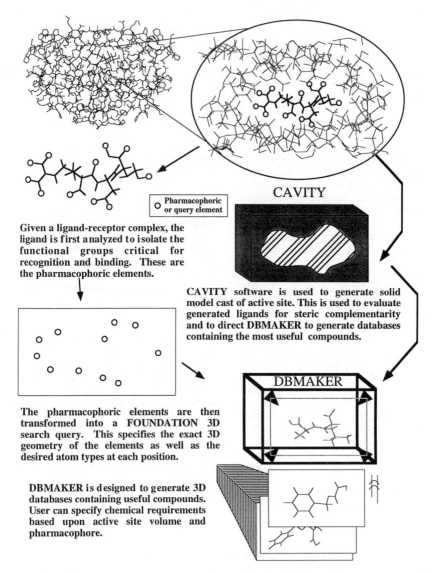

Given a ligand-receptor complex, the ligand is first analyzed to isolate the functional groups critical for recognition and binding. These are the pharmacophoric elements.

CAVITY software is used to generate solid model cast of active site. This is used to evaluate generated ligands for steric complementarity and to direct DBMAKER to generate databases containing the most useful compounds.

The pharmacophoric elements are then transformed into a FOUNDATION 3D search query. This specifies the exact 3D geometry of the elements as well as the desired atom types at each position.

DBMAKER is designed to generate 3D databases containing useful compounds. User can specify chemical requirements based upon active site volume and pharmacophore.

Figure 1. Overview of CAVITY, DBMAKER, FOUNDATION, and SPLICE ligand design system.

FOUNDATION then searches 3D databases and retrieves all fragments bearing a user-specified minimum number of matching query elements. These fragments are collected and analyzed by the EDIT module. Retrieved fragments that collide with the receptor are eliminated. The remaining components are then pruned to ensure steric complementarity with the original receptor.

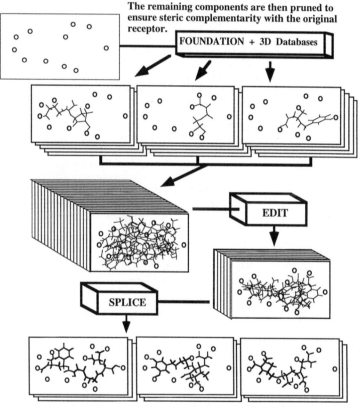

What results is a collection of structures each of which contains several different pharmacophoric elements. SPLICE then analyzes these fragments and assembles the most complete, novel ligands possible by joining the appropriate components.

Figure 1. *Continued*

pharmacophore. Based on the crystal structure or other methods, the pharmacophoric elements are isolated and their 3D relationships identified. Aside from visual inspection, GRID(35) and MCSS(26) could be utilized. DISCO(36) is an automated procedure for pharmacophoric pattern identification used to generate various pharmacophoric maps. In the case where no crystal data is available, pharmacophoric elements can be elucidated from a working 3D-QSAR model(37). The isolated pharmacophoric elements are then transformed into a 3D search query, which specifies both the position of each element relative to one another, as well as acceptable atom types.

In order to search for complementary components, we must have a source of 3D data. The program DBMAKER allows a user to generate databases of 3D structures according to numerous specified parameters. These parameters monitor content, composition, size, and connectivity information, but allow the program to generate random compounds within the scope of these constraints.

FOUNDATION is then used to search and retrieve all chemical components from our databases that contain a specified fraction of the pharmacophoric elements. By accepting structures that match various portions of the pharmacophore, we can retrieve a large number of diverse building blocks. FOUNDATION aligns each hit with the pharmacophore; thus, structures are docked in the active site with the appropriate orientation. FOUNDATION approximates the fit of each component within the active site, and will discard structures that are clearly contacting the receptor. However, a more rigorous screening is required to insure steric compatibility. This is performed by the EDIT module of SPLICE. Structures that require subtle modifications are pruned using an automated, standardized procedure(32). What results is a mass of components residing within the active site that link various elements to one another.

SPLICE takes these structures and forms new ligands that contain a greater number of pharmacophoric elements than any single component. This is accomplished by linking ("splicing") fragments that contain different portions of the pharmacophore with a chemical bond. Through iterative processing, SPLICE determines the largest, most complete ligands through the assembly of appropriate components.

Generation of vector-contact cavity shell - CAVITY

Our first concern is to visualize the active site. To study such a ligand-receptor system, the use of crystallographic coordinates and powerful molecular modeling software allow the visualization of complex steric relationships. Vector-based models have traditionally been the most popular representation due to their simplistic form and ease of display. However, with the inherent lack of molecular volume information, steric contacts are extremely difficult to judge(38).

Often, one must immediately know when two complex molecular entities are in contact. An example is the docking of ligands within receptor cavities. In this task, one is attempting to optimize the fit of the compound by subtle manipulation to improve its interaction with the active site. Another situation is the design of a novel compound with a more complementary binding surface. Although molecular surfaces make these tasks possible, there are drawbacks as well. In most molecular modeling systems, surfaces must be calculated and redisplayed with each ligand modification(38). Furthermore, the docking of surfaces may become visually overwhelming.

To circumvent this problem, Barry calculated the molecular surfaces of the receptor atoms after adding a constant distance to each van der Waals radius (39). By setting this constant distance to the van der Waals radius of a particular atom type (usually hydrogen) steric contacts are revealed where penetration of the receptor molecular surface by the ligand vector model occurs. The advantage of this display technique is that the molecular surface of the ligand need not be calculated. This allows uninterrupted ligand manipulation and modification while maintaining continuous feedback regarding receptor contact.

In previous work, we described a flood-fill algorithm to produce casts of the region enclosed within a molecular cavity using solid modeling techniques(*30*). We perform the same task here. However, by applying Barry's procedure to our cavity isolation algorithms, we generate a display called the **vector-contact cavity shell**.

Components of the cavity-like active site of a receptor comprise a *minute percentage* of the protein, and are scattered throughout the primary sequence. The majority of the protein consists of supporting sequences, densely folded to position the walls of the cavity. Dot surfaces usually include these uninteresting portions of the protein, cluttering the visual scene unless the area of interest is selectively edited. Because the shell is actually a **cast** of the cavity, it isolates all molecular surfaces in **direct contact** with the active site. The entire system is reduced to a single, continuous shell within which the *vector model* of the binding ligand must reside.

Determination of pharmacophoric sites

The next step is to determine the active-site pharmacophore. A simple method is to visually scan the active site for potential hydrogen bond donors or acceptors. One must determine the loci and atom types of the functional groups complementary to the receptor cavity. Hydrogen bonds of the receptor-ligand complex are calculated. The receptor functional groups responsible for inter-molecular hydrogen bonds are isolated, and the receptor is scanned to denote all other potential sites where inter-molecular hydrogen bonding could occur. Appropriate complementary functional groups are placed within the cavity and allowed to seek optimal binding loci by minimization. These are added to the original hydrogen bonding sites to produce the complete receptor active site pharmacophore.

Generation of 3D search query - FOUNDATION

In previous work, we described the three-dimensional database search and retrieval program FOUNDATION(*31*). Its search functionality is common to many such programs, its uniqueness being the use of clique detection algorithms (*40*)to retrieve partial query solutions. The program searches chemical structures containing specific 3D configurations of atoms and/or bonds (in any combination of the user-specified query elements) that match with pharmacophore requirements. One important feature of FOUNDATION is that it determines the *structural complementarity* that a query solution possesses with the ligand-accessible space in which it must reside. To represent the ligand-accessible space, we utilize the same filler lattice of points created above with the CAVITY program that conforms to the internal volume of the active site. Each potential hit is then realigned with the query following an RMS-fit procedure. A ligand atom is considered to be in the active site if it resides within a user-specified distance from any lattice point. A search is then performed to determine how many of the ligand atoms connecting the matching pharmacophoric elements reside within the cavity. This constraint prevents the recovery of structures that would clearly collide with the receptor. FOUNDATION is the heart of our hybrid strategy since it provides the building blocks to construct novel ligands.

Generation of 3D Databases - DBMAKER

DBMAKER (*33*)is a set of programs that allow investigators to generate their own 3D structural databases. User-defined parameters monitor the *content* (rings vs linear or branched), *composition* (elements and hybridization), *size*, and *connectivity information* (size and number of rings, spiro ring fusion, etc) of the database. SMILES strings (*41*) are generated, then converted to 3D-structures with CONCORD (*42*). This assures both the quality of the 3D structures produced, as well as the ability to generate them in a number of proprietary formats. Methods are present to maintain compound registration, thereby preventing duplication. This allows one to continually enlarge old databases by the addition of new compounds as they are needed. An unlimited variety of structures can be generated with DBMAKER. Since the process is user-controlled, compounds can be tailored to the investigator's needs. For example, if sturdy, hydrophobic anchoring segments are required, one can generate aliphatic carbocycles. If flexible chains containing hydrogen-bonding elements are needed, then linear structures with heteroatoms can be created. DBMAKER also contains a powerful filtering feature, allowing the user to easily remove undersirable chemical constructs. For example, if one is attempting to develop peptidomimetics, structures containing amide bonds may be selected against. DBMAKER was developed to allow investigators previously hindered by a lack of 3D-information to employ 3D-searching as a tool for molecular design, and to assist in design of compounds previously not synthesized.

Generation of Fragments - FOUNDATION

The core of our approach is the generation of molecular fragments. Although structures from any source may be satisfactory, those which more effectively complement the target receptor are preferred. By finding fragments that each contain several pharmacophoric elements, we maximize complementarity with minimal structural mass. The 3D database search program FOUNDATION(*31*) retrieves all fragments containing a user-specified minimum number of matching query elements. Each retrieved structure is re-oriented to match the configuration specified in the query; thus, structures are docked in the active site with the appropriate orientation. Appropriately bonded atoms, as well as desired atom types are recovered according to query specification.

Automated Editing Procedures - SPLICE [EDIT module]

FOUNDATION approximates the fit of each structure within the active site, discarding those that are clearly overlapping the receptor. A rigorous screening is required to ensure steric compatibility, especially for data from other sources. The EDIT module of the SPLICE(*32*) program is described below. Each structure is first examined using a depth-first search (*40*) to determine the atoms comprising the shortest paths between all pharmacophoric elements. These atoms are termed *path atoms*. If any path atom collide with the receptor, the structure is rejected since maintaining the pharmacophoric pattern would be impossible. Receptors are not static, some conformational flexibility being observed upon binding, which allows binding for structures that otherwise would contact these atoms. To compensate for atomic motion, SPLICE decreases the van der Waals radii of user-designated receptor atoms. All ligand atoms present in ring systems are then located. If a ring clashes with the receptor and contains any path atom, the parent structure is also rejected. Structures that pass the above criteria are edited with an automated procedure to ensure steric fit. Starting from the most distal atoms and proceeding towards the interior, the structure is systematically checked for receptor contact. Fragments are clipped off until the structure is satisfactory. Atoms that contact the receptor are removed along with all neighbors. Should any part of a non-essential ring collide with the receptor, the entire ring structure is deleted. In the case of receptor contact with fused multi-ring systems, only the atoms that are not stably locked in any ring are deleted. When completed, each structure should reside within the active site, yet retain the necessary elements that maintain its conformation. What results is a mass of components residing within the active site that link various elements to one another.

Component Assembly - SPLICE

Novel ligands are then assembled from the set of overlapping fragments, each containing a subset of the pharmacophore. The union of two fragments requires a precise alignment of atoms to allow the formation of a linking bond (e.g., bond angles and lengths have to be within tolerance levels). To determine which structures can be joined, pairs of structures that both contain a mutual bond whose atoms overlap nearly perfectly are isolated. For example, structure A contains a bond (A1-A2) that overlaps nearly perfectly with bond (B1-B2) in structure B. If the distances between atoms A1 <-> B1 and atoms A2 <-> B2 are within a user-specified range, then structures A and B can be joined by creating the new bond: A1-B2. We define this procedure as **splicing** A and B together at bond A1-B2 to form a new hybrid structure. When processing is completed, SPLICE automatically joins each matched pair of fragments to form a novel structure. These molecular structures are stored as Sybyl mol2 files(9). Depending upon the minimum number of elements specified for an acceptable combination, resulting structures may not satisfy the entire pharmacophore. There may exist several fragments that must be combined to form the most complete ligand. To find these combinations, SPLICE is repeatedly executed using the structures produced in the previous generation. With each run, the number of query elements required is increased. Through iterative processing, SPLICE determines the largest, most complete ligands through the assembly of appropriate components. The resulting ligands are then processed within Sybyl, and a preexisting QSAR model is used to evaluate the binding affinity.

Example of denovo design - HIV-1 protease

Figure 2 details an example of our *de novo* design work. Here, we will attempt to generate ligands capable of binding with the P1/P2 region of the HIV-1 protease active site(43). As detailed in this figure, the P1/P2 region is a crescent shaped structure incorporating both P1 and P2 sidechain binding pockets. A ligand binding pharmacophore deduced from the crystal structure of the protease is shown superimposed upon a CAVITY(30) generated cast of the active site. This region has numerous hydrogen bond donor and acceptor sites scattered throughout. As shown in the side view, this area is also relatively flat.

From our observations of this region, we can conclude that a planar, cyclic structure would best serve as a foundation upon which complementary functional groups could be anchored. Although this portion of the active site is fairly flat, it is also quite extensive in area. Thus, we can use five-membered, six-membered, and fused multi-ring assemblies as a foundation for ligand construction.

This is accomplished by using the three cyclic templates shown in the figure in conjunction with DBMAKER(33). Specific combinations of C=N, N=C, C=C, C(=O), N, and O will produce planar rings when written to SMILES(41) strings with the cyclic templates. All carbon atoms in the backbone must be sp2 hybridized to assure planarity, hence our choice of elements. With these components, the generated rings are planar and contain numerous sites to which substituent groups may be attached.

Two to four sidechains are specified per structure, each containing one to four atoms or functional groups. Four sidechain components are used: "C", "N", "O", and "C(=O)". When used in various combinations, DBMAKER can generate SMILES representations of functional groups that include alcohols "CO", ethers "COC", ketones "CC(=O)C", esters C(=O)OC, aldehydes "CC(=O)", acids "CC(=O)O", amines "CN",

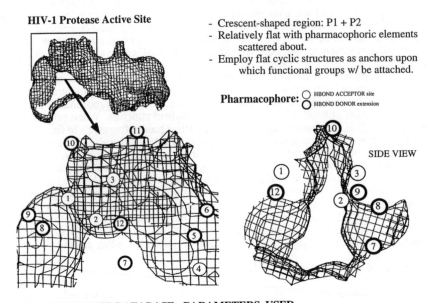

HIV-1 Protease Active Site

- Crescent-shaped region: P1 + P2
- Relatively flat with pharmacophoric elements scattered about.
- Employ flat cyclic structures as anchors upon which functional groups w/ be attached.

Pharmacophore: ○ HBOND ACCEPTOR site
Ⓞ HBOND DONOR extension

SIDE VIEW

HIV "CRESCENT" DATABASE - PARAMETERS USED

Component	Backbone Freq.	Sidechain Freq.	# Atoms	No. of Connections	# Sidechains	Sidechain Lengths
C=N	0.20	0.00	2	3=2		
C=C	0.20	0.00	2	3=3		
N=C	0.20	0.00	2	2=3		
C(=O)	0.20	0.10	1	2	2 (40%)	1 (30%)
C	0.00	0.30	1	4	3 (50%)	2 (30%)
N	0.10	0.30	1	3	4 (10%)	3 (30%)
O	0.10	0.30	1	2		4 (10%)

TEMPLATES

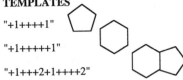

"+1++++1"

"+1+++++1"

"+1+++2+1++++2"

Figure 2. Use of CAVITY, FOUNDATION, and DBMAKER to generate ligands capable of binding with P1/P2 of HIV-1 protease active site.

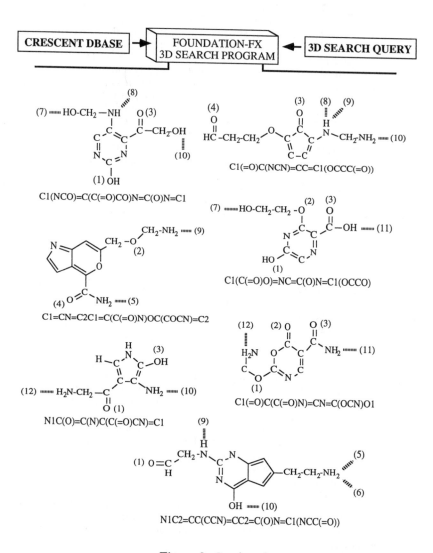

Figure 2. *Continued*

and amides "C(=O)N". These components are listed in the parameters file along with their frequencies of utilization.

An initial group of 5000 SMILES strings was randomly produced within the scope of the defined parameters. Processing and extraction of unique SMILES strings followed, leaving 4125 different structures. These strings were then converted to SYBYL multi_MOL format with CONCORD. The entire process required approximately 15 minutes of CPU time (R4000 Indigo).

To determine how well these structures complemented the active site for which they were designed, a three-dimensional database search was again conducted against the HIV-1 protease pharmacophore using the program FOUNDATION-FX(44). The search was performed to retrieve structures containing any combination of 4 or more query elements. A number of constraints were used to limit the number of hits retrieved. First, a maximum of four hits per structure (different conformations) was specified. Second, the coordinates of each retrieved hit were required to differ from any previous hit by greater then 0.500 angstroms RMS. deviation. Third, a maximum error of 0.500 angstroms RMS deviation from the query was tolerated. Fourth, a maximum number of 4000 search conformations was allowed per database structure. Finally any hit was required to fit within the active site using a reduced atom radius factor of 0.700.

Given the query of 12 hydrogen bond donor and acceptor sites, 25 five query element hits were retrieved (9 different structures) and 140 four query element hits were found. Selected structures are shown in Figure 2 along with the pharmacophoric elements they complement. The group of retrieved compounds is diverse, each engineered to specific criteria in order to complement the P1-P2 region of the HIV-1 protease active site.

Overview of the CoMFA technique

CoMFA(37) (Comparative Molecular Field Analysis) is a 3D quantitative structure-activity relationship (3D-QSAR)(45) method that computes steric and electrostatic interactions for a series of ligands with a regular lattice of probe atoms(37,46). The results are tabulated and multivariate analysis methods yield a QSAR model for the set of compounds in the training set. The recommended statistical technique for CoMFA is partial least squares(47) (PLS), with cross-validation (leave-one-out method) to select, among several PLS models, the one with the highest internally predictive value(48). The most common version of CoMFA is implemented in the QSAR module of Sybyl(9).

The following hypotheses form the basis of the CoMFA approach(46):
• non-covalent receptor-ligand interactions can be simulated by the steric (Lennard-Jones) and electrostatic (charge-charge) interactions of a ligand with a probe atom;
• the unknown receptor can be replaced by a regular grid, and steric and electrostatic interactions of probe atoms with each aligned ligand are calculated at grid points; such calculations generate columns that are tabulated for each molecule (row) in the series; the set of ligands upon which the CoMFA model is derived is termed training set, whereas other ligands predicted (or tested) with the training-set based CoMFA model are included in an external (or test) set;
• multivariate analysis (PLS) of the resulted table yields a regression model which highlights those features of the receptor that are implied by the given structure/activity data set. The resulting QSAR coefficients (fields) are then examined graphically as contours in 3D space.

A successful CoMFA study has to be a model with(49):

• self-consistent properties (robustness, e.g., similar molecular properties for similar ligands occupy the same 3D-space - based on user-defined alignment rules); the cross-validated r^2 of a robust CoMFA model should be higher than 0.5;
• predictive power (e.g., the model is able to predict compounds that were not present in the training set within reasonable limits - i.e. predictive $r^2 \geq 0.5$)
• explanatory power (e.g., the model is able to explain the interactions between the examined series of compounds and their putative binding site).

Flexible compounds are the most difficult case for 3D-QSAR data sets. The active conformation for each molecule and its superimposition on other structures (the **alignment** rules) have to be chosen, either in accordance with available experimental data, or based on hypothetical assumptions. Defining the alignment rule is a key step in 3D-QSAR. For flexible structures, a variety of methods can be used. If crystallographic data are available, the field-fit procedure(50) may prove useful (crystals being used as template molecules). This procedure minimizes the RMS difference between a fixed (steric and electrostatic) template field and the corresponding fields of the structure being aligned, by adjusting atomic coordinates (hence, field values). This procedure has been extensively discussed(51) in conjunction with alignment issues, and applied to determine the alignment of 52 human immunodeficiency virus 1 protease (HIV PR) inhibitor peptides(52) based on 7 experimentally determined structures of inhibitor-enzyme complexes.

Methods that investigate conformational space (e.g., using simulated annealing and cluster analysis(53)) may find the best match between various structurally dissimilar ligands when no crystal data are available. During this procedure(53), low-energy conformers are selected and minimized pairwise - and the best match obtained from all different conformations can be selected. For unknown receptors, the Active Analog Approach may be used in conjunction with (constrained) systematic search(54) (implemented as the RECEPTOR™ module(55) in Sybyl) to generate a set of sterically allowed conformations and to determine the existence of common 3D orientations of specified functional groups (the pharmacophore), in a series of compounds.

Using five different alignment rules and crystallographic data for seven ligands, a CoMFA model for HIV protease inhibitors was developed(52). The CoMFA PLS results for Alignment I yielded a cross-validated r^2 of 0.778 (leave-one-out technique(48)), with six principal components(56), and a conventional r^2 of 0.984. The predictive(37) r^2 (for 18 compounds) was 0.662. The same model was submitted to a HASL (Hypothetical Active Site Lattice) analysis. HASL is a 3D-QSAR method which distributes partial activities at molecular lattice points (defined within the van der Waals volume of each ligand)(57). HASL has a predictive power comparable to CoMFA (conventional r^2 1.00, for 59 compounds, and predictive r^2 0.438, for 18 compounds)(58). Due to methodological differences, the outliers in prediction for the two methods are not identical. By excluding the common outlier (compound m10 in the test set(52)), the predictive power improves for both methods: 0.74 (CoMFA) and 0.53 (HASL). The robustness of the HIV-protease inhibitors model and its predictive power were conserved between two different 3D-QSAR methods.

Evaluation of the predictive and explanatory power of QSARs

The fundamental problem of the above-mentioned (and similar) 3D-QSAR methods is that the proposed model is not uniquely determined. For flexible molecules, many conformers can present a particular pharmacophoric pattern, and the rationale for choosing one (the alignment rule) is usually done on an energetic and/or experimental basis. If the choice of the alignment has no reference to experimentally determined structures, results have to be treated with caution(*51*), because other conformations may in fact bind to the receptor, while the proposed alignment rule compensates inadequacies in considering entropic and enthalpic effects. A similar problem is encountered in choosing test set conformers. Test set compounds (with known activity, external to the model) are used to verify the predictive power of different QSARs. In the case of flexible compounds, the alignment rules are already defined, hence the alignment of test compounds is constrained by the existing model. Applying the same conformational choices and superimposition procedures to generate a test set of single conformers is useful only if insignificant changes exist in the structures present in the test set, compared to molecules in the training set.

If test set molecules have flexible moieties not present in the training set, the appropriate conformation is ambiguous even within the alignment rules. In this case, a (limited) conformational analysis performed on the test set molecules, using the alignment rules as constraints during conformational search, will generate multiple conformers for the same ligand - all consistent with the initial model, yet geometrically different and hence with a range of predicted activities that often spans several log units, instead of a single value (Figure 3). A rational approach to select the active conformation for prediction is required.

We have proposed(*59*) a semi-automated procedure, NewPred, compatible with the Sybyl/CoMFA method, that allows limited conformational analysis based on the alignment rules of an initial CoMFA model, and which automatically selects a single conformation for test set compounds, then predicts activities based on the initial QSAR. All conformers are minimized, either in the average steric and electrostatic field of the CoMFA model (option available in Sybyl), using the field fit procedure, or, preferably, in the receptor-binding site (when available), to optimize individual conformer alignment. The lowest energy conformer found is then chosen to be included in the final test set (which is used to evaluate the predictive power of the model).

NewPred has been applied to a set of 30 HIV-protease inhibitors, used to evaluate the predictive power of a previously described model. The predictive power proved to be significantly better for neutral (as opposed to charged) models(*59*), and for CoMFA fields compared to hydrophobic fields(*49*) (using HINT(*60*)). The resulting proposed conformations(*59*) for 1(S)-amino-2(R)-hydroxyindan-containing peptides(*61*) was used to explain their poor inhibitory activity.

The steps leading to a robust CoMFA model, i.e. choice of compounds, decisions concerning conformer selections and superposition rules, are of paramount importance, and are usually detailed in all CoMFA reports. The aim of any QSAR analysis is to go beyond that point, and make valid conclusions to be used in drug design. However, reports detailing this post-processing step - the analysis of the CoMFA fields, were scarce in the literature.

The statistical significance(*62-64*)and predictive power(*59,65*) of QSAR and CoMFA models is subject to discussions, but there is no indication as to how the explanatory power of such models is tested, and no details of the CoMFA post-processing step have been previously presented. Most CoMFA papers graph (by contribution) the scalar product of the standard deviation and the QSAR coefficient(*66*), for both steric and electrostatic fields, without mentioning use of other information available to the molecular modeler.

For drug design purposes, a detailed examination of other CoMFA fields and their interpretation, has been provided(*49*). Our comparative study of various fields available(*9*) in CoMFA has shown that, to a certain degree, CoMFA fields have structural correspondence in receptor atoms. For the case of HIV protease, different CoMFA fields from our HIV-protease inhibitors model were compared with the binding site crystal structure. The average steric field (at 70% contribution level) matches the

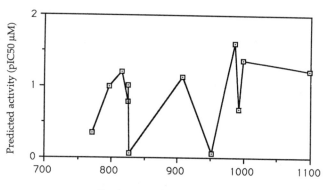

Figure 3. Total energy of active-site and inhibitor complex vs. predicted activity of isolated conformers[57] for compound m3. Twelve local minima, representing geometrically and biologically different conformers, were obtained using NewPred[57].

invagination contours between Ile^{184} and Pro^{181} (S1 to S2'), also between Val^{182} and Ile^{50} (S2). Steric contours overlap with Arg^{108} (S3), Asp^{30} (S2), Ile^{50} and Gly^{49} (S1), and with four residues in S1'-S2' (Pro^{81}, Ile^{150}, Gly^{148} and Gly^{149}). Electrostatic contour regions correspond with immediate contact residues Asp^{30} (S2), Asp^{25} (S1-S1'), Gly^{149} (S1'), and also with long-distance polar contact residues Arg^{187} (S2') and Asp^{129} (S2'-S3'). Some of the residues involved in non-polar (Ile^{184}, Pro^{181}, Val^{182}) and polar (Asp^{29}, Asp^{125} and Gly^{27}) contact with the ligands could not be overlapped at the usual (80%) contribution levels. These limitations are probably due to the reduced diversity of the training set. Using the combined information from CoMFA steric and electrostatic fields, and HINT fields(67) on the same model, a new hydrophobic pocket (between Gly^{49}, Ile^{50}, Gly^{149} and Ile^{150}) has been identified in the binding site, behind the scissile bond. While this pocket is not occupied by any of the ligands, its occupation would lead, according to our results, to an increase in binding affinity. Thus, the explanatory power of QSAR models becomes instrumental in suggesting new targets for drug design.

The HIV-protease inhibitors CoMFA model has been used to predict compounds that were designed using our in-house *de novo* design software. Novel transition-state isostere compounds have been suggested, and their binding affinity was predicted to be in the nanomolar range. While some of the earlier synthesized compounds did not perform as predicted in the in vitro assays, the later compounds show encouraging results(68). Using a good model in terms of statistical significance, internal consistency and predictive power, to evaluate *de novo* designed ligands offers a promise as a tool for drug design.

Literature Cited:
(1)   Gund, P., *Prog. Mol. Subcell. Biol.* **1977**, *11*, 117-143.
(2)   Lesk, A. M., *Communications, A.C.M.* **1979**, *22*, 221-224.
(3)   Jakes, S. E.; Willett, P., *J. Mol. Graphics* **1986**, *4*, 12-20.
(4)   Jakes, S. E.; Watts, N.; Willett, P.; Bawden, D.; Fisher, J. D., *J. Mol. Graphics* **1987**, *5*, 41-48.
(5)   Van Drie, J. H.; Weininger, D.; Martin, Y. C., *J. Comput. -Aided Mol. Des.* **1989**, *3*, 225-251.
(6)   Sheridan, R. P.; Rusinko, A., III; Nilakantan, R.; Venkataraghavan, R., *Proc. Natl. Acad. Sci. USA,* **1989**, *86*, 8165-8169.

(7)   MACCS-3D, Molecular Design Ltd.,*2132 Farallon Dr., San Leandro, CA., 94577,*
(8)   ChemDBS-3D, Chemical Design Ltd., *Suite 120, 200 Route 17 S, Mahwah NJ 07430,*
(9)   SYBYL™, *Tripos Associates, Inc., 1699 S Hanley Rd, St. Louis, MO 63144.*
(10)   Kuntz, I. D.; Blaney, J. M.; Oatley, S. J.; Langridge, R.; Ferrin, T. E., *J. Mol. Biol.,* **1982**, *161*, 269.
(11)   DesJarlais, R. L.; Seibel, G. L.; Dixon, J. S.; Kuntz, I. D., *J. Med. Chem.,* **1990**, *31*, 722-729.
(12)   Bartlett, P. A.; Shea, G. T.; Telfer, S. J.; Waterman, S. In *Molecular Recognition: Chemical and Biological Problems*; S. M. Roberts, Ed.; Royal Society of Chemistry: London, 1989, 1989; Vol. 78; pp 182-196.
(13)   Martin, Y. C.; Bures, M. G.; Willett, P. In *Rev. in Computational Chemistry*; K. Lipkowitz and D. Boyd, Ed.; VCH Publishers, Inc.: New York, 1990; pp 213-263.
(14)   Martin, Y. C., *J. Med. Chem.,* **1992**, *35*, 2145-2154.
(15)   Borman, S., *C&EN* **1992**, *August 1*, 18-26.
(16)   Moon, J. B.; Howe, W. J., *Proteins: Struct., Funct. Genet.* **1991**, *11*, 314-328.
(17)   Nishibata, Y.; Itai, A., *Tetrahedron* **1991**, *47*, 8985-8990.
(18)   Rotstein, S. H.; Murcko, M. A., *J. Med. Chem.* **1993**, *36*, 1700-1710.
(19)   Fersht, A. *Enzyme Structure and Mechanism;* 2 ed.; W.H. Freeman and Company: New York, 1985, pp 475.
(20)   Bohm, H.-J., *J. Comput.-Aided Mol. Design* **1992**, *6*, 61-78.
(21)   Bohm, H.-J., *J. Comput.-Aided Mol. Design* **1992**, *6*, 593-606.
(22)   Lewis, R. A.; Dean, P. M., *Proc. R. Soc. Lond. [Biol.]* **1989**, *236*, 125-140.
(23)   Lewis, R. A.; Dean, P. M., *Proc. R. Soc. Lond. [Biol.]* **1989**, *236*, 141-162.
(24)   Lewis, R. A., *J. Mol. Graphics* **1992**, *10*, 131-139.
(25)   Verlinde, C.; Rudenko, G.; Hol, W., *J. Comput.-Aided Mol. Des.* **1992**, *6*, 131-147.
(26)   Miranker, A.; Karplus, M., *Proteins: Struct. Funct. Genet.* **1991**, *11*, 29-34.
(27)   Chau, P. L.; Dean, P. M., *J. Comput.-Aided. Mol. Des.* **1992**, *6*, 385-396.
(28)   Chau, P. L.; Dean, P. M., *J. Comput.-Aided. Mol. Des.* **1992**, *6*, 397-406.
(29)   Chau, P. L.; Dean, P. M., *J. Comput. -Aided Mol. Des.* **1992**, *6*, 407-426.
(30)   Ho, C. M. W.; Marshall, G. R., *J. Comput. -Aided Mol. Des.* **1990**, *4*, 337-354.
(31)   Ho, C. M. W.; Marshall, G. R., *J. Comput. -Aided Mol. Des.* **1993**, *7*, 3-22.
(32)   Ho, C. M. W.; Marshall, G. R., *J. Comput. -Aided Mol. Des.* **1993**, *7*, 623-647.
(33)   Ho, C. M. W.; Marshall, G. R., *J. Comput.-Aided Mol. Des.* **submitted**.
(34)   Ho, C.; Marshall, G. In *Proceedings of the twenty-seventh annual Hawaii International Conference on system sciences*; L. Hunter, Ed.; IEEE Computer Society Press: Washington, DC, 1994; Vol. 5; pp 213-222.
(35)   Goodford, P. J., *J. Am. Chem. Soc.* **1985**, *28*, 849-56.
(36)   Martin, Y.; Bures, M.; Danaher, E.; DeLazzer, J. In *Trends in QSAR and Molecular Modelling 92*; C. Wermuth, Ed.; ESCOM: Leiden, 1993; pp 20-27.
(37)   Cramer III, R.; Patterson, D.; Bunce, J., *J. Am. Chem. Soc.* **1988**, *110*, 5959-5967.
(38)   Langridge, R.; Klein, T. E. In *Quantitative Drug Design*; C. A. Ramsden, Ed.; Pergamon Press: Oxford, 1990; Vol. 4; pp 413-430.
(39)   Barry, C.; **1979**, personal communication.
(40)   Cormen, T. H.; Leiserson, C. E.; Rivest, R. L. *Introduction to Algorithms*; The MIT Press: Cambridge, Massachusetts, 1990, pp 947-949.
(41)   Weininger, D., *J. Chem. Info. Comp. Sci.* **1988**, *28*, 31.
(42)   CONCORD, 3.01, Pearlman, R., TRIPOS Associates - St. Louis, MO. USA.

(43) Miller, M.; Schneider, J.; Sathyanarayana, B.; Toth, M.; Marshall, G.; Clawson, L.; Selk, L.; Kent, B.; Wlodawer, A., *Science* **1989**, *246*, 1149-1152.

(44) Ho, C. M. W.; Marshall, G. R., *J. Comput.-Aided Mol. Des.* **in preparation**.

(45) Marshall, G.; Cramer III, R., *Trends Pharmacol. Sci.* **1988**, *9*, 285-289.

(46) Cramer III, R.; Bunce, J. In *QSAR in Drug Design and Toxicology* D. Hadzi and B. Jerman-Blazic, Ed.; Elsevier: Amsterdam, 1987; pp 3-12.

(47) Stahle, L.; Wold, S., *J. Chemometrics* **1987**, *1*, 185-196.

(48) Cramer III, R.; Bunce, J.; Patterson, D.; Frank, I., *Quant. Struct.-Act. Relat.* **1988**, *7*, 18-25.

(49) Oprea, T.; Waller, C.; Marshall, G., *Drug Des. Discov.* **in press**,

(50) Clark, M.; Cramer, R. I.; Jones, D.; Patterson, D.; Simeroth, P., *Tetrahedron Computer Methodology* **1990**, *3*, 47-59.

(51) Klebe, G.; Abraham, U., *J. Med. Chem.* **1993**, *36*, 70-80.

(52) Waller, C.; Oprea, T.; Giolitti, A.; Marshall, G., *J. Med. Chem.* **1993**, *36*, 4152-4160.

(53) Perkins, T.; Dean, P., *J. Comput.-Aided Mol. Des.* **1993**, *7*, 155-172.

(54) Dammkoehler, R.; Karasek, S.; Shands, E.; Marshall, G., *J. Comput.-Aided Mol. Des.* **1989**, *3*, 3-21.

(55) RECEPTOR™, 2.4, available from Tripos Associates, 1699 S. Hanley Rd, St. Louis MO 63144, St. Louis, MO.

(56) Wold, S.; Esbensen, K.; Geladi, P., *Chemometrics and Intelligent Laboratory Systems,* **1987**, *2*, 37-52.

(57) Doweyko, A., *J. Med. Chem.* **1988**, *31*, 1396-1406.

(58) Doweyko, A.; **1994**, personal communication.

(59) Oprea, T.; Waller, C.; Marshall, G., *J. Med. Chem.* **in press**,

(60) Kellog, G.; Semus, S.; Abraham, D., *J. Comput.-Aided Mol. Des.* **1991**, *5*, 545-552.

(61) Thaisrivongs, S.; Turner, S.; Strohbach, J.; TenBrink, R.; Tarpley, W.; McQuade, T.; Heinrickson, R.; Tomasselli, A.; Hui, J.; Howe, W., *J. Med. Chem.* **1993**, *36*, 941-952.

(62) Mager, H.; Mager, P., *Quant. Struct.-Act. Relat.*.**1992**, *11*, 518-521.

(63) Wold, S., *Quant. Struct.-Act. Relat.* **1991**, *10*, 191-193.

(64) Clark, M.; Cramer, R. I., *Quant. Struct.-Act. Relat.* **1993**, *12*, 137-145.

(65) Baroni, M.; Costantino, G.; Cruciani, G.; Riganelli, D.; Valigi, R.; Clementi, S., *Quant. Struct.-Act. Relat.* **1993**, *12*, 9-20.

(66) Cramer III, R.; DePriest, S.; Patterson, D.; Hecht, P. In *3D-QSAR in Drug Design*; H. Kubinyi, Ed.; ESCOM: Leiden, 1993; pp 443-485.

(67) HINT!, 1.1, Kellogg, KE., Abraham, DJ, Richmond, VA 23298-0540

(68) Quintana-Morales, E.; Marshall, G.; **1994**, personal communication.

RECEIVED October 11, 1994

Chapter 6

De Novo Design of Highly Diverse Structures Complementary to Enzyme Binding Sites

Application to Thermolysin

Regine S. Bohacek and Colin McMartin

Research Department, Pharmaceuticals Division, Ciba-Geigy Corporation, Summit, NJ 07901

A computer program for *de novo* molecular design was used to explore the diversity of molecules complementary to the binding sites of enzymes. The program, **GrowMol** (*1,2*), generates molecules with spatial and chemical complementarity to the three dimensional structure of a host binding site. The molecules are created in the host binding site one atom or functional group at a time. At each step the position and type of atom to be added is randomly selected using Boltzmann statistics to bias acceptance towards atoms which can form favorable interactions with the binding site.

When applied to thermolysin, the program generated structures which were identical to or closely resembled known inhibitors. In addition, the program rapidly produced tens of thousand of distinct molecules which display a large variety of structural motifs. New methods to analyze the resulting structures have been developed.

An analysis of the diversity of thermolysin inhibitors generated by **GrowMol** will be presented.

The accurate 3-dimensional structure of a binding site, in principle, provides all the information required for the design of high affinity drugs. A major challenge for computer-aided drug design is to develop algorithms which will rapidly produce lists of compounds which have a high probability of binding strongly to a given binding site. Ideally this list will contain highly diverse structures representing all the major structural motifs which satisfy the requirements of the site.

Two main strategies are currently being used for this purpose. One method explores data bases by docking known compounds into the binding site and evaluating their interactions. The second approach generates, *de novo*, molecular structures that fit into a 3-dimensional representation of a binding site. This method is not limited to the molecules registered in a data base and has the potential of revealing a much larger diversity of structures.

A number of *de novo* methods have been developed which differ from each other in the ways the structures are generated and evaluated. One method first identifies "hot" spots in the binding site where a ligand molecule can form a hydrogen bond or fill a hydrophobic pocket and then docks large, predetermined

0097–6156/95/0589–0082$12.00/0
© 1995 American Chemical Society

molecular fragments into these sites. The fragments are then connected to form molecules (*3*). Another method constructs molecules by connecting smaller molecular fragments which are linked to form molecules. At each step the addition of a new group is evaluated using molecular mechanics energy (*4*). A program has recently been reported which first generates fragments to fill a volume and then changes the atoms to satisfy the electrostatic and hydrophobic requirements of the binding site (*5*). Other methods construct structures an atom at a time. One program uses only sp3 carbons (*6*) while another uses a complete set of atom types and evaluates the structures using molecular mechanics energies both in the growth step and for the final ranking of the structures (*7,8*).

In this paper, we present a progress report of the methods we are developing to meet the challenge of *de novo* ligand design. Our program, GrowMol, has been designed to rapidly produce large numbers of diverse structures which are highly complementary to the binding site. GrowMol "grows" structures into a binding site an atom or small fragment at a time (*2*). The program uses a rapid method for evaluating the complementarity of each new atom to the binding site. This provides a fast method to control the generation of atoms during the "growth" process. Complementarity has been found to correlate with potency and is, therefore, also used to rank order the structures in the final evaluation step.

Use of Complementarity to Evaluate Binding Affinity.

To be efficient, a program which rapidly grows potent molecules in a binding site needs to be able to asses the likely effect on potency of atoms or functional groups which are being considered for addition to the growing molecules. We have previously developed an algorithm which accurately describes the complementarity between a ligand and an enzyme binding site (*9*).

The algorithm was developed using high resolution protein crystal structures. Using a molecular modeling program, MACROMODEL (*10*), a buried strand of protein was identified and removed from the structure creating a cavity. The strand was treated as a virtual ligand and the cavity as a virtual binding site. A large number of virtual ligand/binding site pairs were created. The solvent accessible surface was then computed for each virtual binding site. Rules for describing the complementarity properties of this surface were then tested. These rules were based on properties of the binding site atoms including the electrostatic potential and electrostatic gradient. In each case the ability of a set of rules to predict the hydrogen bonding and hydrophobic character of ligand atoms lying close to the surface was analyzed. Several hundred interactions were examined for each set of rules. The most effective rules were based on distances of a point on the accessible surface to the nearest atoms in the binding site. When the distance from the surface to a binding site hydrogen bonding hydrogen is less than 2.6 Angstroms, then a ligand atom close to that point was found to be a hydrogen acceptor (i.e. C=O, OH) 94% of the time. When the distance from the surface to a binding site oxygen was less than 3.0 Angstrom, then the atom closest to that point was found to be a hydrogen bond donor 91% of the time. Points on the surface lying at a distance greater than 2.6 and 3.0 Angstrom to hydrogen bonding hydrogens and oxygens, respectively, indicate that the nearest ligand atom will be hydrophobic 91% of the time.

These rules provide a powerful method for the prediction of complementarity. In a further application of this algorithm, the complementarity scores of a series of thermolysin inhibitors was determined. The interactions between the inhibitors and thermolysin were known from the crystal structures of the thermolysin/inhibitor complexes. Using the above mentioned algorithm, the complementarity scores were found to correlate accurately with potency.

For *de novo* growth we wanted a description of complementarity encoded into a three dimensional grid representation of the binding site. A grid map of this sort can then be used to very rapidly evaluate the potential contribution of each atom in the grown structure. The rules for the accessible surface were, therefore, modified to meet this requirement.

Instead of considering only the accessible surface, a 3-dimensional representation of the entire volume of the binding site cavity is generated. This volume is divided into a grid at 0.25 Angstrom intervals. Each grid point is encoded with information about the nearest binding site atom. The binding site is divided into three major zones: forbidden, contact, and neutral. A grid point is in the forbidden zone if it is too close to one of the binding site atoms. The contact zone is further divided into hydrogen bond donor, hydrogen bond acceptor, and hydrophobic zones. The neutral zone includes all grid points which are in none of the above zones and less than the van der Waals radius plus 3.5 Angstroms from binding site atoms. The exact definitions of these zones have been described (2). Generation of the binding site grid map and assignment of all the values is performed by a computer program called GRIDBOX.

The cutoff distances used to construct the different zones for this grid map are designed to allow extra space for the generation of structures which upon further optimization may adopt conformations that are spatially and energetically compatible to the binding site. A second grid map is generated to evaluate the structures after they have been subjected to energy minimization in the active site. More stringent cut-off distances are used for this grid map.

The relationship between potency and favorable enzyme/inhibitor contacts, was evaluated using the data for thermolysin inhibitors which had been used for the accessible surface study. Nine potent thermolysin inhibitors, for which the binding mode to thermolysin had been determined by X-ray crystallography, were placed in the grid representation of the active site of thermolysin. The number of hydrophobic contacts (defined as the number of ligand carbons other than carbonyl carbons which occupy the hydrophobic zone) and the number of hydrogen bonds (i.e., the number of ligand hydrogens in the hydrogen acceptor zone plus the number of ligand oxygens found in the hydrogen bond donor zone) was determined for each ligand. These interactions were determined using a computer program called EVAL which uses the grid map to determine the binding site zone occupied by each inhibitor atom. Table I gives the results of this analysis. A multiple linear regression was carried out to correlate the potency of each inhibitor with the number of hydrophobic contacts and the number of hydrogen bonds each inhibitor makes with the enzyme. Using Grid Map 2, the following results were obtained with the structures determined by X-ray crystallography:

$$\log (K_i) = \qquad 3.16 \qquad -0.42 \,(\text{PHOB}) \qquad -0.39 \,(\text{HBOND})$$

| significance (p) | 0.0003 | 0.0049 | 0.0091 |
| $r^2 = 0.94$ | n=9 | sd=0.40 | p=0.0002 |

where PHOB = the number of hydrophobic contacts
       HBOND = the number of hydrogen bonds

Generated structures which satisfy the preliminary complementarity requirements are energy minimized in the active site and then re-evaluated. In this final evaluation potency is estimated using a linear regression equation relating potency to complementarity. Therefore, it was necessary to carry out the linear regression analysis using the complementarity scores obtained from known <u>energy minimized</u> structures. The nine thermolysin inhibitors were each energy

Table I

Complementary between Ligand and Enzyme in Thermolysin/inhibitor complexes determined by x-ray crystallography

Ligand[a]	Potency[b] Ki, μM	Hydrophobic Contacts		Hydrogen Bonds[c]		Estimated[d] Potency Ki, μM X-ray Data	Estimated[e] Potency Ki, μM Energy Minimized
		X-ray	Minimized	X-ray	Minimized		
ZF^{P}LA	0.000068	9	12	8	8	0.00018	0.0035
ZG^{P}LL	0.0091	7	7	6	7	0.0076	0.054
phosphoramidon	0.028	5	5	6	8	0.053	0.12
CLT	0.05	6	12	7	7	0.0082	0.0313
HONH-BAGN	0.43	5	6	4	5	0.032	0.839
BAG	0.75	5	5	3	3	0.78	2.19
P-Leu-NH2	1.3^f	2	3	5	5	2.0	3.31
THIO	1.8^g	4	5	3	3	2.0	3.31
RETRO	2.3^g	4	5	3	3	2.3	1.11

[a]X-ray structures of the ligand/thermolysin complexes were used. Cbz = carbobenzoxy; ZF^{P}LA = Cbz-Phe^P-L-Leu-L-Ala (11); ZG^{P}LL = Cbz-Gly^P-L-Leu-L-Leu (12,13); phosphoramidon = N-[(a-L-rhamnopyranosyloxy)hydroxyphosphinyl]-L-Leu-L-Trp (14); CLT = N-(1-carboxy-3-phenylpropyl)-L-Leu-L-Trp (15); HONH-BAGN = HONH-(benzylmalonyl)-L-Ala-Gly-p-niroanilide (16); BAG = (2-benzyl-3-mercaptopropanoyl)-L-alanylglycinamide (17); P-Leu-NH2 = N-phosphoryl-L-leucinamide (14); THIO = thiorphan, (N-[(S)-2-(mercaptomethyl)-1-oxo-3-phenylpropyl]glycine)(18); RETRO= retro-thiorphan (((R)-1-(mercaptomethyl)-2-phenylethyl)amino)-3-oxopropanoic acid (18). [b]From data compiled by Matthews (19). [c]Bifurcated hydrogen bonds, i.e. between ligand carbonyl and Arg 203, are counted only once. [d]Ki computed using the equation resulting from the linear regression between the logarithm of potency, the number of hydrophobic contacts, and the number of hydrogen bonds using structures determined by x-ray crystallography. [e]Ki computed using the equation resulting from the linear regression between the logarithm of potency, the number of hydrophobic contacts, and the number of hydrogen bonds using structures energy minimized in the active site of thermolysin. [f]Because there is a discrepancy between the data compiled by Matthews (19) and the original data, the original data of Powers, et al (20) is used. [g]See reference 21.

minimized in the active site of thermolysin using the GROWMIN program (McMartin, C. unpublished). The correlation between biological potency and complementarity for these structures gave the following results:

$$\log (K_i) = \qquad 2.49 \qquad -0.18 \text{ (PHOB)} \qquad -0.42 \text{ (HBOND)} \qquad (1)$$

significance (p)   0.0175   0.1381   0.0395
$r^2 = 0.80$ n=9   sd=0.77   p=0.0086

The energy minimized structures do not exhibit the same high degree of correlation between complementarity and potency as the structures in the conformations determined by X-ray. This decrease of correlation is not entirely surprising as the zones of the accessible surface upon which the grid boxes were based were parameterized using high resolution x-ray diffraction data of proteins. In this study, the linear regression equation (1) obtained using the energy minimized compounds was used for estimating the potency and ranking GrowMol generated structures.

We do not claim that this method can accurately predict potency. However, these results do indicate that there is a relationship between potency and the number of favorable enzyme/inhibitor contacts.

Generation of Molecular Structures.

The generation of molecular structures by GrowMol can be divided into three distinct steps: A) initiation, B) "growth" of each new atom, and C) acceptance or rejection of each new atom.

A) Initiation. The user selects a root atom, i.e. a point where "growth" is initiated. The root atom can be any atom present in the binding site structure from which a molecule is to be "grown". The atom can belong to the enzyme or to an inhibitor fragment. In the example shown below the root atom is the carbon atom attached to sulfur.

The computer program determines growth points for the root atom. Growth points indicate all the positions available for new atoms which might be added to the growing structure. The position of the growth point for a given atom is obtained using a lookup table of rotational isomeric states appropriate for that atom.

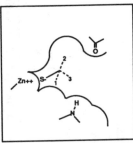

B) To "Grow" a New Atom. The position and atom type are randomly selected using Monte Carlo sampling:

    1.) One of the growth points is randomly selected

    2.) An atom of a functional group is randomly chosen from the functional group library:

C, N, -N=, O, O-, H, C=O, N-H, benzene or a five-membered aromatic ring with a nitrogen in one of the five positions

3.) The coordinates of the new atom are computed using the bond length, bond angles and dihedral angles associated with the growth point. These geometrical parameters are based on the MM2 force field (22).

C) Evaluation of New Atom. The atom is accepted or rejected depending on its degree of complementarity to the binding site using Boltzmann statistics (for more details describing this method see reference 2):

1.) The binding site zone occupied by the new atom is identified. If the atom is in a forbidden zone, i.e., too close to a binding site atom (see (a) in the diagram below), the atom and the associated growth point are erased and the program returns to step B).

2.) If the atom is in one of the allowed zones, a complementarity score is assigned.

A metropolis-like sampling criterion is used to decide if the new atom will be retained and added to the growing chain. The probability of retaining the new atom is given by the Boltzmann factor:

$$BF = exp \, (- \, complementary \, score/RT)$$

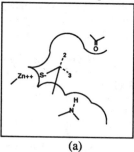
(a)

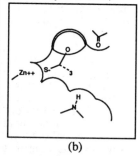
(b)

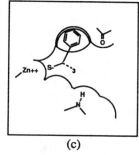
(c)

3) If the atom is accepted, growth points are computed, and it is connected to the growing chain, otherwise it is erased. If the atom, for example an oxygen, is in a non-complementary zone, such as a hydrophobic zone, then the probability is high that it will be rejected (example b above). If the atom or group is complementary to the binding site, such as a benzene in a hydrophobic zone, then the probability is high that it will be retained (example c above).

4) The process continues until the user specified number of atoms has been reached and the molecule is saved.

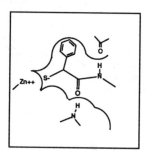

If, however, in the course of the generation all of the growth points have been used and the user specified number of atoms or complementary contacts have not been reached, the molecule is not saved. In either case the program proceeds to step 5 below.

5.) The growth point arrays are re-initialized and the program returns to the beginning to "grow" a new molecule until the user specified number of molecules have been generated.

A few additional rules are used to ensure chemically realistic structures. For example, each sp^2 must be connected to at least one other sp^2 atom, and chemically unstable groups such as peroxides are not permitted.

Evaluation and Ranking of Structures.

GrowMol generates structures rapidly. Structures with an average size of 14 atoms were generated at a rate of approximately 17 structures per minute using a VAX 6410 computer. Therefore, large number of structures can be generated in a relatively short time. Although all the structures are generated to be spatially and chemically complementary to the binding site and to have low conformational energy, they do not all form equally good interactions with the binding site. To rank order the structures, two criteria were used: a) favorable interactions with the enzyme atoms (expressed either as the number of favorable contacts or the estimated K_i) and b) the internal molecular mechanics energy of the bound structures.

The procedure used is outlined in Table II. The first step is the rejection of all duplicate structures. Next, those structures with a minimal degree of complementarity are selected for further optimization. Each of these structures is subjected to energy minimization in the active site and energy minimization in the absence of the active site. The difference between the energy of the bound conformation and the energy of the conformation minimized outside of the binding site is taken as an initial measure of the ligand strain energy.

The ligand strain energy is not used as an estimate of potency (as we have found no correlation between potency and ligand strain energy) rather as a measure of how well the structures fit into the binding site cavity. Structures with a strain energy larger than a user specified strain energy are rejected.

After energy minimization, a second more stringent degree of complementarity is applied. Grid Map 2 is used to determine the number of hydrophobic contacts and the number of hydrogen bonds. If the binding site is one for which sufficient data is available, the correlation between potency and favorable contacts can be determined. The coefficients of the regression equation correlating potency with complementary contacts, such as equation (1), are then used to compute an estimated potency. This allows for the estimation of the K_i from the number of favorable contacts. The user specifies a minimal threshold and retains only the best structures.

The remaining structures can be clustered into families based on similarity. Two structures are taken to be in the same family if 60% of the atoms of the larger structure are within 0.5 Angstrom of an atom in the other structure. A representative structure with the lowest K_i and the lowest strain energy is selected from each cluster for visual inspection.

Application.

Enzymes are "designed" by nature to bind to the transition state of a substrate. We were intrigued by the question: how different can a molecule be from a substrate

Table II. Results for evaluation of 22,000 structures generated in the S1' and S2' subsites of thermolysin

Total number of grown structures	22,000
Unique structures	12,654
Criterion: chemical complementarity (at least 2 hydrogen bonds and 3 hydrophobic contacts)	5,000
Energy minimize in the active site and in vacuum; compute the strain energy	
Unique structures	4,449
Criterion: **strain energy** (less than 35 kJ/mol)	3,937
Criterion: **complementarity** (estimated Ki of < 2.0 mM)	1,373
Cluster into families	
Select representative structures from each family	308

and still bind to the enzyme? The question can be partially explored by the systematic, synthetic modification of inhibitors based on known substrates. However, this strategy is unlikely to sample the full range of possible structures. Since GrowMol uses only the 3-dimensional structure of the binding site as a design template, it is an ideal tool to explore the structural diversity of molecules complementary to a known binding site.

GrowMol has been applied to a number of different enzymes including serine proteases, such as thrombin; aspartic proteases, i.e., pepsin and HIV protease; kinases; and purine nucleoside phosphorylase (PNP). For this study we chose the zinc metallo protease thermolysin for the following reasons: high resolution data of a variety of inhibitors complexed to thermolysin is readily available (19), the thermolysin binding site does not change appreciably when complexed with numerous different inhibitors, in the crystalline state thermolysin is enzymatically active (16), and there is considerable interest in zinc metallo protease inhibitors as therapeutic targets.

For this study only the S1' and S2' subsites of the thermolysin binding site were used. The binding site atoms were obtained from the structure of thermolysin bound with ZG^{P}LL (13). The structure, labeled 5TMN was obtained from the Brookhaven Protein Data Bank (23,24). This part of the binding site has the best defined pockets, and X-ray crystal structures of small inhibitors which only use this part of the binding site are available. Known inhibitors display a variety of different zinc chelating groups. For this study a sulfur atom which binds to the catalytic zinc ion was selected to be the root atom. The position of the sulfur atom as well as the initial growth point, which defines the dihedral angle that an atom connected to the sulfur will adopt, was obtained from the coordinates of thiorphan complexed with thermolysin determined by X-ray crystallography (18). The structure of thiorphan is shown in Table IV. In this study, we decided to probe the site using aliphatic thiols, and, therefore, a carbon was selected for the first guest atom.

To study diversity, three large sets of structures were generated: 50,000 with an average of 12.0 atoms, 50,000 with an average of 13.8 atoms and 72,000 with an average of 18.7 atoms. The number of unique structures in each of these sets was determined. Table III summarizes the results.

Table III. Effect of Molecular Size on the Number of Unique Structures Generated in the Binding Site of Thermolysin

Average no. of atoms per structure	No. of structures generated	No. of unique structures	Replication rate[a]	α[b]
12.0	50,000	3,938	12.7	1.99
13.8	50,000	10,543	4.7	1.96
18.7	72,000	39,680	1.8	1.76

SOURCE: Reprinted with permission from ref. 2. Copyright 1994.
[a]The number of structures divided by the number of unique structures found
[b]Measure of diversity: Number of unique structures = α (number of atoms)

To estimate if the sample size was large enough to adequately sample all the possible structures which could be generated, the average replication rate for each set was determined. For the smallest set, each unique structure was replicated on average 12.7 times. This indicates that most of the unique structures of this size that can be generated by this version of GrowMol were found. For the larger set, the replication rate was only 1.8 even though 72,000 structures were generated. We conclude, that the number of diverse structures obtained will almost certainly underestimate the true diversity of molecules in this size range.

The increase in the number of diverse structures with increasing molecular size is due to a combinatorial explosion. Therefore, the number of unique structures might be expected to increase as an exponential function of the number of atoms in the molecule:

$$N_{unique} = \alpha^{N_{atom}}$$

This equation expresses diversity in terms of a single parameter, α. The value found for α was close to 2 (1.99 and 1.96) for the first two sets and somewhat lower (1.76) for the third set. This lower value is very likely due to insufficient sampling. The parameter, α, is useful as it can be used to predict the number of unique structures GrowMol can generate for a given size molecule. For example, there should be approximately one billion unique structures containing 30 atoms!

An additional set of molecules with 25 to 35 atoms was also generated in order to investigate features of large molecules. 7000 structures were generated; of these 4706 were unique.

Results

A set of 22,000 structures with an average size of 18.7 atoms was evaluated. The results are summarized in Table II. The entire analysis resulted with 308 structures which were inspected visually at the graphics terminal. These structures were classified in a number of different ways: the way in which the backbone chain extends thorough the binding site; the type of functional group occupying a specific area of the binding site; structures which have a constant feature, e.g. a phenyl group in the S1' pocket. In addition, the structures were compared to known inhibitors to provide an experimental validation of the method.

Figure 1 gives examples of some of the different binding modes exhibited by the GrowMol structures. The ellipses represent hydrophobic groups, e.g. branched alkane chains, cyclohexane or benzene rings. The dotted lines show hydrogen bonding to enzyme residues. The first example shows the binding mode proposed for the substrate. Most known inhibitors are based on this binding mode. These examples represent only a fraction of the total number of different binding modes found in this study.

Figure 2 shows examples of novel ways in which the S1' and S2' binding pockets of thermolysin can be occupied. Structures A-1 through A-3 occupy the S1' pocket; structures B1 to B-3 occupy the S2' pocket; and the last structures are macrocycles which bridge the two pockets. These examples show the ability of the *de novo* growth algorithm to generate conformationally restricted structures which are chemically complementary and fill the complex volume of the binding site.

To assess the ability of GrowMol combined with the evaluation procedure to produce potent inhibitors, the reported thiol inhibitors of thermolysin were compared to structures generated by GrowMol. A literature search revealed fourteen thiol thermolysin inhibitors. Six of these compounds are alpha thiols. At

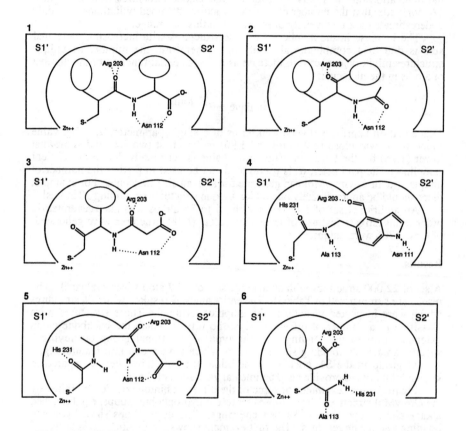

Figure 1. Examples of different ways in which the backbone of grown structures extend through the S1' and S2' sub sites of thermolysin. The ellipses represent hydrophobic groups. The dotted lines show hydrogen bonding to enzyme residues. The first example shows the binding mode proposed for the substrate. Most known inhibitors are based on this binding mode. The remaining examples show binding motifs which differ appreciably from this mode and provide novel solutions to the hydrogen-bonding requirements of the enzyme. (Reproduced with permission from ref. 1. Copyright 1994 American Chemical Society)

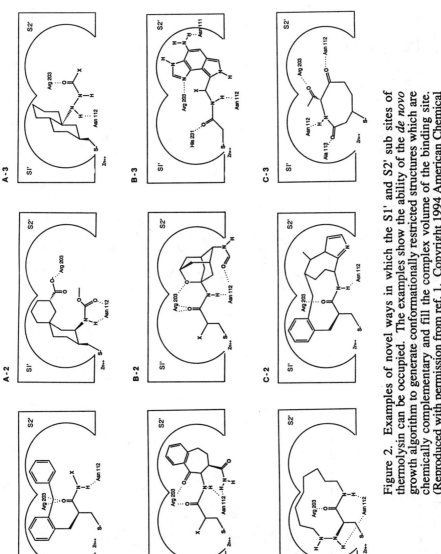

Figure 2.   Examples of novel ways in which the S1' and S2' sub sites of thermolysin can be occupied. The examples show the ability of the *de novo* growth algorithm to generate conformationally restricted structures which are chemically complementary and fill the complex volume of the binding site. (Reproduced with permission from ref. 1. Copyright 1994 American Chemical Society)

Table IV. Comparison between thiol inhibitors of known potency and structures generated by GrowMol.

Known Inhibitors	Experimental Potency K_i, μM	Estimated Potency K_i, μM	GROWMOL Structures	Estimated Potency K_i, μM	Estimated Potency K_i, μM Before Energy Minimization In Active Site
CGS 26670 (structure)	0.019a	0.63	(structure)	1.4	2.2
			(structure)	1.4	3.3
(structure)	3.8b	5.0	(structure)	1.4	3.3
(structure)	0.75c	2.2	(structure)	0.55	0.74

Compound				
(S) Thiorphan	1.8d	0.95	0.11	5.0
(R) Thiorphan	3.0d	2.2	0.77	17.4
Retro-thiorphan	2.3d	2.2	0.11	5.0
	5.3e	51.5	22.5	34.0
	52f	77.98	none	

[a]Ksander, G.; Bohacek, R. S.; de Jesus, R.; Yuan, A.; Sakane, Y.; Berry, C.; Ghai, R.; Trapani, A. J.; manuscript in preparation; [b]See reference 26; [c]See reference 17; [d]See reference 18; [e]This compound binds to thermolysin in a noncompetitive manner and can, therefore, not be compared with the rest of the potency data (25). [f]See reference 25.

present no experimental data is available concerning the binding mode of alpha thiols to the zinc ion in thermolysin. It has been postulated that alpha thiols bind to thermolysin in a bidendate fashion (25). The structures generated by GrowMol in the present application can only bind in a monodendate manner, and, therefore, the published alpha thiols cannot be used as a comparison. The (S) isomer of retro-thiorphan has a K_i of 94 μM, and was, therefore, also excluded. This left seven inhibitors. We added an eighth compound to the list, a potent benzofused macromolecule, CGS 26670 (Ksander, G.; Bohacek, R. S.; de Jesus, R.; Yuan, A.; Sakane, Y.; Berry, C.; Ghai, R.; Trapani, A. J.; manuscript in preparation). Table IV shows a comparison between these compounds and similar structures generated by GrowMol. Table IV also gives the experimental and estimated potencies for the structures.

GrowMol generated structures identical to both (S) and (R) thiorphan. In addition GrowMol generated structures similar to five of the six remaining known inhibitors. The inhibitor which was not found is not very potent (experimental K_i of 52μm). It forms few interactions with the enzyme and, therefore, it is not surprising that no structure similar to this inhibitor was generated.

The estimated potencies are all within approximately one order of magnitude of the experimentally determined K_i s.

Conclusions

A combinatorial, *de novo* growth algorithm has proved to be a useful tool for exploring the diversity of potential ligands to an enzyme binding site.

When applied to the thermolysin binding site a large number of diverse structures with steric and chemical complementarity were found. Molecules identical to and similar to known inhibitors were found in the set of computer generated structures.

The large size of the resulting sets of molecules could be considered a liability, however, we believe that the ability to reveal such a large set of diverse structures makes the method very powerful. Large data-bases of complementary structures can be created and then searched to reveal structures with various desired properties. We have also found that using molecular mechanics energies and clustering of similar structures, it is possible to identify a relatively small set of highly diverse structures having properties believed to lead to good binding.

The present version of GrowMol does have limitations. In its present form, GrowMol limits the number of structures that can be generated by fixing the bond lengths, bond angles and dihedral angles to ideal values corresponding to the rotational isomeric states of each torsion bond. The small number of atoms and functional groups in the current library is another limitation. Refinements are presently under way which will remove these limitations.

In conclusion, the *de novo* growth program we have developed is currently able to provide medicinal chemists with a list of novel structures which are highly complementary to a known binding site. With further refinement in methods for potency prediction and structure generation, it is expected that in the future this list will contain nearly all the structures which can bind with high affinity to a known conformation of the site.

Acknowledgments

We thank our colleagues Drs. Frank Clark, Wayne Guida and Jeff Watthey for the numerous suggestions and stimulating discussions.

Literature Cited

1. Preliminary results presented at the XIIth International Symposium on Medicinal Chemistry, Basel, Switzerland, **1992**.
2. Bohacek, R. S.; McMartin, C. *J. Am. Chem. Soc.* **1994**, *116*, 5560-5571.
3. Bohm, H.-J.; *J. Comput.-Aid.Mol.Des.* **1992**, *6*, 61-78.
4. Rotstein, S.H.; Murcko, M.A. *J. Med. Chem.***1993**, *36*, 1700-1710.
5. Gillet, V.; Johnson, A. P.; Mata, P.; Sike, S.; Williams, P. *J. of Comput.-Aided Molec. Design* **1993**, *7*, 127-153.
6. Rotstein, S.H.; Murcko, M.A. *J. of Comput.-Aided Molec. Design* **1993**, *7*, 23-43.
7. Nishibata, Y.; Itai,A. *Tetrahedron* **1991**, *47*, 8985-8990.
8. Nishibata, Y.; Itai,A. *J. Med. Chem.* **1993**, *36*, 2921-2928.
9. Bohacek, R. S.; McMartin, C. *J. Med. Chem.* **1992**, *35*, 1671-1684.
10. Mohamadi, F.; Richards, N.G.; Guida, W.C.; Liskamp, R.; Lipton,M.; Caufiled, C.; Chang, G.; Hendirkson, T.; Still, C. *J. Comput. Chem.* **1990**, 11, 440-467.
11. Holden, H.M., Tronrud, D.E., Monzingo, A.F., Weaver, L.H.; Matthews, B.W. *Biochemistry* **1987**, *26*, 8542-8552.
12. Bartlett, P.A.; Marlowe, C. K. *Science* **1987**, *235*, 569-571.
13. Tronrud, D.E.; Holden, H.M.; Matthews, B.W. *Science* **1987**, *235*, 571-574.
14. Tronrud, D.E., Monzingo, A.F.; Matthews, B.W. *Eur. J. Biochem.* **1986**, *157*, 261-268.
15. Monzingo, A.F.; Matthews, B.W. *Biochemistry* **1984**, *23*, 5724-5729.
16. Holmes, M.A.; Matthews, B.W. *Biochemistry* **1981**, *20*, 6912-6920.
17. Monzingo, A.F.; Matthews, B.W. *Biochemistry* **1982**, *21*, 3390-3394.
18. Roderick, S.L., Fournie-Zaluski, M.C.; Roques, B.P.; Matthews,B.W. *Biochemistry* **1989**, *28*, 1493-1497.
19. Matthews, B.W. *Acc. Chem. Res.* **1988**, *21*, 333-340.
20. Kam, C.; Nishino, N.; Powers, J.C. *Biochemistry* **1979**, *18*, 3032-3038.
21. Benchetrit, T., Fournie-Zaluski, M.C.; Roques, B.P.*Biochem. and Biophys. Res. Comm.* **1987**, 8127-8140.
22. Allinger, N.L. *J. Am. Chem.Soc.* **1977**, *99*, 8127-8140.
23. Berstein, F.C.;Koetzle, G.J.B.; Williams, G.J.B.;Meyer, Jr., E.F.; Brice, M.D.; Rodgers, J.R.; Kennard, O.; Shimanouchi, R; Tasumi,M. *J.Mol.Bio.* **1977**, *112*, 535-542.
24. Abol, E. E.; Bernstein, F. C.; Bryant, S. H.; Koetzle, T. F.; Weng, J. In*Crystallographic Database-Information Content, Software Systems, Scientific Applications*; Allen, F.H.; Bergerhoff, G.; Suievers, R., Eds.; Data Commission of the International Union of Crystallography: Bonn/Cambridge/Chester, 1987; 107-132.
25. Pickering, D.S.; Krishna, M.V.; Miller, D.C.; Chan, W. W. *Archives of Biochem. and Biophys.* **1985**, *239*, 368-374.
26. MacPherson, L. J., Bayburt, E. K., Capparelli, M. P., Bohacek, R. S., Clarke, F. H., Ghai, R. D., Sakane, Y., Berry, C. J. Peppard, J. V., Simke, J. P., Trapani, A. J. *J. Med. Chem.* **1993**, *36*, 3821-3828.

RECEIVED October 11, 1994

Chapter 7

Computer-Aided Design of New Drugs Based on Retrometabolic Concepts

Nicholas Bodor and Ming-Ju Huang

Center for Drug Discovery, University of Florida, P.O. Box 100497,
Health Science Center, Gainesville, FL 32610-0497

Traditional methods of drug design have relied on maximization of drug activity, which, however, often produce highly potent derivatives with equally elevated toxicities, resulting in no change in the therapeutic index. Inclusion of metabolic and toxicological considerations in the drug design process is embodied in retrometabolic concepts. This systemic methodology employs several rules for design of safe drugs, either by metabolic activation (chemical delivery systems) or strategic enzymatic deactivation processes (soft drugs). Automation of these drug design processes was accomplished by artificial intelligence-based computer programs. Thus, new structures are generated from any lead compound, using the "soft analog" and the "inactive metabolite" approach strategies. All structures are conformationally optimized. Then they are ranked based on calculated properties: solubility and partition characteristics, isosteric-isoelectronic comparison to the lead, and estimated metabolic rates of the predicted enzymatic degradation.

The main objective of drug design is to develop drugs on as rational a basis as possible and to reduce the trial-and-error factor in this process to the absolute minimum. In the classical drug design process, one starts with a lead compound, which is a well-defined structure having known biological action. The objective then is to modulate the biological action-activity, assuming that structure-activity relationships (SAR) exist. Accordingly, random and then some systematic modifications of the structure will allow one to learn as much as possible about the structure-activity relationships which then will lead to maximization of the drug action. Most of the time, in this way one can generate new compounds with increased activity. However, in most cases the ratio of the activity to undesired side-effects (toxicity), that is, the drug therapeutic index does not change. There are many reasons for this. Most of the time it is because the side effects are

0097–6156/95/0589–0098$12.00/0
© 1995 American Chemical Society

related to the intrinsic receptor affinity responsible for the desired activity and that the drug pharmacodynamic and pharmacokinetic properties will change parallel to the activity changes. On the other hand, when introduced in the body, drugs will undergo multiple enzymatic conversion (metabolism), as the body is trying to get rid of these foreign chemicals. However, many of these metabolic conversions will lead to modified compounds, still closely related to the drug, which can have enhanced or a different type of biological activity, or which could be quite toxic (epoxides, radicals, etc.). Most drugs will generate multiple metabolites which can affect drug action and toxicity, even if the drug, that is, the new drug, has better receptor binding or other desired properties. The bottom line is that most of the time the therapeutic index does not improve.

All data accumulated on the metabolic activation-toxication and/or deactivation of drugs and biologically active chemicals clearly indicate the necessity of including metabolic considerations in the general drug design process. That is, rather than waiting on studies on the drug metabolism after selection of the best drug candidate, structure-activity relationships (SAR) should be combined from the beginning with structure-metabolism relationships (SMR) throughout the drug design process. The combination of SAR and SMR is incorporated into the retrometabolic drug design (RMDD) concept. If one considers a drug (D) as the lead compound, then the retrometabolic drug design could cover two directions. In one case the drug is converted chemically into an inactive form by covalently attaching to it bioremovable moieties, of two general types. The more important one is called targetor (T), while the others are modifiers (F), which will optimize physico-chemical properties of the molecule. All these combinations lead to a molecule which we can call a chemical delivery system (CDS), which by design will sequentially undergo metabolic conversions, removing the modifiers after the targetor fulfills its site- or organ-targeting role. That is, the CDS by designed metabolic conversion will be activated to the desired drug at the site of action. This forms one part or one side of the retrometabolic drug design loop, as shown in Fig. 1.

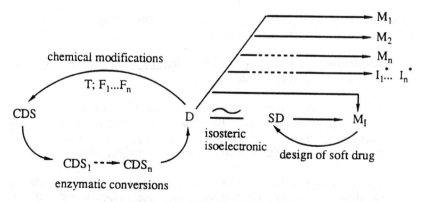

Fig. 1. The retrometabolic drug design loop.

On the other side of the loop are newly designed drugs with desired intrinsic activity. However, these are drugs of a special kind. These so-called "soft drugs" are designed in such a way that they will be metabolized in a predictable and controllable way to an inactive metabolite (M_I) after they achieve their therapeutic role. As shown on the right side of the retrometabolic loop, the soft drug (SD) is isosteric and/or isoelectronic, with the drug, but it will be metabolized in one step to an inactive metabolite (M_I), which in general is one of the inactive metabolites of the drug as well. The design of the soft drug takes place with well-defined strategic modifications involving the drug metabolism. It is evident that the two classes of compounds involving the retrometabolic design, that is, CDS and SD, are opposite to each other. While the CDS is inactive and metabolically activated via a designed route to the drug, the SD is active itself, which then is metabolically deactivated via a designed route.

The concept of soft drugs in the present sense was introduced in 1980 (*1-3*), and in a relatively short time a number of different classes of soft drugs were identified, which were classified (*4,5*) in five distinct groups as follows:
1)   Soft Analogs
2)   Activated Soft Compounds
3)   Active Metabolite Based Drugs
4)   Controlled Release Endogenous Agents
5)   The Inactive Metabolite Approach

Examples for practical use of each of these classes were provided in the literature (*6-13*), but since the early 1980's the two classes which have emerged as the most important ones are the soft analogs and the inactive metabolite approach. Common to these two and actually all soft drug classes is the fact that the soft drug after being delivered to perform its therapeutic role is metabolized singularly to one inactive metabolite. This inactive metabolite can undergo further modifications, but, from the point of view of activity-toxicity, this is irrelevant.

The Inactive Metabolite Approach

The working hypothesis and strategies to be used in developing new drugs based on the inactive metabolite approach are as follows:

1)   The design process starts with a known inactive metabolite of a drug that is used as the lead compound.

2)   The new soft drug structures are designed from this inactive metabolite by performing chemical modifications to obtain structures that resemble (isosterically and/or isoelectronically) the drug from which the lead inactive metabolite was derived. This is the activation stage.

3)   The new structure of the soft analog is designed in such a way that its metabolism will yield the starting inactive metabolite in one step and without going through toxic intermediates (predictable metabolism).

4)   The transport and binding properties, as well as the rate of metabolism and

pharmacokinetics of the soft drug, are <u>controlled</u> by molecular manipulations in the first activation stage (<u>controllable metabolism</u>).

Some additional considerations are the following:

If a useful inactive metabolite of a drug is not known, based on our current knowledge of drug metabolism, inactive metabolites can be predicted. Second, there are two major classes of soft analogs based on inactive metabolites. In <u>one class</u>, where the inactive metabolite formation <u>involves the pharmacophore</u>, generally the requirements for isosteric-isoelectronic analogs are much more restricted. The new structures have to be very close analogs of the lead compound. On the other hand, if the inactive metabolite is formed by enzymatic modification of a part of the molecule which is <u>not involved in the activity</u>, generally there is more freedom in modifying this kind of inactive metabolite in the activation stage.

The soft drug, like the lead compound or its active metabolites have good affinity to the receptor and will trigger the desired pharmacological action at the site(s) it has reached. The inactive metabolite by definition is void of receptor binding activity.

The soft drug, whose activity is similar to that of the lead drug, has, however, a distinct property of being converted to the inactive metabolite in a <u>separate, independent process</u> (not related to the active site). In the general soft drug design process, <u>one of the basic principles</u> involves avoiding oxidative metabolism as much as possible and basing the deactivation on hydrolytic enzymes, such as non-specific esterases, in order to achieve the predictable, controllable, and directed drug metabolism. The inactive metabolite which is used in the drug design process, is, on the other hand, normally produced by oxidative metabolism in the body, when the lead drug is administered.

One good example is provided by corticosteroids. A typical corticosteroid, such as hydrocortisone (1), undergoes generally multiple metabolic conversion, as illustrated in the following figure.

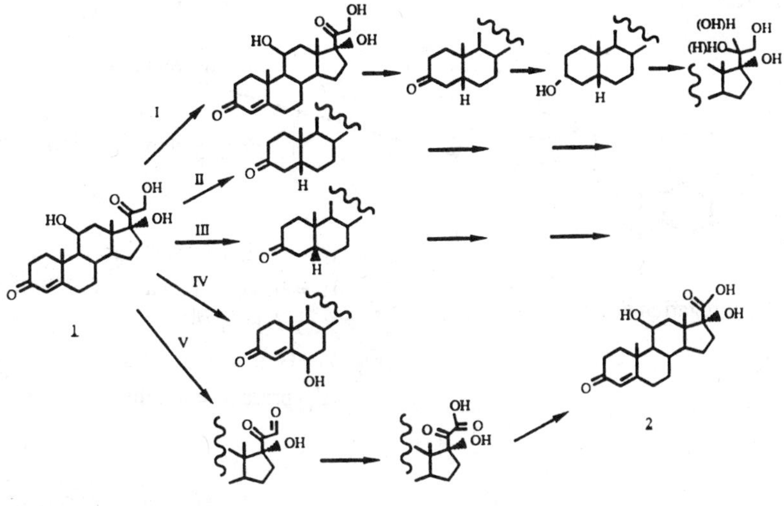

Fig. 2. The major routes of metabolism of hydrocortisone.

Enzymatic conversions of rings A or B, of the 11-hydroxy function, and the degradation of the dihydroxy acetone side chain, all take place. These are all oxidative or reductive conversions, which generally occur very easily if the compound, that is, the natural corticosteroid, is present in normal physiological concentrations. If, however, the concentrations are higher and/or if analogs which do not metabolize easily are introduced, the activity-toxicity profile changes substantially. Significant side effects, most of which are related to the ubiquitous corticosteroid receptors, will be present and sometimes prohibit use of even therapeutic doses of these steroids. The oxidation of the dihydroxy acetone side chain (process V) takes place step-wise, first the 20-keto-21 aldehyde, then the 20 keto-21-oic acid is formed, and finally the cortienic acid (2), in which the C-21 carbon is not present anymore. This is a common, inactive metabolite found in human urine after administration of corticosteroids. It is thus an ideal lead compound for applying the inactive metabolite approach. Modification of the 17-β-carboxyl function, in addition to the modifiers introduced in the 17-α and the usual substituents in 6,9 and 16 positions of the steroid ring, will lead to a host of more or less active analogs (3) of the basic corticosteroid. The activity will depend on the specific functions, but they should all have one common property, that is, hydrolysis of the 17-β-carboxylate to the corresponding substituted cortienic acid leading to inactivation. As shown from receptor binding studies (14), all substituted cortienic derivatives (4) have essentially zero binding to glucocorticoid receptors. On the other hand, the strategic modifications of the ester function and appropriate substitution of other positions as represented by 3 can lead to extremely potent corticosteroids, as indicated in Table I.

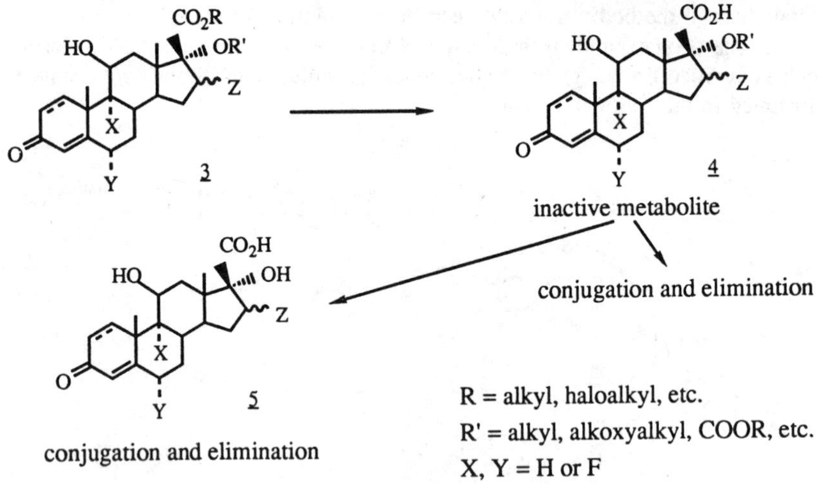

inactive metabolite

conjugation and elimination

conjugation and elimination

R = alkyl, haloalkyl, etc.

R' = alkyl, alkoxyalkyl, COOR, etc.

X, Y = H or F

Z = H or α- or β-CH_3

Δ^1 - present or absent

Table I. Binding of Selected Soft Glucocorticoids to the Glucocorticoid Receptor of Rat Lung

Compound 3	R_1	R_2	X_1	X_2	X_3	RBA[a]
a	i-C$_3$H$_7$	α-CH$_3$	Cl	F	H	560
b	n-C$_3$H$_7$	α-CH$_3$	Cl	F	H	870
c		α-CH$_3$	Cl	F	H	840
d	i-C$_3$H$_7$	β-CH$_3$	-CH-Cl[c] CH$_3$	H	F	11
e	C$_2$H$_5$	α-CH$_3$	CH$_2$Cl	F	H	19
f	C$_2$H$_5$	α-CH$_3$	Cl	F	H	740
g[b]	C$_2$H$_5$	α-CH$_3$	Cl	F	H	16
h	i-C$_3$H$_7$	α-CH$_3$	Cl	F	F	1,100
i	n-C$_3$H$_7$	α-CH$_3$	Cl	F	F	1,000
j	n-C$_3$H$_7$	α-CH$_3$	Cl	H	F	1,000
k	CH$_3$	α-CH$_3$	Cl	H	F	1,200
l	CH$_3$	β-CH$_3$	Cl	F	H	990
m	n-C$_3$H$_7$	β-CH$_3$	Cl	F	H	1,460
n	i-C$_3$H$_7$	α-CH$_3$	F	F	H	820
o	n-C$_3$H$_7$	α-CH$_3$	F	F	H	990
p	C$_2$H$_5$	H	F	H	H	200
r	i-C$_3$H$_7$	H	F	H	H	70
s	C$_2$H$_5$	α-CH$_3$	Cl	F	F	2,100
t	CH$_3$	H	Cl	H	H	180
u	C$_2$H$_5$	H	Cl	H	H	490
v	n-C$_3$H$_7$	H	Cl	H	H	540
w	H^3	β-CH$_3$	Cl	H	H	3
z	H^3	α-CH$_3$	Cl	F	H	7

[a] RBA dexamethasone = 100.

[b] 11-keto.

[c] Note branching:

One of the compounds, the one derived from unsubstituted prednisolone, which has a unique 17-α-ethyl carbonate function together with the chloromethyl ester in the 17-β position, was selected for development. This compound, loteprednol etabonate (3u), shows a good glucocorticoid receptor-binding activity of around 500 as compared to dexamethasone having 100, while its hydrolytic cleavage to the corresponding cortienic acid 17-α etabonate shows again no activity whatsoever. However, while the activity shown on the receptor binding is transformed into significant *in vivo* activity when applied topically, both as a dermatological agent or as an ophthalmic anti-inflammatory agent, or even to control allergic reactions by inhalation, while the typical corticosteroid side effects are significantly and dramatically reduced. The molecule is essentially metabolized during absorption, and thus no free active corticosteroid can reach the systemic receptors, including the ones in the thymus, and so it does not produce thymolysis, adrenal-and-immunosuppression. In animal studies, the therapeutic index of LE (3u) was found to be 24 as compared to 1 for betamethasone valerate, (clobetabol-17-α-propionate (1.5) or hydrocortisone 17-α-butyrate (1.3)). Human vasoconstrictor activity and topical anti-inflammatory activity in a variety of ophthalmic inflammatory diseases confirmed LE to be a very useful therapeutic agent, lacking however, the important side effects observed in humans, such as the extremely restrictive one of elevating intraocular pressure, that causes glaucoma.

Calculation at the AM1 level of the optimal conformers of various known active steroids, as compared to Loteprednol Etabonate, confirmed that the activity is associated primarily within the exposed 21-hetero atom, which in general is an oxygen (OH in hydrocortisone) or a Cl in clobetasol, while the "pseudo" 21-atom in LE is also Cl. Analogous esters which do not have this heteroatom placed in the right position, do not show any significant activity. Thus, this class of soft drug belongs to the first group, where the inactive metabolite essentially destroys the important component in the pharmacophore, which then has to be rebuilt in the way shown above, to get the soft drug.

Developing a soft β-blocker (7) starting from the well-known β-adrenergic antagonists metaprolol (6) and atenolol (7) involves a common inactive metabolite, again a carboxylic acid type, however, which is formed by oxidation of a part of the molecule which is not involved with the pharmacophore. The corresponding phenylaclic acid derivative (8), an isolated inactive metabolite, can be reactivated to soft β-blockers (9) using a wide variety of alcohols, which ultimately will affect the molecular lipophilicity and the rate of cleavage to the inactive metabolite. Here, we have much more freedom to modify the molecule without affecting the intrinsic activity too much. What is affected are the distributional and binding properties.

Accordingly, one of the lipophilic ester analogs, the adamantane ethyl derivative (9b), is being developed as an ophthalmic antiglaucoma antagonist, due to its enhanced lipophilicity, which, however, does not prevent the compound from being deactivated to the very inactive metabolite (8) during absorption through the gastrointestinal mucosa.

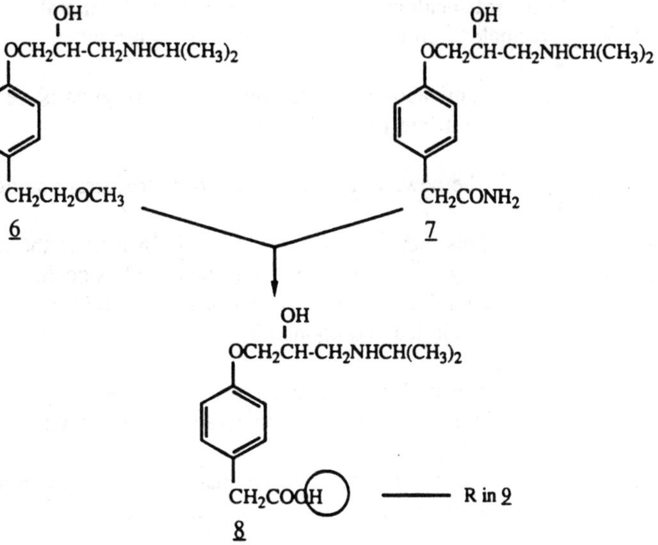

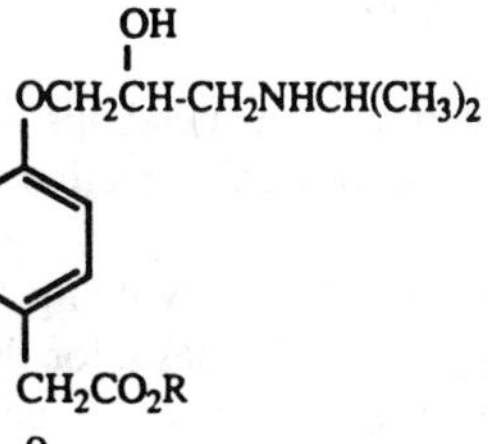

OH
|
$OCH_2CH\text{-}CH_2NHCH(CH_3)_2$

$CH_2CH_2OCH_3$

6

OH
|
$OCH_2CH\text{-}CH_2NHCH(CH_3)_2$

CH_2CONH_2

7

OH
|
$OCH_2CH\text{-}CH_2NHCH(CH_3)_2$

CH_2COOH ⃝ ——— R in **9**

8

OH
|
$OCH_2CH\text{-}CH_2NHCH(CH_3)_2$

CH_2CO_2R

9

R

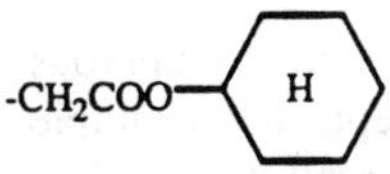

$-CH_2COO$⃝ H   **9a**

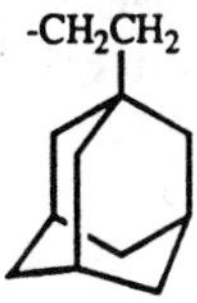

$-CH_2CH_2$ **9b**

R

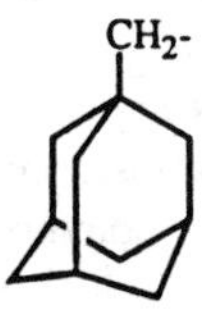

$CH_2\text{-}$ **9c**

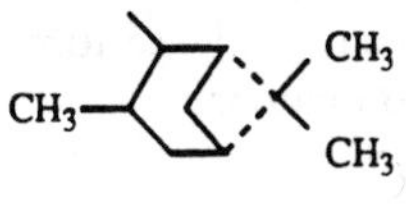

CH_3 CH_3 CH_3 **9d**

There are numerous practical examples, which were developed through the years based on empirical rules, and which can very easily be generalized and applied to essentially any drug class. In addition, the well-defined rules involved in designing soft analogs can also be described in general terms. According to this, a soft analog can be designed in the following way:

1)   A metabolically, preferentially hydrolytically, sensitive part is to be built into the lead molecule.

2)   The new, soft analog is a close structural analog.

3)   This metabolically weak spot is located in the molecule in such a way that overall physical, physicochemical, steric, and complementary properties of the soft analog are very close to those of the lead compound.

4)   The built-in metabolism is the major, or preferentially the only, metabolic route for deactivation of the drug.

5)   The rate of predictable metabolism can be controlled by structural modifications.

6)   The products resulting from the metabolism are nontoxic and have no significant biological activity.

7)   The predicted metabolism does not require enzymatic processes leading to highly reactive intermediates.

The combined rules involving soft analog design and the inactive metabolite approach were the basis for developing an expert computer program, which is general in nature and can be used for designing soft drugs starting with essentially any lead compound. Accordingly, the flow chart of the soft drug design, as part of the overall retrometabolic drug design approach, is shown below (Fig 3).

As shown on the flow chart, the computer program contains numerous specific transformation rules based on the soft drug concept, for example generating common oxidative metabolites by oxidations of methyl function, oxidation of hydroxymethyl function, or β-oxidation of various alkyl chains. Accordingly, corresponding carboxylic acid type metabolites are generated by the computer and then converted into various esters and reversed esters. The computer will perform the $-CH_3$ or $-CH_2OH \rightarrow -COOH \rightarrow -COOR$ transformations.

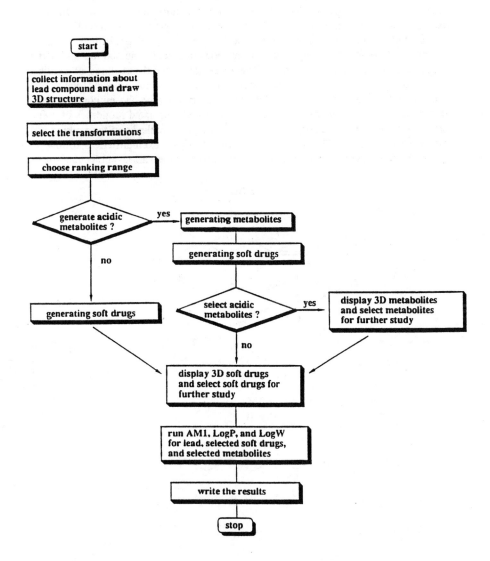

Fig. 3. Flow chart of the soft drug design.

There are a number of other ways to produce soft analogs: by strategic replacement, for example, of neighboring methylene groups by -O-CO or -CO-O- functions or other metabolically sensitive groups like thioesters or amides. In this example, we consider only the simple esters and perform the modifications exhaustively. That is, all possible modifications will be performed by the computer. Following the chosen transformations, the acidic metabolites and soft drugs are automatically generated. The three-dimensional structures of acidic metabolites and soft drugs are displayed on the screen and selected ones are chosen. AM1, LogP, and LogW calculations are run for the chosen compounds. Based on the overall order of closeness of combined calculated properties of soft drugs to those of the lead compound, the soft drugs are ranked and the ranking is displayed in a table. The use of this program can be exemplified as follows. Prostaglandins undergo facile oxidative metabolism at the various molecular sites (Fig 4).

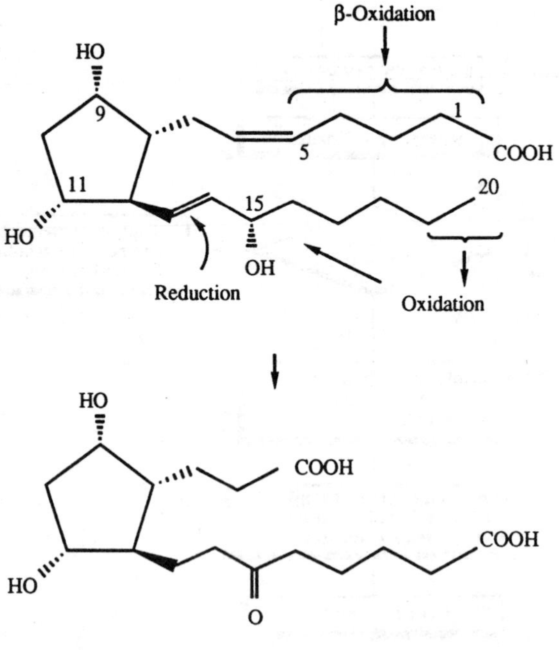

Fig. 4. General metabolism of prostaglandins in humans

Many times the metabolism of prostaglandins is too fast or too slow to produce selectively the desired activity. But most of the transformations are oxidative. By replacing them with hydrolytic processes, the activity could be better controlled and targeting could be achieved. In this particular case, starting from PGE_1, this exercise will lead to sixteen different soft drug analogs.

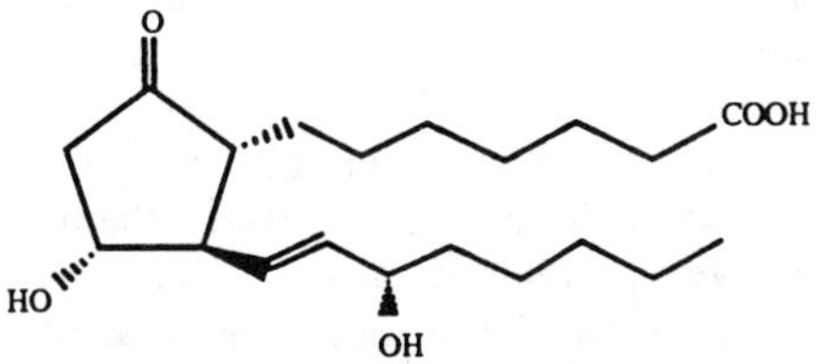

Prostaglandin E₁ (PGE1)

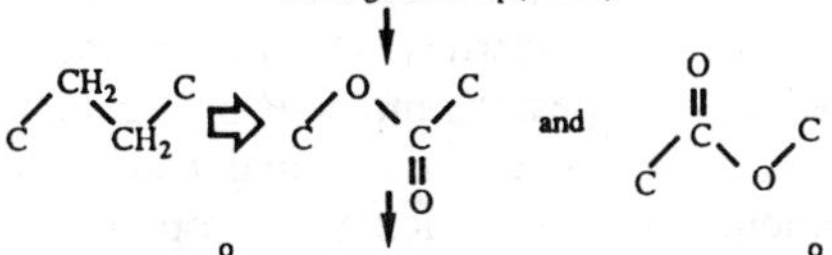

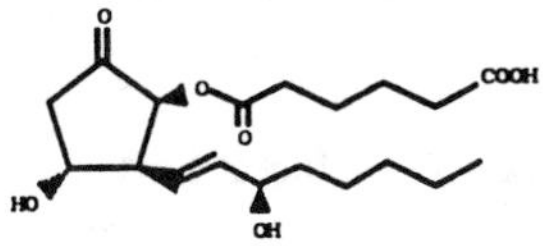

PGE1-1

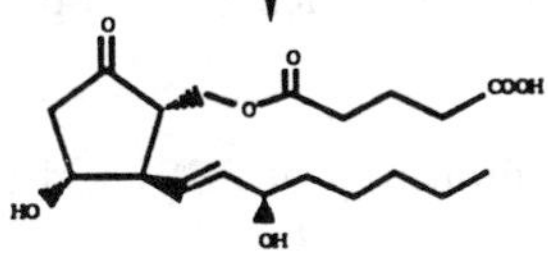

PGE1-2

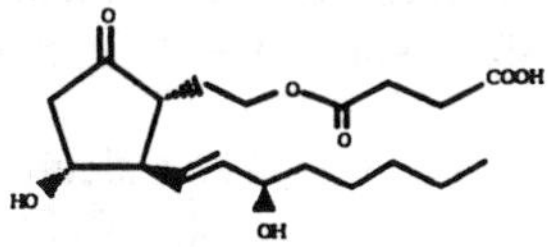

PGE1-3

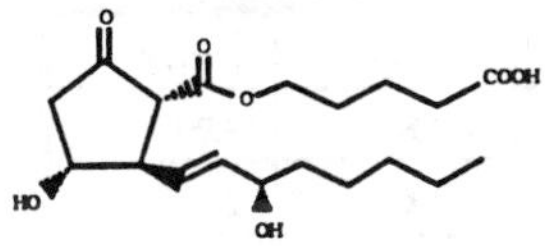

PGE1-4

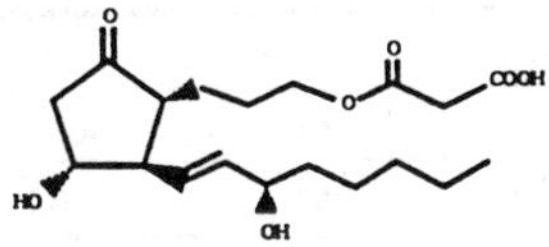

PGE1-5

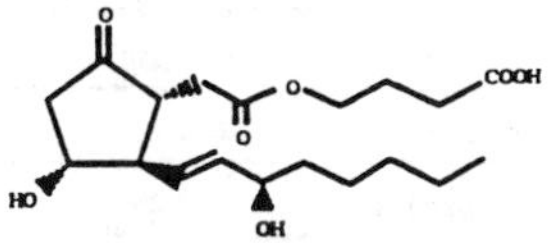

PGE1-6

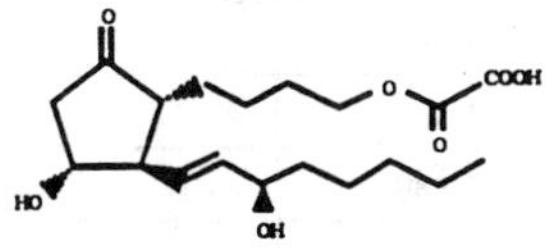

PGE1-7

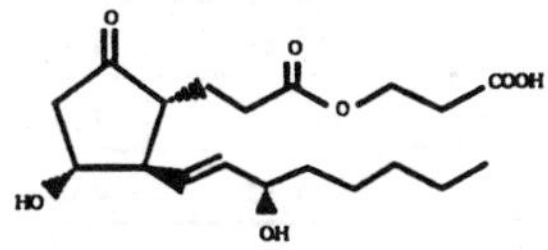

PGE1-8

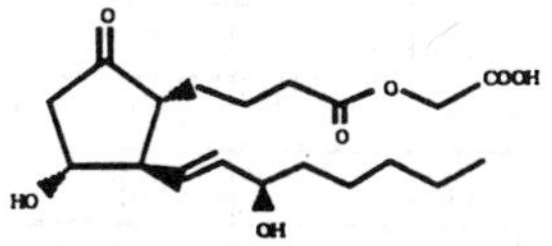

PGE1-9

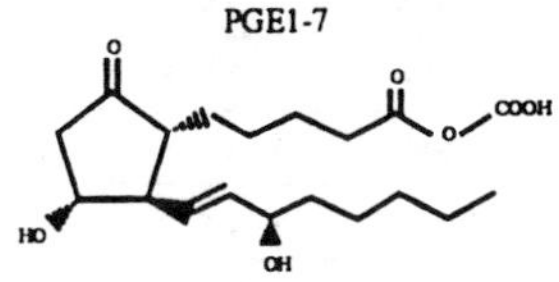

PGE1-10

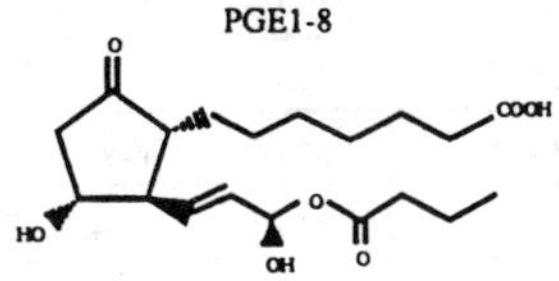

PGE1-11

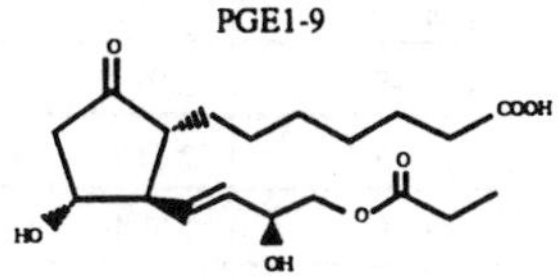

PGE1-12

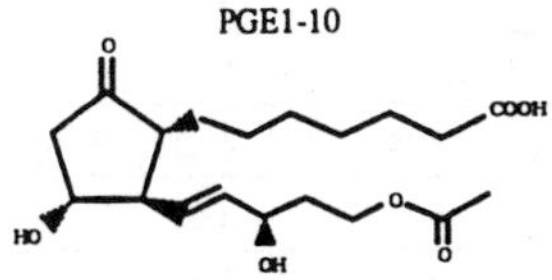

PGE1-13

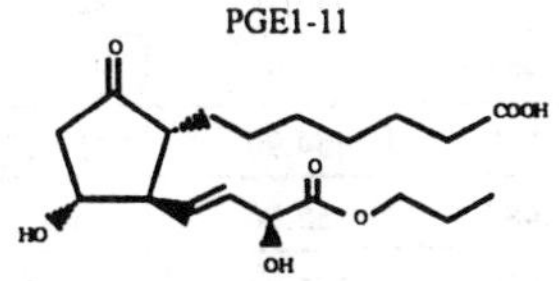

PGE1-14

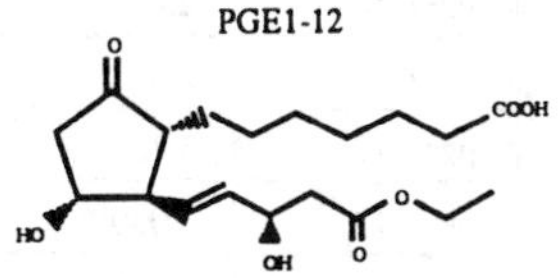

PGE1-15

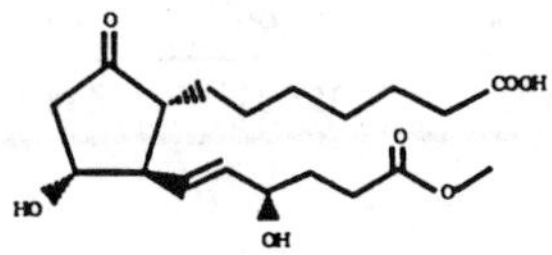

PGE1-16

These structures are distinctly different, but they have to be differentiated somehow quantitatively, as one certainly would not want to synthesize and test all of them. Accordingly, specific selection rules need to be introduced. The basic rules should be related to the closeness of the isosteric-isoelectronic, transport, and/or binding (and, ultimately, rate of metabolism) properties of the new drugs.

In order to be able to compare some of these properties, new methods of estimating lipo- and hydrophilicity (as expressed by the partition coefficients and aqueous solubilities) were developed (16-19), based on correlations between molecular descriptors calculated by quantum chemical semi-empirical methods, such as AM1. In this way the optimized structures (isomers) can be effectively compared, and the important properties which enter into consideration involve, besides the lipophilicity properties, the molecular volume, molecular surface area, molecular ovality (the ratio between the actual and minimum, that is, spheric surface), dipole moment, etc. The calculated properties are summarized in the following Table II:

Table II. The Calculated Properties of PGE1 and Its Sixteen Different Soft Drug Analogs.

cpds	LogP	LogW	Volume	Surface	Ovality	Dipole
pge1	4.231	2.732	361.01	468.02	1.909	3.902
pge1-1	3.290	5.283	354.92	460.71	1.900	2.403
pge1-2	3.026	6.410	355.07	459.24	1.894	4.875
pge1-3	2.695	5.879	354.21	450.19	1.860	2.126
pge1-4	3.093	5.903	355.13	460.35	1.898	4.809
pge1-5	2.909	5.472	355.15	458.12	1.889	4.416
pge1-6	3.169	6.118	355.09	459.40	1.894	2.952
pge1-7	2.994	3.455	355.05	459.24	1.894	3.078
pge1-8	2.867	6.458	355.03	457.50	1.887	5.215
pge1-9	2.771	5.111	354.46	451.58	1.864	1.512
pge1-10	2.800	4.177	354.70	458.16	1.891	4.909
pge1-11	2.832	7.981	354.59	455.96	1.882	5.474
pge1-12	3.105	6.528	354.90	459.22	1.894	4.053
pge1-13	3.050	6.489	355.06	458.35	1.890	3.629
pge1-14	3.029	6.070	355.04	458.60	1.891	4.230
pge1-15	3.027	6.606	354.90	458.01	1.889	4.006
pge1-16	3.033	6.538	355.05	458.03	1.889	3.805

As one can see, the partition coefficient, that is, the Log P, varies between 2.7 and 4.2, a 1.5 log-unit spread. The dipole varies between 1.5 and 5.5, again, a very significant difference. The dipole, and generally the charge distribution, should also affect binding properties of the molecule. The other calculated properties also reflect the molecular variability, although most of these are exact isomers. In the final step, the combined closeness of properties to the lead compound is considered, and the next figure shows the order obtained, while the structures of the closest analogs are depicted. The second most preferred compound, structure PGE 1-6, is identical to a compound recently patented (*20*).

Overall order of closeness of combined calculated properties of soft drugs to those of the lead compound, based on equally weighted contributions of LogP, LogW, volume, surface and dipole.

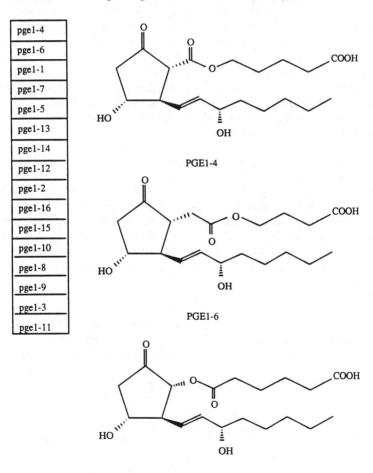

| pge1-4 |
| pge1-6 |
| pge1-1 |
| pge1-7 |
| pge1-5 |
| pge1-13 |
| pge1-14 |
| pge1-12 |
| pge1-2 |
| pge1-16 |
| pge1-15 |
| pge1-10 |
| pge1-8 |
| pge1-9 |
| pge1-3 |
| pge1-11 |

PGE1-4

PGE1-6

PGE1-1

One of the final selections involves the ease with which the soft drugs will be converted to the inactive metabolite. Depending on the specific use, this needs to be sometimes faster, sometimes slower. Depending on the class of drugs and specific use, the user is going to require certain properties from the soft drugs. A method of determining or estimating the rate of hydrolytic cleavage of the various ester-type soft drugs was developed, using a pattern-recognition technique involving the partial volumes around the ester functions of various known compounds.

The usefulness of the retrometabolic drug design concept was demonstrated in practice by developing a number of new drug candidates. Some of them are in the final stages of clinical development. For example the inactive metabolite based Loteprednol Etabonate (LE, 3u in Table 1) is currently completing Phase III clinical trial program in the US, for the treatment of ophthalmic inflammations and allergies. LE was found to be efficacious and to have a superior safety profile based (21) on the low incidence of observed side effects, (most noteworthly, essential lack of elevation of intraocular pressure which normally occurs in 15-25 precent of the patients) as compared to current ophthalmic steroid treatments. Some 740 patients with giant papilary conjunctivitis (GPC) and seasonal allergic conjunctivitis (SAC) were studied at 46 sites. LE is currently being developed for other uses, like asthma, dermatitis and inflammatory GI conditions.

Other soft drugs in clinical development include the soft β-blocker Adaprolol Maleate (9b) and the soft anticholinergic tematropium methyl sulfate (12).

These are general concepts, and the availability of an expert computer program which can be applied to various other classes of drugs will make the use of the concept much easier. The program is open to additions and modifications. Specific transformations can be added, the underlying semi-empirical method used to calculate molecular properties can be replaced, and the methods used to calculate molecular properties can also be upgraded.

Finally, the program and the general concepts can be extended to other classes of important organic chemicals, like pesticides, agrochemicals, etc.

Literature Cited

1.   Bodor, N.; Kaminski, J. *J. Med. Chem.* **1980**, *23*, 469-474.
2.   Bodor, N.; Kaminski, J. *J. Med. Chem.* **1980**, *23*, 566-569.
3.   Bodor, N.; Wood, R.; Raper, C.; Kearney, P.; Kaminski, J. *J. Med. Chem.* **1980**, *23*, 474-480.
4.   Bodor, N. In *Strategy in Drug Research*; Buisman, J.A.K., Ed.: Elsevier Scientific Publishing Company: Amsterdam, 1982.
5.   Bodor, N. *Chemtech* **1984**, 28-38.
6.   Bodor, N. *Med. Res. Reviews* **1984**, *3(4)*, 449-469.
7.   Bodor, N.; Oshiro, Y.; Loftsson, T.; Katovich, M.; Caldwell, W. *Pharm. Res.* **1984**, *3*, 120-125.
8.   Bodor, N.; Sloan K.; Little, R.; Selk, S.; Caldwell, L. *Int. J. Pharm.* **1982**, *10*, 307-321.
9.   Bodor, N.; Sloan, K. *J. Pharm. Sci.* **1982**, *71(5)*, 514-520.

10. Bodor, N. In *Topical Corticosteroid Therapy: A Novel Approach to Safer Drugs*; Christopers, E., et al., Eds.; Raven Press Ltd.: New York, 1988.
11. Bodor, N.; El-Koussi, A.; Kano, M.; Khalifa, M. *J. Med. Chem.* **1988**, *31*, 1651-1656.
12. Hammer, R.; Amin, K.; Gunes, Z.; Brouillette, G.; Bodor, N. *Drug Des. & Del.* **1988**, *2*, 207-219.
13. Bodor, N. In *Enc. of Human Biol.*: Academic Press: San Diego, 1991.
14. Bodor, N. In *Topical Glucocorticoids with Increased Benefit-Risk Ratio*; Korting, H., Ed.: A. G. Karger: Basel, 1993.
15. Bodor, N. and El-Koussi, A. *Cur. Eye Res.* **1988**, *7(4)*, 369-374.
16. Bodor, N.; Gabanyi, Z.; Wong, C. *J. Am. Chem. Soc.* **1989**, *111*, 3783-3787.
17. Bodor, N.; Huang, M. *J. Pharm. Sci.* **1992**, *81(9)*, 954-960.
18. Bodor, N.; Huang, M. *J. Pharm. Sci.* **1992**, *81(3)*, 272-281.
19. Bodor, N.; Harget, A.; Huang, M. *J. Amer. Chem. Soc.* **1991**, *113*, 9480-9483.
20. EPO Pat. 386901/Dec. 12, 1990 by ONO Pharmaceutical Co., Japan.
21. Pharmos Corporation, Phase III Clinical Trials Show Loteprednol Etabonate Efficacious for Eye Inflammation and Allergy Press release, New York, April 27, 1994.

RECEIVED November 7, 1994

Chapter 8

Molecular Mechanics and Dynamics Studies on Amide-Modified Backbones in Antisense Oligodeoxynucleotides

R. M. Wolf, V. Fritsch, A. De Mesmaeker, J. Lebreton, and A. Waldner

Central Research Laboratories, Ciba-Geigy Ltd., 4002 Basel, Switzerland

Replacing one or several phosphodiester linkages $-O-PO_2^--O-CH_2-$ in oligodeoxynucleotides by five structural isomers of an amide bond $-NH-CO-$ and two $-CH_2-$ groups yields antisense oligonucleotides that retain the ability to form duplexes with complementary mRNA. Molecular mechanics and dynamics simulations reveal that the modified sequences can assume various conformations which allow for standard Watson-Crick base pairing with a complementary RNA strand without major strain or steric hindrance. The overall structural features and dynamical behavior of the modified RNA·DNA hybrid duplexes are comparable although not identical to those of the wild-type RNA·DNA duplexes.

Antisense oligonucleotides represent a new class of potential therapeutical drugs. Their action is based on the repression of a defined protein by blocking specifically a portion of the corresponding mRNA (1-3). To be used as antisense agents, modified synthetic oligo(deoxy)nucleotides must retain the specificity towards complementary mRNA (given by the base sequence) and should have similar or higher binding constants with mRNA as compared to their wild-type analogues. Furthermore, they should be stable against degradation by nucleases and they should have an increased cell permeability with respect to natural oligonucleotides. Among the various possible modifications of natural oligonucleotides, backbone-modified nucleic acids look the most promising (4-6). Various approaches may be used to substitute the phosphodiester backbone which is exposed to the attack by nucleases. However, in order to allow for sequence-specific duplex formation with complementary RNA, the modified backbone must be able to adopt conformations which orient the bases in the best possible way for Watson-Crick base pairing. Although not necessarily required, this may be achieved most easily by keeping the number of bonds in the backbone as well as towards the nitrogen of the bases the same as in natural DNA, i.e., six bonds in the backbone and three bonds to the base nitrogen. Even for drastic changes in the backbone, like for example in

0097–6156/95/0589–0114$12.00/0
© 1995 American Chemical Society

"peptide nucleic acids" (PNAs) in which the sugar portion is also substituted, the numbers of bonds are respected (7).

The amide backbone modifications (see Figure 1) analyzed by molecular mechanics and dynamics in the scope of this work keep the sugar portions completely unaltered. Thus the modified backbone segment must be able to fold into a conformation which either resembles that of the natural phosphodiester or at least which leaves the orientation of the furanoses (and hence of the bases) essentially unchanged.

The structures shown in Figure 1 were found to satisfy various criteria in order to be used in antisense compounds (8-14). Details about synthesis and experimental data are published elsewhere (9-14). Melting points T_m of duplexes with complementary RNA were obtained for different sequences with various numbers of substituted phosphate linkages (see Table I).

Table I. Melting point differences between RNA · DNA duplexes with amide-modified DNA strands and the corresponding wild-type RNA · DNA hybrid duplexes. Average values over various sequences.

modification	amide **1**	amide **2**	amide **3**	amide **4**	amide **5**
ΔT_m per modification (°C)	-2.8	-1.6	+0.4	0.0	-3.5

Amide modifications **3** and **4** have the highest T_m values (superior to the corresponding wild-type RNA · DNA hybrid duplexes in some cases). The other three amide modifications increasingly destabilize the duplexes in the order amide **2**, **1**, and **5**. Although the experimental data clearly show differences between the various backbone modifications, these differences are not as spectacular as anticipated. Above all, an interesting similarity in the T_m values was observed between those amide modifications which can be regarded as isomers of the same *trans* double bond, namely amides **1** and **5**, and amides **3** and **4**, respectively. Without further proof, this may be taken as a first hint that the actual geometry of backbone modifications might be more relevant than the detailed electrostatics. The latter clearly change when reversing the orientation of the amide group, as is the case when passing from amide **1** to amide **5** or from amide **3** to amide **4**. Also, it turns out that modifications directly connected to one of the sugars (amide modifications **1**, **2**, and **5**) have a negative influence on the stability of the duplexes formed with complementary RNA.

In order to understand in more detail the structural features of the various amide backbone modifications, molecular mechanics (MM) and molecular dynamics (MD) studies were undertaken.

Computational Methods

MM and MD simulations were carried out with the AMBER all-atom force field (15)

$$
E_{total} = \sum_{bonds} \frac{k_l}{2} [l - l_0]^2 + \sum_{angles} \frac{k_\theta}{2} [\theta - \theta_0]^2 +
$$

$$
\sum_{torsions} \frac{V_n}{2} [1 + \cos(n\phi - \phi_0)] +
$$

$$
\sum_{i=1}^{N}\sum_{j>i}^{N} 4\varepsilon_{ij} \left[\left(\frac{\sigma_{ij}}{r_{ij}} \right)^{12} - \left(\frac{\sigma_{ij}}{r_{ij}} \right)^{6} \right] + \tag{1}
$$

$$
\sum_{i=1}^{N}\sum_{j>i}^{N} \left[\frac{332 \, \delta_i \, \delta_j}{\varepsilon \, r_{ij}} \right] + \sum_{H-bonds} \left[\frac{C_{ij}}{r_{ij}^{12}} - \frac{D_{ij}}{r_{ij}^{10}} \right]
$$

as incorporated in the software package *InsightII 2.2.0*/*Discover 2.9* from BIOSYM Technologies, San Diego, USA.

Electrostatic energy contributions were evaluated by using partial charges and applying Coulomb's law. Partial charges were assigned by an acceptor-donor scheme which reproduces as closely as possible the original AMBER charges (15) in the unmodified portions of the structures (unpublished work by Thacher, T., BIOSYM Technologies, San Diego). The permittivity was adjusted by a distance-dependent dielectric function $\varepsilon = 4 \cdot r_{ij}$, where r_{ij} is the distance separating two charges δ_i and δ_j. The use of a distance-dependent dielectric function was found appropriate to account for the absence of explicit solvent molecules and counterions (16-18). Specific 1-4 nonbonded interaction energy terms were reduced to 50% (19). No cut-offs were used.

Conformational Analysis. For the conformational analysis, starting structures were generated from an initial A-form RNA· DNA octamer duplex r(GA$_6$G)· d(CT$_6$C) (see Figure 2) by making the appropriate changes to introduce the desired backbone modification between the middle residues in the DNA strand. Note that "A-form" in this respect refers to the helical parameters, the backbone conformation, and the sugar puckering, i.e., all furanose units were initially in C3'-endo puckering mode. Obviously however, the sugars were free to adopt any energetically accessible puckering mode in subsequent MD simulations. Different conformers were generated by enforcing chosen backbone torsion angles incrementally by 30°, followed by a complete relaxation by conjugate gradient until the maximum derivative was < 0.1 kcal· mol^{-1}·Å^{-1}. The enforcement of a torsion angle ϕ to a predefined value ϕ_{cstr} was performed by applying an additional harmonic energy term E_{cstr} of the form

$$
E_{cstr} = k_{cstr} (\phi - \phi_{cstr})^2 \tag{2}
$$

where the force constant k_{cstr} was set to 1000 kcal· mol^{-1}· rad^{-2}. This procedure allows a scan through conformational space without an actual disruption of the duplex.

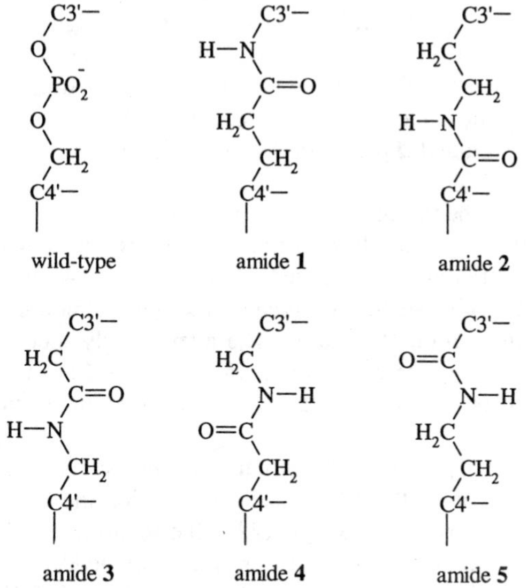

Figure 1. Amide-modified backbones replacing the wild-type phosphodiester (top left) in antisense oligo(deoxy)nucleotides.

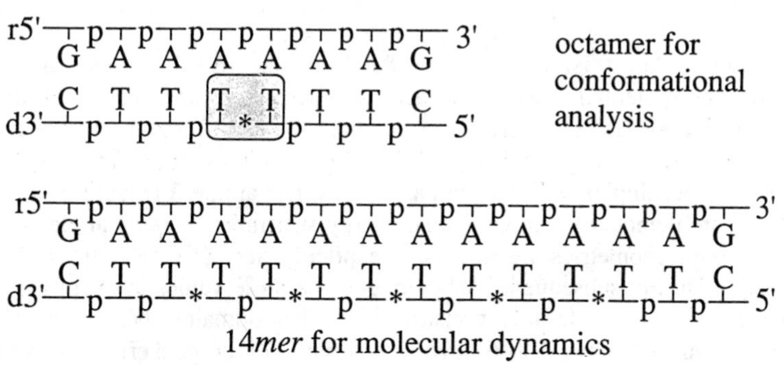

octamer for conformational analysis

14mer for molecular dynamics

Figure 2. Structures used for conformational analysis (top) and molecular dynamics (bottom). * designates the replacement of a phosphodiester linkage by an amide modification.

Although the method cannot ensure the finding of all possible local minima, it yields a representative range of low-energy conformers to be considered further.

Torsion angles around bonds directly connected to one of the sugar rings, i.e., either to C3' in the case of residue *i* or to C4' for residue *i+1* in Figure 3, were not enforced. The enforced torsion angles for the amide modifications were α and β for amides **1** and **5**, ζ and α for amide **2**, ζ and β for amides **3** and **4** (see Figure 3 and reference (20) for definitions). Note that the amide bond was always preset to either *trans* (180°) or *cis* (0°) and then let free to relax, i.e. deviating moderately from the planar structure. Although the *cis* amide is energetically less favorable in the isolated form, its occurrence was not excluded *a priori* in the modified duplex structures.

Molecular Dynamics. The various local-energy-minimum conformers found during the conformational analysis were used as starting geometries for the MD runs, which were carried out on alternatingly modified RNA·DNA 14*mer* duplex structures r(GA$_{12}$G)·d(CT(T*T)$_5$TC), where * designates a specific amide-modified linkage in a defined starting conformation (Figure 2). The alternatingly modified structures were chosen for simulations because of the experimental feasibility of such backbones, for which modified dimers can be connected by standard nucleotide oligomerization techniques (21).

Prior to molecular dynamics, the structures were completely minimized. MD simulations were then run in the *NVT* ensemble, keeping the temperature constant at 300 K by coupling to an external heat bath (22). One-femtosecond time steps were used for the numerical integration. The system was preconditioned by heating stepwise from 0 to 300 K over a period of 24 picoseconds: 2 ps at 50 K and 100 K, 4 ps at 150 K and 200 K, 5 ps at 250 K, 7 ps at 300 K. Trajectories were then recorded for 100 ps. Instantaneous coordinates were saved every 0.5 ps for subsequent analysis.

Results and Discussion

Conformational Analysis. For each of the backbone modifications various local minima on the potential energy surface were found, sometimes differing only marginally in energy. Lowest-energy conformations for the different backbone-modified structures are shown in Figures 4 to 6. These figures represent the T*T sequence cut out of the octamer duplex without the base atoms. For amide **3** (Figure 5) and amide **4** (Figure 6) are depicted the three lowest-energy structures. Torsion angles for local-energy-minimum geometries are not listed explicitly here (see references (9-13) for more details). During subsequent MD computations these values are subject to either moderate oscillations or definite transitions to other domains. Note that all amide structures shown in Figures 4 to 6 have sugars with N-type puckering (between C3'-endo and O4'-endo), i.e., the torsion enforcement on the modified backbone linkages did normally not alter considerably the puckering mode from the C3'-endo starting geometry. A more detailed discussion of backbone torsion angle transitions and of the sugar puckering follows in the subsequent sections dealing with MD simulations.

A comparison between the lowest-energy conformations of amide **1** and amide **5** (Figure 4), respectively of amide **3** and amide **4** (Figures 5 and 6) reveals a close geometrical resemblance between these structures that are the respective isomers of the same hypothetical *trans* double bond. The small differences can be attributed to the

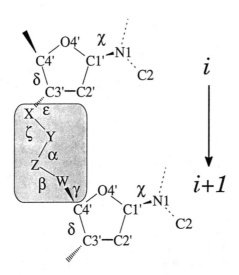

Figure 3. Definition of torsion angles in backbone-modified oligonucleotides: $X-Y-^{\alpha}-Z-^{\beta}-W-^{\gamma}-C4'-^{\delta}-C3'-^{\varepsilon}-X-^{\zeta}-Y-Z$; $O4'-C1'-^{\chi}-N1-C2$ (according to reference (20).

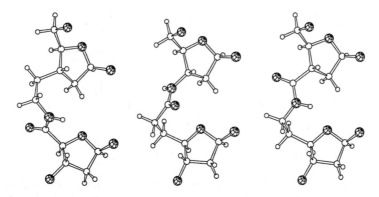

Figure 4. Lowest-energy geometries for the amide backbone modifications 2 (left), 1 (center), and 5 (right). The structures were cut out of the octamer duplex. Base atoms are not shown for clarity.

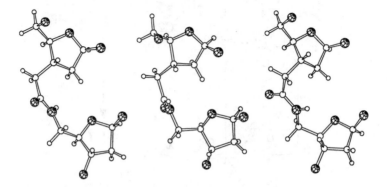

Figure 5. Three lowest-energy geometries for the amide **3** backbone modification: **3a** (left), **3b** +0.5 kcal·mol⁻¹ (center), **3c** +2.9 kcal·mol⁻¹ (right). Energy values refer to the entire octamer duplex as shown in Figure 2. See also legends of Figure 4.

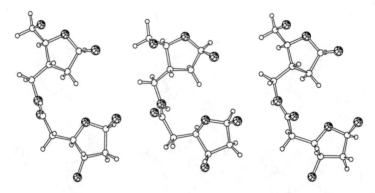

Figure 6. Three lowest-energy geometries for the amide **4** backbone modification: **4a** (left), **4b** +1.6 kcal·mol⁻¹ (center), **4c** +3.8 kcal·mol⁻¹ (right). See also legends of Figures 4 and 5.

repulsion between the amide oxygen and the furanose ring oxygen (O4') in amides **1** and **4**, as compared to the corresponding attraction of the amide hydrogen to the O4' atom in amides **5** and **3**, respectively. Overall, the similar lowest-energy backbone conformations corroborate the interpretation of the experimental results that geometrical considerations seem to govern the relative stability of the modified duplexes.

In amides **1** and **5**, the amide bond corresponds to the backbone torsion angle ζ which is thus forced to roughly *trans*. The other backbone torsion angles have to adjust to this "unnatural" conformation, resulting in an overall backbone arrangement not commonly found in nucleic acids, e.g., α and β in *gauche plus* (g^+).

In the amide **2** modification, β coincides with the amide bond. The lowest-energy conformer is found to adopt a backbone conformation similar to that in wild-type A-DNA, i.e, all backbone torsion angles automatically adjust to the standard conformational ranges found in DNA once β is fixed to the "natural" *trans* range. In that sense, the amide **2** modification is the perfect geometrical match of a natural DNA backbone.

In both amide **3** and amide **4**, the amide bond corresponds to α. In standard DNA duplexes, α is generally in the *gauche minus* (g^-) range and the torsion angle γ is in the *gauche plus* (g^+) range. However, in another low-energy geometry, found also experimentally in A-DNA crystals (*23-25*), both α and γ are in the *trans* range. The lowest-energy conformations induced by the modifications amide **3** and amide **4** are found to have [α,γ] in [*t,t*] with the other torsion angles adopting the same conformational ranges as in the alternative A-DNA structures mentioned above. Thus, both amide **3** and amide **4** modifications induce a backbone conformation found experimentally in X-ray studies on DNA and hence virtually free of strain.

Molecular Dynamics. It is experimentally established that in RNA·DNA hybrid duplexes, the RNA strand riboses stay in the C3'-endo puckering domain, the DNA strand deoxyriboses adopt an average puckering mode between O4'-endo and C1'-exo, and the global helical parameters of the hybrid duplexes are closer to A-form than to B-form (*26-30*). Considering furthermore the low-energy barriers for deoxyribose puckering transitions, the A-form starting geometry seems appropriate for MD simulations. This was also verified in MD simulations of a wild-type RNA·DNA hybrid duplex r(GA$_{12}$G)·d(CT$_{12}$C) for which the experimental results were correctly reproduced (*31*). The various modifications in different low-energy conformations were introduced alternatingly in the DNA strand of the 14*mer* RNA·DNA duplex (Figure 2). For the amide modifications, the resulting energy differences in the 14*mer* duplex between the distinct conformers were roughly proportional to the differences observed for a single modification in the octamer duplexes investigated in conformational analysis. This finding may be explained by a compensation of geometrical changes by the alternating wild-type phosphodiesters in such a way that consecutive amide modifications do not "feel" each other when separated by a natural linkage.

MD results were analyzed with respect to backbone conformational transitions and to the behavior of the sugar puckering in the modified strands. Detailed helical parameters were not considered at this stage of the investigation. For a global overview, Figures 7 and 8 depict the 14*mer* duplex starting structures and the average dynamics geometries, with coordinates averaged over the 100 ps trajectories. These figures visualize qualitatively the fact that the amide modifications lead to stable duplexes in which base pairing is more or less conserved on average, although momentary disruptions were noticed during the MD trajectories, an observation made also for the wild-type reference structure (structures on the left in Figures 7 and 8).

In Table II are listed time-averaged values (with standard deviations in parentheses) for the backbone torsion angles, the glycosidic torsion angle χ and the sugar puckering parameters P and τ_m (defined according to reference (32)) for the middle residues **T*T** of the amide-modified 14mer duplexes $r(GA_{12}G) \cdot d(CT(T*T)_5 TC)$. Note that there was no substantial difference between the behavior of these middle dimers and the adjacent modification-linked dimers. All data reported in Table II were obtained starting from the lowest-energy conformations as shown in Figures 4 to 6. The corresponding data for an unmodified RNA·DNA duplex, obtained under identical conditions, are listed for comparison.

Backbone Transition. DNA strands with amide modifications **1** and **3** oscillated around the lowest-minimum-energy conformation shown in Figures 4 to 6. This is evidenced by the small standard deviations for the backbone angles reported in Table II. For amide **3**, starting from the second-lowest-energy conformer (**3b** in Figure 5), four out of five modified residues converted into the lowest-energy geometry (**3a** in Figure 5) during the 100 ps dynamics run. During this simulation period, no transition from **3a** to **3b** was observed. Since the conversion from **3b** to **3a** is readily observed, the involved barrier can be overcome on the 100 ps time scale. Also, the difference in potential energy between the two conformations is rather small (~ 0.5 kcal·mol^{-1}). The absence of **3a** $\rightarrow$ **3b** transitions has been attributed to the larger puckering amplitude observed in the **3a** conformation, i.e., the overall entropy of the system increases when passing from **3b** to **3a**. Thus, the **3a** conformation would be preferred over **3b** for free energy reasons. A more detailed analysis on these transitions and their possible reasons will be given elsewhere (Wolf et al., *submitted for publication*).

In amide **2** and amide **5** modified DNA strands, transitions in the modified backbone portions were observed even when starting from the lowest-energy geometry as depicted in Figure 4. The amide **2** modified strand underwent transitions [α,γ] from [g^-,g^+] to [t,t] in various unmodified portions, but also in the amide-modified residues. This type of transition is commonly observed in molecular dynamics simulations on wild-type DNA·DNA and RNA·DNA duplexes (see e.g. references (31,33,34). Its occurrence in the simulation of the amide **2** modified backbone portions underlines the geometrical similarity of the amide **2** modification and the natural phosphodiester linkage, already observed in the conformational analysis. In amide **5** modified strands α and ε in the modified part oscillated coherently between two conformational domains, the lowest-energy domain as depicted in Figure 4 and the next-lowest-energy domain found in the conformational analysis. Similarly, the amide **4** modified portions showed transitions between the low-energy structures **4a** and **4b** depicted in Figure 6. These transitions are related to changes in the torsion angles β and ζ, as seen also in the higher values for the MD standard deviation of these angles in Table II.

Sugar Puckering. The alternating character of the amide-modified backbones has an interesting effect on the overall sugar puckering scheme. Indeed, there are two different types of deoxyriboses, one having the modified backbone part attached to the C3' carbon (residue *i* in Figure 3) and one having the modified backbone sequence

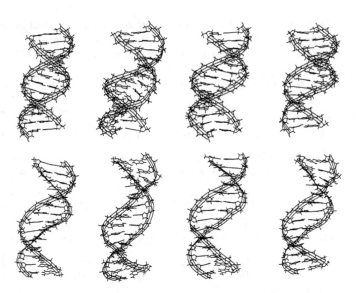

Figure 7. Starting structures (top) and average dynamics geometries (bottom) of 14mer duplexes. From left to right: wild-type RNA·DNA, amide 1, 2, and 3a modified duplexes. The 5'-end of DNA or modified DNA strands is always at the left top of each duplex. The ribbons go through C3' as trace atom with C2' being the plane atom.

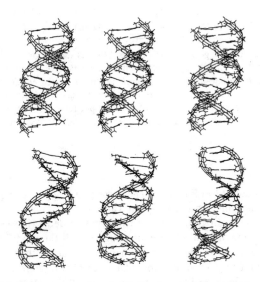

Figure 8. Starting structures (top) and average dynamics geometries (bottom) of 14mer duplexes. From left to right: wild-type RNA·DNA, amide 4, and amide 5 modified duplexes. See also legend of Figure 7.

Table II. Time averages and standard deviations (in parentheses) of torsion angles and puckering parameters (in degrees) for the central dimer in the amide-modified DNA strand in 14*mer* duplexes r(GA$_{12}$G)·d(CT(T*T)$_5$TC).

	wild-type	amide 1	amide 2	amide 3a[a]	amide 4a[a]	amide 5
α[b]	-70 (11)[c]	-73 (12)	-71 (12)	-74 (11)	-78 (11)	-75 (11)
	-70 (12)	**60 (14)**[d]	**-140 (64)**	**171 (8)**	**171 (9)**	**105 (40)**
β	175 (10)	172 (9)	175 (10)	179 (8)	174 (9)	171 (9)
	176 (10)	**64 (7)**	**179 (10)**	**-149 (24)**	**-94 (21)**	**69 (9)**
γ	63 (9)	59 (9)	60 (10)	58 (10)	59 (10)	61 (10)
	62 (11)	**62 (10)**	**127 (58)**	**173 (9)**	**178 (9)**	**67 (10)**
δ	108 (18)	98 (23)	83 (14)	94 (19)	84 (16)	88 (18)
	108 (17)	**90 (14)**	**104 (27)**	**107 (25)**	**100 (23)**	**90 (18)**
ϵ	-176 (8)	**-178 (14)**	**-175 (8)**	**176 (9)**	**169 (9)**	**155 (44)**
	-176 (9)	-175 (10)	-170 (10)	-166 (9)	-166 (10)	-172 (9)
ζ	-89 (12)	**180 (11)**	**-71 (13)**	**-100 (18)**	**-137 (24)**	**-176 (10)**
	-88 (11)	-83 (12)	-78 (14)	-79 (13)	-76 (13)	-79 (12)
χ	-139 (16)	-134 (19)	-156 (12)	-156 (13)	-154 (11)	-145 (19)
	-140 (15)	-139 (16)	-146 (15)	-154 (17)	-156 (15)	-139 (16)
P[e]	110 (27)	82 (41)	66 (28)	86 (35)	61 (33)	63 (40)
	112 (26)	93 (23)	88 (55)	103 (43)	90 (40)	87 (31)
τ_m[e]	42 (5)	44 (6)	41 (6)	42 (6)	42 (6)	43 (6)
	42 (6)	46 (5)	40 (6)	42 (6)	42 (6)	43 (6)

[a] Using the lowest-energy geometry in Figures 4 to 6 as starting points. [b] See Figure 3 and reference (20) for torsion angle definitions; [c] the first line for each value refers to the residue with the amide-modified backbone portion attached at C3' (*i* in Figure 3), the second one to the residue with the modification attached to C4' (*i+1* in Figure 3); [d] values in bold correspond to modified backbone portions; [e] P is the phase angle of pseudorotation and τ_m the maximum degree of puckering (see reference (32) for definitions).

attached to C4' (residue *i+1* in Figure 3). The corresponding puckering modes are found on the first and second line, respectively, in Table II.

For all five amide modifications, these two types of deoxyriboses have a different average puckering behavior. In all cases the sugars with the modification attached at C3' (*i*) have smaller average values for P than the sugars with the modified portion bonded to C4' (*i+1*). Considering that the difference of two degrees observed between the P values of the middle residues in the wild-type RNA·DNA duplex reflects the random character of molecular dynamics, the corresponding differences in the amide-modified strands are significant (e.g. almost 30° for amide **4**).

Furthermore, the sugar puckering scheme was found to depend also on the actual backbone conformation, and not only on the nature of the amide modification. Indeed, starting the dynamics trajectories from other low-energy conformers (not shown in Table II) leads to a different puckering scheme in some cases, as already reported in the previous section. The backbone conformation controls both the amplitude and the mean value of the phase angle of pseudorotation P. Furthermore, the puckering state of the sugar determines to some extent the torsion angle to the base χ. Thus the various substructures in the modified DNA strands are strongly interrelated and modifications to one part inevitably affect other portions. Details about this structural interdependence will be given elsewhere (Wolf et al., *submitted for publication*).

The (unmodified) complementary RNA strands showed no unusual behavior in any of the amide-modified duplexes. All riboses remained confined to the C3'-endo puckering mode as has also been found for RNA strands during MD simulations of wild-type RNA·DNA hybrid duplexes (*31*).

Conclusions

The simulation results of alternatingly amide-modified DNA strands paired to complementary RNA sustain the concept that such modifications do not introduce considerable strain or steric hindrance.

The amide-modified duplex structures behave quite similarly to the natural RNA·DNA hybrids (*26-31*) with all riboses in the RNA strand adopting the standard A-type puckering C3'-endo, whereas the deoxyriboses oscillate between the classical C3'-endo and C2'-endo puckering modes, with an average value concentrating around O4'-endo. Some modifications can adopt "natural" DNA conformations because the amide bond corresponds to a torsion angle which is *trans* (like β in the canonical A-DNA, or α in the other low-energy conformation found in standard DNA (*23-25*)). Thus, amide **2**, with the amide bond corresponding to β, adopts all torsion angles in the range of wild-type DNA. Furthermore, the transition $[\alpha,\gamma]$ from $[g^-,g^+]$ to $[t,t]$, observed in MD simulations of wild-type DNA·DNA (*33,34*) or RNA·DNA (*31*) is also noticed for this modification. Both amide **3** and amide **4**, with the amide bond corresponding to α, adopt the lowest-energy conformation with $[\alpha,\gamma] = [t,t]$.

Obviously, the simulations alone cannot explain the T_m differences observed between the five amides reported. These differences would have to be analyzed from the point of view of free energy, i.e., including entropy considerations. Such an approach is currently excluded considering the very large conformational space available to single strands. Still, the simulations have revealed structural details concerning possible backbone conformations and sugar puckering schemes which may be considered as useful hints for the design of further backbone modifications of this type.

Literature Cited

1. Uhlmann, E. and Peyman, A. *Chem. Rev.* **1990**, *90*, 543.
2. Crooke, S.T. *Annu. Rev. Pharmacol. Toxycol* **1992**, *32*, 329.
3. Cook, P.D. *Anti-Cancer Drug Design* **1991**, *6*, 585.
4. Quaedflieg, P.J.L.M.; van der Marel, G.A.; Kuyl-Yeheskiely, E.; van Boom, J.H. *Recl. Trav. Chim. Pays-Bas* **1991**, *110*, 435.
5. Vasseur, J.-J.; Debart, F.; Sanghvi, Y.S.; Cook, P.D. *J. Am. Chem. Soc.* **1992**, *114*, 4006.

6. Jones, R.J.; Lin, K.Y.; Milligan, J.F.; Wadwani, S.; Matteucci, M.D. *J. Org. Chem.* **1993**, *58*, 2983.
7. Nielsen, P.E.; Egholm, M.; Berg, R.H.; Buchardt, O. *Science* **1991**, *254*, 1497.
8. De Mesmaeker, A.; Lebreton, J.; Waldner, A.; Cook, P.D. *International Patent WO 92/20823*, **1992**.
9. Lebreton, J.; De Mesmaeker, A.; Waldner, A.; Fritsch, V.; Wolf, R.M.; Freier, S.M. *Tetrahedron Lett.* **1993**, *34*, 6383.
10. De Mesmaeker, A.; Waldner, A.; Lebreton, J.; Hoffmann, P.; Wolf, R.M.; Freier, S.M. *Angew. Chem. Int. Ed. Engl.* **1994**, *33*, 226.
11. De Mesmaeker, A.; Lebreton, J.; Waldner, A.; Fritsch, V.; Wolf, R.M.; Freier, S.M. *Synlett* **1993**, 733.
12. De Mesmaeker, A.; Lebreton, J.; Waldner, A.; Fritsch, V.; Wolf, R.M. *Bioorg. Med. Chem. Lett.* **1994**, *4*, 873.
13. De Mesmaeker, A.; Waldner, A.; Fritsch, V.; Wolf, R.M.; Lebreton, J. *Tetrahedron Lett.* **1994**, *35*, 5225.
14. Lebreton, J.; Waldner, A.; Lesueur, C.; De Mesmaeker, A. *Synlett* **1994**, 137.
15. Weiner, S.J.; Kollman, P.A.; Nguyen, D.T.; Case, D.A. *J. Comp. Chem.* **1986**, *7*, 230.
16. Whitlow, M. and Teeter, M.M. *J. Am. Chem. Soc.* **1986**, *108*, 7163.
17. Orozco, M.; Laughton, C.A.; Herzyk, P.; Neidle, S. *J. Biomol. Struct. Dyn.* **1990**, *8*, 359.
18. Ferentz, A.E.; Wiorkiewicz-Kuczera, J.; Karplus, M.; Verdine, G.L. *J. Am. Chem. Soc.* **1993**, *115*, 7569.
19. Weiner, S.J.; Kollman, P.A.; Case, D.A.; Singh, U.C.; Ghio, C.; Alagona, G.; Profeta, S.J.; Weiner, P. *J. Am. Chem. Soc.* **1984**, *106*, 765.
20. IUPAC-IUB, Joint Commission on Biochemical Nomenclature. *Eur. J. Biochem.* **1983**, *131*, 9.
21. Gait, M.J. *Oligonucleotide Synthesis: A Practical Approach;* IRL Press: Oxford, 1984.
22. Berendsen, H.J.C.; Postma, J.P.M.; van Gunsteren, W.F.; DiNola, A.; Haak, J.R. *J. Chem. Phys.* **1984**, *81*, 3684.
23. Frederick, C.A.; Quigley, G.J.; Teng, M.-K.; Coll, M.; van der Marel, G.A.; van Boom, J.H.; Rich, A.; Wang, A.H.-J. *Eur. J. Biochem.* **1989**, *181*, 295.
24. Ramakrishnan, B. and Sundaralingam, M. *Biochemistry* **1993**, *32*, 11458.
25. Ramakrishnan, B. and Sundaralingam, M. *J. Mol. Biol.* **1993**, *231*, 431.
26. Fedoroff, O.Y.; Salazar, M.; Reid, B.R. *J. Mol. Biol.* **1993**, *233*, 509.
27. Chou, S.-H.; Flynn, P.; Reid, B. *Biochemistry* **1989**, *28*, 2435.
28. Salazar, M.; Fedoroff, O.Y.; Miller, J.M.; Ribeiro, N.S.; Reid, B.R. *Biochemistry* **1993**, *32*, 4207.
29. Hall, K.B. *Curr. Op. Struct. Biol.* **1993**, *3*, 336.
30. Lane, A.N.; Ebel, S.; Brown, T. *Eur. J. Biochem.* **1993**, *215*, 297.
31. Fritsch, V. and Wolf, R.M. *J. Biomol. Struct. Dyn.* **1994**, *11*, 1.
32. Rao, S.T.; Westhof, E.; Sundaralingam, M. *Acta Cryst.* **1981**, *A37*, 421.
33. Brahms, S.; Fritsch, V.; Brahms, J.G.; Westhof, E. *J. Mol. Biol.* **1992**, *223*, 455.
34. Fritsch, V.; Ravishanker, G.; Beveridge, D.L.; Westhof, E. *Biopolymers* **1993**, *33*, 1537.

RECEIVED November 2, 1994

Chapter 9

Simulations of Drug Diffusion in Biomembranes

Terry R. Stouch, Howard E. Alper[1], and Donna Bassolino

Department of Macromolecular Structure, Bristol-Myers Squibb
Pharmaceutical Research Institute, Princeton, NJ 08654

Bioavailability, largely due to membrane permeation, is an important step in the drug delivery process and therefore drug design. Here we discuss, over 10nsec of molecular dynamics simulation of small molecules in lipid bilayer membranes that were used to elucidate the mechanism of diffusion of drugs within biomembranes.

These simulations accurately reproduce many experimentally observed parameters. The simulations also agree with theory that indicates the lipids have internal structure that influences the diffusion process. These simulations show that even within the hydrocarbon chains there are distinct regions between which the rate and mechanism of diffusion varies. These variations are linked to the frequency and size of spontaneously arising voids within the bilayer as well as the rate of torsional isomerization of the hydrocarbon chains.

Bioavailability is an important component of drug efficacy. Many promising drug candidates have been abandoned because of their inability to travel from their site of administration to their site of action. Needless to say, this process is the subject of much research.

Often, bioavailability is dependent on passive transport across biological membranes. Correlations are often drawn between a molecule's bioavailability and its partition coefficient between water and organic phases (1). These coefficients are a thermodynamic measure of the relative solubility of a compound between the phases. Membrane

---

[1]Current address: Moldyn Corporation, Cambridge, MA 02138

0097–6156/95/0589–0127$12.00/0
© 1995 American Chemical Society

permeation, however, is a measure of the flux of a compound from one side of a membrane to another and includes, in addition to relative solubility, the rate of movement within the membrane. Correlations have also been drawn between permeability itself and partition coefficients. Although these correlations are good for many molecules, they do not universally hold. Small molecules (MW ~<100 amu) show especially anomalous behavior (2-4).

This is actually consistent with earlier theories of Lieb and Stein (5, 6) who speculated that the mechanism of diffusion within membranes may be similar to that for solutes within soft polymers. There, diffusion is thought to occur by intermittent "jumps" between spontaneously arising voids within the polymer. This suggests that molecules which are small enough to fit within the voids present in biomembranes show enhanced rates of permeation because they have enhanced rates of diffusion due to this jumping mechanism.

The observed high correlations between permeation and partition coefficients for many compounds has led to the incorrect-but-common assumption that the water/organic interface is a good representation of a biomembrane. Yet, due to the amphiphilic nature of the lipid molecules comprising biomembranes, biomembrane properties (including the ordering of the hydrocarbon region) are different from those of bulk hydrocarbon. This is further supported by a growing body of theoretical and experimental data. Statistical mechanical studies of Dill and coworkers clearly demonstrate the interfacial nature of a membrane. The anomalous behavior of small membrane permeants, mentioned above, also suggest differences. In addition, White and coworkers (7, 8) have shown this to be true in experimental studies of membrane/solute interactions.

Here we review atomic-level molecular dynamics simulations done in our laboratory which support both the concepts on diffusion proposed by Lieb and Stein as well as the concept of internal biomembrane structure proposed by Dill and coworkers (9, 10) and by White and coworkers (7, 8). We find that small solutes do indeed travel by jumping between available voids in the bilayer and that this motion is size dependent. We also show that this movement is not homogeneous throughout, but varies with distance from the membrane/water interface which suggests an internal membrane structure unlike that of bulk alkane.

METHODOLOGY

Although much useful understanding can often be gained from static structures of some biomolecules, such as proteins, biology is a dynamic process that is dependent on molecular motion. This is perhaps nowhere as true as for biomembranes whose properties are completely dependent on their fluidity. The magnitude of this fluidity is of such importance that nature maintains tight control of it through the content of cholesterol and unsaturated lipid molecules. Because of this, we use molecular dynamics (MD) simulations as a tool to study biomembranes.

Previously, we studied the ability of empirical force fields and MD methods to duplicate experimenatally-known properties of both the crystal structures of lipid molecules as well as those of the physiologically-relevant Lα phase of lipid bilayers. These simulations have successfully reproduced a wide range of structural and dynamical properties (11-16). Confident in the physical relevance of these simulations, we extended them to study the diffusion of solutes within the bilayers. Much of our discussion here will be drawn from our early studies of benzene as a diffusant (at concentrations equivalent to those used clinically for anesthetics) (17) although it also includes results obtained from studies of several other molecules of varying size and complexity. Perhaps the first thing to be examined in any simulation is its ability to duplicate known, experimentally determined, properties. As noted above, comparison between simulation and experiment for the bilayer itself was quite good. As we have also noted (17), calculated properties of the diffusing solutes (diffusion, rotation) were also in good agreement with values observed experimentally as were the responses of the bilayer itself to the presence of the solutes.

All atoms were included in the simulations, including the hydrogens of the lipid hydrocarbon chains. Progressively more studies demonstrate the need for this level of detail in order to adequately reproduce physical quantities, particularly dynamical quantities such as diffusion coefficients (18-21). Each monolayer contained 18 lipid molecules and almost 500 water molecules. This resulted in a bilayer of over 7000 atoms and 34.5Å on a side. Simulations were done with several concentrations of benzene (1-4 benzene molecules) all of which were at about the concentrations of anesthetics used clinically.

The details of the simulations were presented previously. In brief, using a modified version of Discover version 2.6 (Biosym Technologies, Inc. San Diego, CA) the trajectory was integrated using the Verlet (22) algorithm with a timestep of 1.0 fsec. The temperatures for the lipids and waters were separately maintained at a target temperature through coupling to an external bath (23). Two-dimensional periodic boundary conditions were used in the plane of the membrane to effectively simulate an infinite bilayer plane. A repulsive "wall" was used in the third dimension to contain the waters and maintain the proper overall density of the system. No other constraints or restraints were used. All of the molecules, lipid, water, and solute, were allowed to evolve according to the classical dynamics of the system which was governed by the force fields.

All of the results are for simulations of a minimum of 1 nsec duration. Our studies show that lengths of this magnitude are required in order for many properties to converge (14). This makes sense from a physical standpoint. For these simulations, the more important motions are those of the hydrocarbon chains, determined primarily by rotation about single bonds. For an individual rotation, this torsional interconversion is known to require from several tens to several hundreds of psec (confirmed both by experiment and by simulation). Long simulations are obviously required in order to adequately sample this motion.

DISCUSSION

During over 7 nsec of simulation, benzene molecules diffused freely throughout the bilayer from positions near to the glycerol portions of the lipids and the polar lipid headgroup region to positions in the center of the bilayer. Often these molecules resided for long periods at particular locations relative to the perpendicular to the plane of the bilayer. The Einstein relationship is commonly used to calculate diffusion coefficients, D, from the slope of a plot of the mean-squared movement versus the time period of that movement. We find that molecules that spend more time near the water/lipid interface have D several times smaller than those near the center. Molecules in intermediate regions diffuse at an intermediate rate. Interestingly, this intermediate rate is similar to that which we see in simulations of benzene in bulk alkane.

Diffusion coefficients are derived from linear fits to time-average data. They are dependent on the degree of sampling during the simulation and we find that they require long simulations to converge to stable values. Also, although some of the benzene molecules tended to stay in particular regions of the bilayer, they were free to move and so comparison of the values of D between molecules was qualitative. In order to gain further quantitative insight into the details of the process of diffusion, we examined the movements that occurred within 1Å slices parallel to the membrane plane. The amount of movement rather than actual diffusion in these slices was used because the amount of time spent in the slices by the benzene molecules was insufficient to provide enough sampling for reliable estimates of diffusion coefficients. Figure 1 is a plot of the average distance traveled within these slices over different periods of time versus the distance of the slice from the bilayer center (the results of both monolayers of the bilayer were merged). At very short times (100 fsec) little difference is seen in the motion in different regions. However, at progressively longer times, a substantial difference occurs, as was suggested by the values of D. Close to the bilayer center, the movements are larger than closer to the headgroup region. The difference develops early, in less than five psec.

Of course, Figure 1 shows averaged quantities, movement averaged over time as well as over the benzene molecules. Insight into the genesis of this effect can be gained through a closer look at the discrete movements that were used to derive that plot. Figure 2 shows for three individual molecules the distance traveled in all 5 psec increments through their entire trajectories for one simulation. One of the molecules

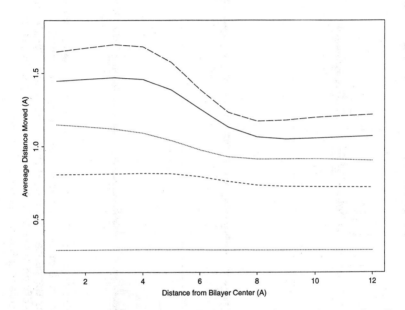

Figure 1. Average distance traveled (averaged over time, molecules and monolayers) by the benzene molecules as a function of distance from the center of the bilayer.

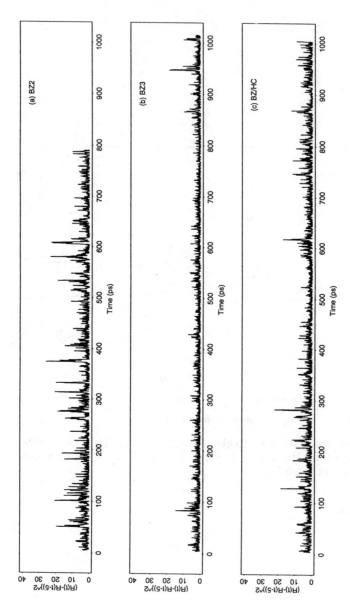

Figure 2. Distance traveled in each 5 psec interval by select benzene molecules. (a) a benzene molecule which spends much of the trajectory near the center of the bilayer (b)a benzene molecule which spends much of the trajectory near the water/lipid interface (c) a benzene molecule in a box of unoriented tetradecane.

(part a) spent most of the simulation near the bilayer center. The second was usually closer to the headgroup region. The third is for a simulation of benzene molecule in bulk alkane.

The plots are noticeably different. That for the molecule near the center has a number of sharp spikes, signifying "jumps," larger-than-average movements for that time period. That for the molecule near the headgroup has few jumps, and these are much smaller and less frequent. Note that the plot for benzene in the bulk alkane demonstrates essentially intermediate behavior, as was true for the D, also. We find that plots such as these are so distinctive that they can be used to predict the location of the solutes.

The increased size and frequency of the jumps are no doubt responsible for the increase rates of movement and diffusion seen for the benzene molecules in the bilayer center. The short time motions, as shown in Figure 1, appear essentially the same regardless of location. The difference in movement starts to occur at about the time that jumps manifest themselves. These jumps typically occur within about 2.5 psec and are of 6-8Å in magnitude.

All of the data presented above shows that the solutes move differently in different regions of the bilayer and suggest that, as hypothesized, the bilayer has an internal order unlike that of bulk alkane. But how does it occur and how does it affect the diffusion of the solutes? The atomic-level nature of the simulations allow us to probe the details of the bilayer structure and how it affects the processes of diffusion. There seem to be two particular features of the bilayer that affect diffusion: the distribution of free volume within the bilayer and the rate of torsional isomerization of the hydrocarbon chains.

It is well-known that lipid bilayers in the Lα phase contain a substantial amount of free volume. The progression from the close-packed crystalline phase of lipid molecules to the physiologically-relevant Lα (or liquid-crystalline) phase is accompanied by a substantial increase in volume.

Figure 3 shows a time composite plot of the available free volume versus position transverse to the bilayer. It shows that the available free volume is greatest near the bilayer center. We have also found that the voids that comprise this volume are substantially larger near the center than closer to the headgroup/water interface. That the rate of diffusion is greatest in the region with the most free volume confirms part of Lieb and Stein's hypothesis that the process of diffusion is mediated by the occurrence of voids between which the solutes can move.

The size and shape of these spontaneously-arising voids are determined by the positions and movements of the lipid molecules. The pertinent regions of the bilayer for this discussion are composed

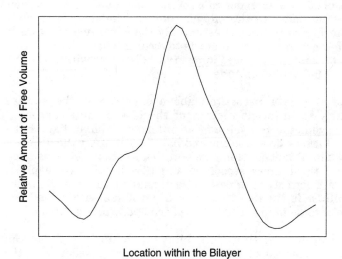

Figure 3. Relative amount of available free volume in the bilayer plotted against location within the bilayer.

primarily of the hydrocarbon chains. For the nsec timescale of these simulations, the most pertinent motions are the rates of rotation about the hydrocarbon single bonds. The residence time of a particular rotamer ranges from several tens of psec to hundreds of psec. Significantly, these times vary with position in the bilayer. Experiment and simulation both show that the faster rates of torsional interconversion occur toward the methyl terminal end of the hydrocarbon chains. This end of the chains is most commonly found near the bilayer center. The torsions near the ester linkage, always near the interface, experience the longer residence times.

These torsional interconversions are important to the process of diffusion. Many instances have occurred where the jumps noted above are mediated by these changes. Figure 4 shows such an instance. The benzene molecule occupied a void near the lipid molecule. One wall of the void was composed of the hydrocarbon chain of the lipid which also served to separate that void from another. In the course of less than 1 psec, one torsion within that chain changed its state from gauche to trans, straightened the chain, and removed it as a barrier between the voids. Within a few psec, the benzene molecule moved 8Å to occupy the new void. During this period, very little else happened in that region of the bilayer. For an equivalent period of time, the average movement for a benzene molecule was less than 2Å.

The increase rate of torsional isomerization at the bilayer center means that these torsion-gated jumps can occur more frequently in this region. This supports the second half of Lieb and Stein's hypothesis, that increased diffusion is due to jumps between voids. Diffusion is fastest at the bilayer center, where, in addition to larger and more frequent voids, the rate of interconversion is larger and jumps between the voids occur more frequently.

CONCLUSION

The simulations provide us with information both about the process of diffusion and about the structure of the bilayer. They confirm Lieb and Stein's hypothesis that increased rate of diffusion occurs when molecules can fit within spontaneously-arising voids in the bilayer and can jump between those voids. They also demonstrate that this process is not the same in all regions of the bilayer which demonstrates a non-homogeneous structure, as proposed by Dill and coworkers. The distribution of voids and the frequency of the torsional interconversions vary with position in the bilayer and result in differences in the rate of diffusion.

In addition, the simulations show, as proposed by both Dill and coworkers and White and coworkers that the bilayer/water interface is not equivalent to a bulk organic/water interface. Bulk organic would be

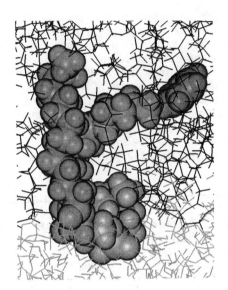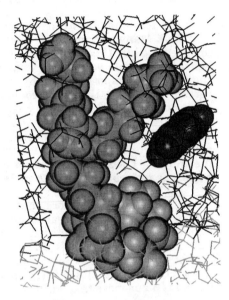

Figure 4. Illustration of a "jump" of a benzene molecule (shown side-on as a space filling figure) which was permitted by isomerization of just one torsion in one hydrocarbon chain in one lipid molecule (also shown as a space filling figure). The rest of the lipid and water molecules are shown as stick figures (a) Prior to the jump. The benzene molecule (upper right) occupied one void for tens of psec. This void was separated from another void by the hydrocarbon chain of a lipid molecule (b) After the jump.   After the change in one torsion, the benzene moved 8Å in about 2.5psec to occupy another void. (Reproduced from reference (17).

expected to be homogenous throughout. The hydrocarbon region of a lipid bilayer close to the headgroups shows substantially more order and tighter packing and, hence, slower diffusion of solutes than does bulk hydrocarbon. Toward the bilayer center, where the terminal methyl groups of the hydrocarbon chains are located, the bilayer is substantially more fluid than bulk hydrocarbon (such "end-effects" are observed in studies of polymer fluidity (24)). Only near the midpoint of the *mono*layer does the hydrocarbon of the bilayer resemble that of bulk as reflected by the rate of movement of the benzene molecules. Other simulations of lipid bilayers have reached similar conclusions regarding the relationship between the properties of hydrocarbons and those of the hydrocarbon portion of the bilayer (25, 26).

Lieb and Stein (5, 6) stated that the magnitude of these effects is dependent on solute size. Our studies confirm this. We have now simulated the diffusion of a number of solutes of varying size. Large solutes show less differential in diffusion rate between different regions of the bilayer. Eventually, a size is reached where no differential is seen. Presumably this results from a change in the mechanism of movement, since these larger solutes no longer fit within the spontaneously arising voids. For solutes of this size, anomalously high rates of permeation would no longer be caused by anomalously high rates of diffusion and permeation could be more directly tied to partitioning, alone.

These studies are additional instances that show that simulation can provide physically realistic results. They serve to help us to gain insight into biochemical events and provide atomic level explanation of experimental results and detailed description and verification of high-level theory.

Acknowledgments:

The authors would like to thank Malcolm Davis for his assistance in preparing the figures and reviewing this manuscript. The authors are also indebted to R. Shaginaw, J. Stringer, R. Gobstein, and G. Burnham (BMS High Performing Computer Center) and S. Samuels (Dept. of Macromolecular Modeling) for orchestrating the CRAY Y-MP and Silicon Graphics computing network and for providing essential computer support. We also thank J. Novotny and J. J. Villafranca for encouraging these studies.

References

1. Overton, E., *Vierteljahrsschr. Naturforsch. Ges. Zuerich,* **1899**. 44 88-135.
2. Cohen, B.E., *J. Membrane Biol,* **1975**. 20 235-268.
3. Finkelstein, A., *J. Gen. Physiol,* **1976**. 68 137-143.
4. Walter, A. and Gutknecht, J., *J. Membrane Biol.,* **1984**. 77 255-264.
5. Lieb, W.R. and Stein, W.D., *Nature,* **1969**. 224 240-243.
6. Lieb, W.R. and Stein, W.D., *Curr. Top. Membr. Transp.,* **1971**. 2 (1-39).
7. White, S.H., Glen, K.I., and Cain, J.E., *Nature,* **1981** 290 161-163.
8. Wimley, W.C. and White, S.H., *Biochem.,* **1993**. 32(25) 6307-6312.
9. Marqussee, J.A. and Dill, K.A., *J. Chem. Phys.,* **1986**. 85(1) 434-444.
10. DeYoung, L.R. and Dill, K.A., *Biochemistry,* **1988**. 27 5281-5289.
11. Alper, H.E., Bassolino, D. A., Stouch, T. R., *J. Chem. Phys,* **1993a**. 98(12) 9798-9807.
12. Alper, H.A., Bassolino, D., Stouch, T.R., *J. Chem. Phys,* **1993b**. 99(7) 5547-5559.
13. Stouch, T.R., *Molec. Simulations,* **1993**. 10(2-6) 335-362.
14. Stouch, T., Alper, H., and Bassolino, D., *Intl. J. Supercomp. Appl.,* **1994**. 8(1) 6-23.
15. Williams, D.E. and Stouch, T.R., *J. Comp. Chem.,* **1993**. 14(7) 1066-1076.
16. Stouch, T.R., Ward, K.B., Altieri, A., and Hagler, A.T., *J. Comp. Chem.,* **1991** 12 1033.
17. Bassolino-Klimas, D., Alper, H. E.,Stouch, T. R., *Biochem.,* **1993**. (32) 12624.
18. Müller-Plathe, F., Rogers, S.C., and van Gunsteren, W.F., *Chem. Phys. Letts.,* **1992b** 199(3,4) 237-243.
19. Bareman, J.P., Reid, R.I., Hrymak, A.N., and Kavassalis, T.A., *Molec. Simul.,* **1993**. 11((204)) 242-250.
20. Pant, P.V.K. and Boyd, R.H., *Macromolecules,* **1993**. (26) 679-686.
21. Yoon, D.Y., Smith, G.D., and Matsuda, T., *J,. Chem. Phys.,* **1993**. 98(12), 10037-10043.
22. Verlet, L., *Phys. Rev.,* **1967**. 159(1) 98.
23. Berendsen, H.J.C., Straatsma, J.P.M., van Gunsteren, W.F., DiNola, A., and Haak, J.R., *J. Chem. Phys,* **1984**. 79 926.
24. Dill, K.A., Flory, P. J., *Proc. Natl. Acad. Sci. USA,* **1980**. 77(6) 3115-3119.
25. Venable, R.M., Zhang, Y., Hardy, B.J., and Pastor, R.W., *Science,* **1993**. 262 223-226.
26. Marrink, S.-J. and Berendsen, H.J.C., *J. Phys. Chem,* **1994**. 98(15) 4155-4168.

RECEIVED October 11, 1994

Chapter 10

Genetic Algorithm Based Method To Design a Primary Screen for Antirhinovirus Agents

E. P. Jaeger[1], D. C. Pevear, P. J. Felock[2], G. R. Russo[3],
and A. M. Treasurywala

Sterling Winthrop, Inc., 1250 South Collegeville Road, P.O. Box 5000,
Collegeville, PA 19426–0900

The human rhinoviruses are the major etiological agents of the common cold in man. Because there are over 100 serologically distinct human rhinoviruses, the design of a common cold treatment requires activity against a broad spectrum of these viruses. It is impractical to test all molecular candidates against all rhinovirus serotypes. Computer methodologies were employed to select a subset of serotypes whose sensitivity to antiviral compounds correlated with the sensitivity of the larger group of viruses. The process involved three steps: i) defining a set of existing molecules which spanned the structural diversity of known human rhinovirus inhibitors; ii) obtaining experimental observations of activity of each of the selected molecules against the larger set of rhinovirus serotypes; and iii) using a genetic algorithm to sort through the resulting biological data and select the serotypes for the new primary screen. The selected subset of viruses provided significantly improved sensitivity prediction over an existing subset of viruses. The method is generally applicable to cases where spectrum or specificity of activity against macromolecular targets is desired.

One of the most common questions asked of computational chemistry groups in industry concerns measures of their usefulness in the mainstream activity of designing more specific, potent and useful agents. These questions are basically centered around the issue of impact; i.e., how much impact is the computational effort having on the project. We offer here an example of a rather unique way in which computational efforts have had impact on an ongoing project: the design of potent broad-spectrum antirhinoviral agents. In this case, a result or activity is said to have impact if it causes the project team to do or plan something differently than what it would in the absence of that result or activity. This impact can be manifested in an explicit project team decision to follow or not to follow a specific course of action.

[1]Current address: 3-Dimensional Pharmaceuticals, 3700 Market Street,
Philadelphia, PA 19104
[2]Current address: Merck Research Laboratories, Division of Merck and Company, Inc.,
West Point, PA 19486
[3]Current address: Virogenetics Corporation, 465 Jordan Road, Troy, NY 12180

0097–6156/95/0589–0139$12.00/0
© 1995 American Chemical Society

There has been an ongoing interest at Sterling Winthrop to design potent agents with a broad spectrum of activity against the major etiologic agents of the common cold, the human rhinoviruses (HRVs). To date more than a hundred strains of HRVs have been isolated and cultured. This large number of viruses made it impractical to screen all new compounds against all of the known serotypes even with high capacity robotic assays. The problem therefore, was to find a smaller set of serotypes that could form the primary screen. The qualities of this subset were: (i) it should be small enough to accomodate the screening of all of the newly synthesized compounds; (ii) it had to be picked in such a way that the results from this subset formed a reliable predictor of the activity that would be found in more complete secondary testing; and (iii) it had to work for the range of structural diversity of compounds that had been synthesized during the project history and it must be able to work for relevant chemical series that might be synthesized in the future. The issue here is not uncommon in the design of primary and secondary biological assays. The screen should not be so specialized to a given chemical series that it would miss good leads from a divergent series of compounds.

The approach that we have devised to address this problem is a generic one that holds significant promise not only for the design of more potent HRV inhibitors, but also for any system which would benefit from subset screening. In this report the methods used to address this problem are detailed so as to bring focus on a novel impact that computational methods have made to the progress of a team and also to inspire the wider use of these techniques.

The series of compounds that have been synthesized in this project bind in a hydrophobic pocket in the HRV capsid and inhibit the attachment of the virus to the cellular receptor and/or the release of the viral RNA into the cell cytoplasm (1-5). The difficulty of designing broad-spectrum active compounds hinges on the fact that these pockets have slightly different sizes and shapes for each of the serotypes (6). Thus, the problem of design becomes one of finding the molecule that will most completely fill the pockets of the largest number of serotypes while at the same time having offending steric interactions with the smallest number of them. The primary screen that was in place at the time of this study was composed of 15 serotypes. The secondary screen contained an additional 39 to make a total of 54 serotypes. Table I lists the 54 human rhinovirus serotypes.

The problem at hand was to find a near-optimal set of 10 or 15 serotypes that best predicted the activity against the 54 viruses. The scope of the problem can be illustrated by recognizing that there are 1.1×10^{25} ways to form a set of 15 serotypes from a pool of 54 serotypes and 8.7×10^{16} ways that a pool of 10 serotypes can be selected from the same pool of 54 viruses.

Genetic algorithms (GAs) have been shown to be very efficient at optimizing functions which are known in the computer science community as being nearly NP-complete (i.e. for which an analytical solution is nearly impossible). This method has been shown empirically to find near optimal solutions to such problems with remarkable efficiency (7). It was therefore resolved to test the GA method to find this (these) set(s). GAs attempt to apply the laws of population evolution (i.e., selection, crossover, and mutation) to sets of "solutions" to a given function (8-11). In our case an individual in this population would be a set of serotypes. The population would be made up of many such sets. For each set, a comparison would be made of how well the set represented the viral sensitivity of the 54 serotypes using a standard collection of compounds. The details of the compounds and methods used for this comparison will be described in the next section. In other words, starting from a set of randomly chosen serotypes, the GA would be asked to evolve another set that was a near optimal predictor of the overall activity of any given compound against the 54 serotypes. Based on the previous reports of the performance of GAs on other unrelated problems it was expected that a near optimal solution would be identified.

Materials and Methods.

Compound Selection. If the GA application was to succeed, it was necessary to select a set of compounds that best represented the structural diversity of the 769 rhinovirus compounds that had been synthesized and, to the best of our abilities, also accomodated types of functionality within the general class that had not yet been synthesized. This smaller set of "representative" compounds would be tested against the 54 serotypes. This data would then be used for the GA optimization. The selection of this smaller set of compounds was extremely important as it could significantly affect the outcome of the experiment. Due to time constraints, it was determined that this set of compounds should not exceed 30. It was decided that a rational method would be developed and applied to the selection of these compounds. Once they were selected by the team through the application of the developed method, they would be expressly tested against the 54 serotypes to generate the data that would be used in the GA-based part of this experiment.

All of the molecules were initially classified according to the presence or absence of molecular keys as defined in MACCS (*12*). Thus each molecule was translated into a bit string. This string was as long as the set of available keys (147 in our case). Each key was represented in this string at a given unique position. The presence of a "0" in a particular position in this string indicated the absence of that key (or functionality) in the molecule while the presence of a "1" indicated its presence in that molecule. The similarity of these bit strings to one another was measured pairwise for all pairs of molecules using the Tanimoto similarity index. This generated a (769 x 769) matrix of similarity indices. The matrix was analyzed by two clustering methods: the Jarvis-Patrick (*13*) and by hierarchical clustering methods available in the SAS (*14*) and S (*15-16*) data analysis packages. This divided the molecules into families. The classification at one extreme consisted of only one family for all molecules. At the other extreme it consisted of 769 families, each composed of only one molecule. The classification level, or resolution, chosen for this study was 30 since the objective was to choose 30 molecules that spanned the property space of the whole set as completely as possible.

Another classification scheme was also used. Here the molecules were described by atom pair(*17-18*) lists. Each pair of atoms in a molecule was classified by two pieces of information: (i) the atom information at each end and (ii) the "distance" between the atoms in terms of the bonded path. This information was converted into a string for each molecule. These strings were compared for all pairs of molecules using the same Tanimoto similarity index. This resulted in another (769 x 769) matrix that was similarly analyzed by clustering methods to define 30 families of compounds. Again the classification scheme used was that of Jarvis-Patrick nearest neighbor and hierarchical clustering.

From these analyses a set of some 30 families would be identified and representatives of these families would be chosen in consultation with the project team. These compounds were tested against the 54 serotypes of the secondary screen in a high capacity tissue culture infectious dose-50% (TCID$_{50}$) assay that has been described previously (*19*).

The project team had also tested 28 other compounds on the basis of their interest to the project against the 54 serotypes used in the secondary assay. This set formed a good test bed which was separate from the so called learning set mentioned above. This set was used to test the quality of the selection of the final primary set of serotypes.

Genetic Algorithm & Analysis. The program used for the GA was GENESIS which was obtained from the public domain (*20*). Default values for most of the parameters were used except for the population size, the convergence threshold, and the number of evaluations. In order to apply GENESIS to a given problem two functions must be provided: (i) an encoding scheme to allow for the representation of the parameters of a problem as a binary string and (ii) a function which measures the performance of a given set of parameters. This customized version of GENESIS is known as GASSER (Genetic Algorithm Selection of SERotypes).

The encoding scheme employed in GASSER had the gene constructed so that the choice of a serotype was accounted for by a six bit codon through a lookup translation. In the case where, for example, a set of 10 serotypes was to be chosen by this method, an individual coded for the genetic algorithm consisted of a bit string of 60 bits. Each 6 bit fragment coded for a particular serotype. Having six bits to represent the identity of the serotype allows a choice of 64 serotypes (2^6). Due to the numerics of translating a 6 bit number to the range from 1 to 54, 10 serotypes were represented by two, 6-bit patterns. More elaborate encoding schemes were not better in arriving at the optimal solutions.

GASSER's performance function used the minimal inhibitory concentration - 80% (MIC_{80}) value to characterize the effectiveness of a compound against a panel of serotypes. The MIC_{80} value is that concentration required to inhibit 80% of the serotypes in a given screen. This value can be calculated for the set of 54 serotypes and for any given subset. With MIC_{80} values computed for 30 compounds against the set of 54 viruses and against a subset of serotypes, it is possible to measure how well the subset of serotypes represent the larger set by calculating a correlation coefficient (r^2). Given a trial set of viruses, the GASSER performance function would calculate the MIC_{80} value for each of the 30 representative compounds against this set. The MIC_{80} values for the 30 compounds would then be calculated against the 54 serotypes. The correlation coefficient of these two sets of MIC_{80} values would be computed and returned to the GA as a measure of the performance of that trial set of viruses. It is ultimately this correlation that GASSER is trying to optimize; what subset of viruses will best characterize the performance of the set of 54 viruses as measured by the correlation of the MIC_{80} values for the 30 representative compounds.

A sample input file for GASSER is included in the experimental section.

Results.

Part I: Compound Selection. The general structure of the molecules under consideration is shown in Figure 1. They consisted of two heterocyclic rings in general connected by an arylalkyl linker of varying lengths. The substituents on the phenyl ring had been varied, along with the length of the linker chain and the type of heterocycle at each end. As previously mentioned, at the start of this analysis 769 compounds of this general class had been synthesized. The similarity of these compounds was computed using both the MACCS keys and atom pairs description methods. The resulting similarity matrices were then analyzed by the Jarvis-Patrick and hierarchical clustering methods. A representative dendrogram showing the classification of this diverse set of molecules based on MACCS keys is shown in Figure 2 and the corresponding dendrogram describing the classification based on atom pairs is shown in Figure 3. These dendrograms show different profiles. This difference may be indicative of the level of information that is provided in the molecular description. The atom pairs-based classification showed a more even distribution of the number of compounds per family and as a result was viewed as a better representation of the subtle differences in chemical structure. The clustering of the atom pairs similarity by the Jarvis-Patrick method led to the identification of 28 families from which 30 representative compounds were selected; two families were represented by two compounds. The selected compounds are shown in Figure 4.

Table I

Human Rhinovirus (HRV) Serotypes Used for Screening

1A	1B	2	3	4	5	6	7	8	9
10	11	12	13	14	15	16	17	18	19
20	21	22	23	24	25	26	27	28	29
30	31	32	33	36	38	39	41	44	47
49	50	54	61	62	66	67	72	75	78
86	87	89	T-39						

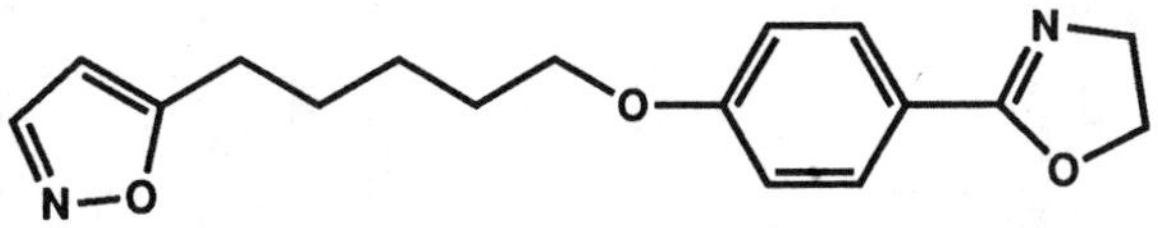

Figure 1. The prototypical structure synthesized for antirhinovirus activity.

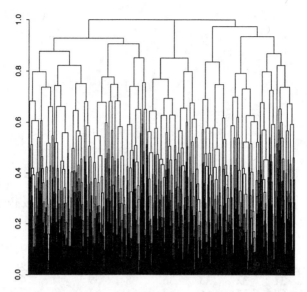

Figure 2. A dendrogram showing the hierarchical clustering of the 769 compounds based on a MACCS keys description.

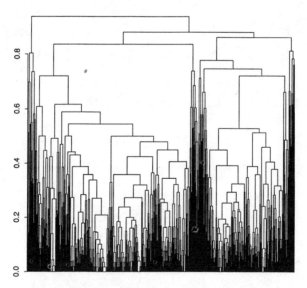

Figure 3. A dendrogram showing the hierarchical clustering of the 769 compounds based on a atom pairs description.

These 30 compounds were tested by the TCID50 assay against 54 serotypes generating 1620 values. The test values are shown in Table II. The decision to perform this amount of biological testing in support of a computational effort is noteworthy and represents in itself a significant impact on the project. However the decision of the team to provide the experiment with an additional 1512 data points (28 compounds x 54 serotypes) is even more noteworthy (Table III). This data was stored in electronic tables from which all GA-based selection analyses were run.

Part II: Serotype Selection. The first issue that was explored was the size of the set of viruses that would be selected; would results differ if there were different numbers of serotypes in this selected set? To test the sensitivity of the MIC_{80} correlation to the number of serotypes in the 'primary' set, GASSER runs were performed to choose sets of 7, 10, 13, and 15 serotypes. All of the runs resulted in serotypes sets which had MIC_{80} correlations greater than 0.9999. The results of this experiment are shown in Figure 5. Clearly, there is not a significant dependence of the results on the number of serotypes within the range examined. The project team then decided that a target of 10 serotypes in the primary screen was suitable.

Once the decision was made that there should be 10 serotypes in the primary screen the production runs of GASSER could begin. Genetic algorithms use random number generators to select the makeup of the initial population and to influence the action of the selection, crossover, and mating operations. As is typical with algorithms that depend on random number generators, multiple runs were made of GASSER with different random seeds. In total, 50 production runs were performed each with a different random seed to ensure that a variety of possible solutions were explored.

Many solution sets were found with high correlations of the MIC_{80}s. In fact, of the 12,870,175 serotype sets that were generated and tested during the 50 production runs of GASSER a total of 583 unique solution sets were identified which had correlation coefficients greater than 0.9999. Hence, GASSER did what was requested of it, namely, to find sets of serotypes which would predict the MIC_{80} performance of the 54 serotypes. The correlation result of one of the 583 solutions is plotted in Figure 6. This particular set of serotypes (HRVs 3, 4, 5, 9, 16, 18, 38, 66, 67, and 75) had an MIC80 correlation coefficient of 0.9999524. The corresponding correlation coefficient for the original set of 15 primary serotypes was 0.313. Thus the GA method had easily found a set of serotypes that was much better at predicting the spectrum of activity of a given compound than the set that was in use.

The predictive ability of these solution sets of viruses was assessed by using them to predict the MIC_{80}s for a different group of compounds. These results, if positive, would provide evidence that this method could be extended beyond the "training set" of compounds and was of general utility. When tested against the 28 compounds of the test set, this same solution set showed much better performance (r^2 = 0.766, Figure 7) than did the original 15 serotypes ($r^2 = 0.02$, Figure 8). This set of viruses therefore showed a 38 fold better correlation with the MIC_{80}s of the 54 serotypes than did the original set of 15 screening viruses. None of these molecules had been used in the so-called training set and as a result these plots showed that the method did indeed succeed in a test of its real predictive ability.

Discussion.

It is important to note that a single run does not typically produce a single suggested set of serotypes to use. The runs produced a number of suggested sets that were all similar in their predictive power. This is a real advantage of this method in that it

Table II: Minimal Inhibitory Concentration for the 30 Representative Compounds Against the 54 HRV Serotypes as Measured by the TCID50 Assay

HRV	Cp1	Cp2	Cp3	Cp4	Cp5	Cp6	Cp7	Cp8	Cp9	Cp10	Cp11	Cp12	Cp13	Cp14	Cp15
1A	1.298	1.568	1.048	11.36	0.553	0.373	0.485	1.157	0.599	2.554	1.590	0.627	0.259	4.264	0.664
89	0.031	0.401	0.296	0.646	0.061	0.177	0.034	0.074	0.024	0.525	25.00	0.055	0.106	0.140	0.112
2	1.043	0.088	25.00	25.00	0.283	1.114	0.318	25.00	25.00	1.293	0.606	0.931	25.00	25.00	0.510
50	0.377	0.696	0.078	0.114	0.651	0.389	0.278	0.233	0.379	2.641	25.00	0.694	1.244	0.859	0.468
22	0.298	0.064	0.118	0.219	0.094	0.304	0.010	0.197	0.269	3.046	3.053	0.230	0.541	0.604	0.152
41	5.648	25.00	25.00	25.00	2.495	2.485	25.00	25.00	1.949	25.00	25.00	2.270	2.542	0.701	2.310
1B	2.551	0.042	0.217	25.00	0.050	0.099	0.117	3.101	0.269	25.00	5.891	1.123	0.630	0.691	0.411
6	0.052	0.550	0.018	0.053	0.844	0.127	25.00	0.036	0.337	0.427	25.00	0.336	0.035	1.623	0.090
14	0.549	1.265	0.122	1.700	0.904	1.225	25.00	0.258	0.236	0.699	25.00	0.661	0.594	2.136	1.064
15	0.339	0.597	0.978	0.554	0.403	0.677	0.458	0.918	0.762	4.136	25.00	0.633	0.784	1.841	0.725
21	0.150	0.005	0.009	0.047	0.028	0.124	0.056	0.067	1.066	1.190	25.00	0.079	0.422	3.100	0.079
25	3.960	0.393	0.209	25.00	1.381	1.411	0.366	0.080	0.640	25.00	25.00	1.312	0.173	0.286	0.262
30	0.189	0.034	0.080	0.330	0.071	0.238	0.050	0.083	0.141	1.979	0.031	0.040	4.457	0.769	0.939
67	0.668	0.663	1.871	2.628	0.331	1.257	0.340	1.075	0.763	25.00	0.252	1.093	0.067	0.577	1.090
86	0.130	0.672	0.037	0.087	0.710	1.031	2.659	0.123	0.343	0.550	25.00	0.322	0.996	3.170	0.551
3	0.709	1.510	0.739	0.694	1.309	1.348	1.794	0.287	0.540	1.987	25.00	1.501	0.322	0.577	1.088
4	2.813	25.00	5.853	1.087	25.00	25.00	25.00	0.325	3.097	2.272	25.00	0.322	2.709	25.00	25.00
5	25.00	0.162	0.261	25.00	25.00	25.00	25.00	25.00	25.00	0.752	25.00	25.00	0.177	0.449	0.104
7	0.067	0.261	25.00	25.00	0.545	0.128	25.00	0.037	0.055	25.00	25.00	0.071	25.00	25.00	1.038
8	25.00	0.747	2.732	9.637	1.276	1.361	0.432	2.737	0.386	25.00	25.00	0.726	2.809	4.796	0.946
9	4.207	25.00	5.918	11.80	0.040	1.568	1.054	2.506	0.558	0.589	25.00	1.197	1.412	0.081	0.125
10	3.043	0.067	0.087	0.360	0.536	0.071	0.031	0.025	0.039	2.017	1.509	0.036	0.058	0.565	0.640
11	0.043	0.067	0.860	0.342	1.358	1.062	0.254	0.542	0.519	0.136	1.414	0.131	1.087	25.00	25.00
12	0.134	2.359	5.384	7.476	25.00	2.440	0.478	4.618	0.160	2.017	3.089	1.250	1.792	0.070	0.363
13	1.174	0.063	0.125	0.237	0.132	0.283	0.044	0.025	1.522	0.136	1.402	0.075	0.162	25.00	25.00
16	0.035	25.00	0.298	3.686	1.521	25.00	25.00	0.558	2.150	0.226	25.00	1.517	1.289	25.00	0.268
17	0.831	0.298	2.010	3.663	0.201	0.380	0.179	3.100	0.179	25.00	25.00	0.605	1.408	3.828	0.329
18	25.00	25.00	0.660	1.668	0.146	0.284	0.112	0.154	0.650	5.832	25.00	0.150	1.433	0.741	0.300
19	1.373	0.093	0.130	1.477	0.134	0.601	0.089	0.079	0.151	2.062	25.00	0.296	0.489	4.838	0.590
20	0.172	0.582	2.010	25.00	0.290	0.612	0.397	1.068	0.650	0.661	25.00	1.430	0.643	2.804	25.00
23	3.401	0.159	0.076	0.361	0.607	0.062	0.683	0.131	0.306	0.475	25.00	0.076	1.362	0.086	0.274
24	0.132	2.689	0.025	0.051	0.065	1.469	0.420	0.880	0.365	0.053	0.536	0.442	0.109	5.222	0.599
26	0.140	2.095	2.376	2.706	0.572	1.243	0.688	0.175	0.306	12.05	0.974	0.076	0.093	25.00	0.268
27	0.035	0.118	2.529	2.367	0.528	0.062	0.255	1.050	0.365	2.062	2.392	1.224	0.902	25.00	0.329
28	1.442	4.738	25.00	3.686	1.521	1.314	0.688	0.880	0.399	5.095	25.00	0.442	1.111	3.828	0.300
29	2.638	25.00	2.010	3.663	25.00	5.079	1.314	0.385	0.365	0.661	25.00	1.224	5.074	5.222	0.268
31	1.407	0.252	0.660	0.569	0.258	0.274	0.043	0.479	0.399	0.593	25.00	0.442	0.583	25.00	0.599
32	0.098	0.199	0.623	1.056	0.072	0.253	0.053	0.125	0.109	5.095	25.00	1.216	0.402	25.00	0.274
33	0.479	0.586	0.429	1.503	0.136	0.506	0.100	0.172	1.336	0.593	25.00	0.170	3.147	0.248	0.230
36	1.366	0.176	0.674	0.668	0.146	0.099	0.566	1.042	0.196	4.801	2.993	0.602	4.941	0.966	0.251
38	0.289	0.482	0.654	1.371	0.503	1.350	0.604	0.275	0.554	5.069	25.00	0.715	25.00	25.00	1.325
39	0.096	0.618	0.367	1.440	0.619	1.555	0.172	0.156	0.192	1.969	25.00	0.659	0.747	0.578	0.358
44	4.198	0.693	1.975	1.841	0.508	1.337	0.303	2.023	0.625	25.00	25.00	0.329	1.943	2.979	0.624
47	0.624	0.559	2.885	8.710	0.461	1.433	0.553	1.050	0.625	1.969	25.00	0.824	1.496	25.00	25.00
49	25.00	8.330	25.00	25.00	1.495	0.199	0.303	2.023	0.192	25.00	25.00	2.604	0.685	25.00	25.00
54	2.431	0.172	0.576	0.895	1.338	2.757	0.101	1.807	0.734	0.546	25.00	0.824	1.766	25.00	0.624
61	25.00	25.00	0.422	25.00	0.143	0.199	1.445	25.00	25.00	25.00	25.00	1.016	1.472	25.00	25.00
66	2.431	25.00	0.075	9.391	1.338	2.757	1.445	3.121	25.00	25.00	5.660	0.677	1.593	25.00	1.365
72	25.00	25.00	25.00	25.00	1.287	25.00	1.272	2.322	1.277	25.00	25.00	25.00	2.567	4.353	25.00
75	25.00	2.219	0.075	0.163	1.287	0.044	2.387	1.419	1.277	0.066	5.660	9.177	2.567	0.300	1.153
78	0.047	0.158	0.075	25.00	0.033	0.044	0.022	2.322	1.359	25.00	0.425	0.355	2.452	25.00	0.452
87	25.00	25.00	25.00	25.00	2.693	25.00	25.00	25.00	25.00	25.00	25.00	25.00	2.452	25.00	25.00
62	25.00	25.00	25.00	25.00	2.693	2.428	1.535	25.00	1.359	25.00	25.00	25.00	2.452	25.00	1.333
T-39	0.266	0.459	0.460	0.916	0.387	0.722	0.237	0.164	0.444	4.875	25.00	0.075	0.375	1.405	0.543

Table II continued

HRV	Cp16	Cp17	Cp18	Cp19	Cp20	Cp21	Cp22	Cp23	Cp24	Cp25	Cp26	Cp27	Cp28	Cp29	Cp30
1A	1.055	0.357	0.316	0.650	4.888	0.183	0.259	2.659	2.745	1.451	0.368	0.060	0.155	0.069	0.0630
89	0.199	0.121	0.042	0.083	1.083	1.807	0.446	0.119	0.149	25.00	0.238	0.008	0.014	0.013	0.0270
2	0.488	0.613	0.073	9.289	0.963	0.787	25.00	25.00	25.00	25.00	2.203	0.009	0.029	0.020	0.0840
50	1.036	0.567	0.160	0.529	1.482	0.510	0.633	1.210	1.081	2.910	0.541	0.064	0.062	0.064	0.0490
41	0.423	0.230	0.023	0.299	0.945	0.335	0.302	1.305	2.020	25.00	0.118	0.018	0.041	0.042	0.0280
1B	25.00	1.248	1.044	2.811	11.11	2.627	25.00	3.909	1.036	3.071	25.00	0.445	1.323	0.424	0.5220
6	0.608	0.593	0.253	1.232	1.250	0.016	0.721	25.00	2.203	25.00	0.341	0.024	0.067	0.018	0.0330
14	2.302	4.682	0.158	0.026	3.888	12.30	0.199	4.439	4.423	25.00	25.00	0.078	0.835	0.246	0.0960
15	1.422	25.00	0.340	0.164	3.424	25.00	25.00	25.00	3.503	25.00	1.363	0.076	0.145	0.374	1.1280
21	0.373	1.198	0.306	0.516	0.360	1.609	0.605	4.674	0.555	4.302	0.149	0.234	0.290	0.271	0.1940
25	0.468	0.138	0.066	0.173	3.191	0.117	0.009	0.141	3.231	4.487	0.131	0.011	0.016	0.015	0.0110
30	0.378	0.573	0.120	2.959	0.589	0.045	25.00	0.422	3.505	2.149	0.659	0.094	0.415	0.026	0.4660
67	4.333	0.289	0.126	0.187	3.259	0.783	0.236	0.587	25.00	1.629	25.00	0.047	0.070	0.066	0.0770
86	25.00	1.117	0.291	0.888	4.696	25.00	0.735	2.756	1.803	25.00	25.00	0.106	0.130	0.117	1.1880
3	25.00	25.00	0.202	0.205	25.00	1.437	0.985	25.00	25.00	25.00	25.00	0.078	0.172	0.357	0.7860
4	25.00	5.449	1.215	0.274	8.728	11.71	25.00	2.934	0.078	25.00	0.758	0.054	0.086	0.336	1.1400
5	0.196	3.052	25.00	0.203	0.755	25.00	25.00	2.904	4.435	5.885	0.009	25.00	25.00	1.686	25.000
7	1.346	25.00	0.081	0.084	25.00	25.00	0.262	0.041	25.00	25.00	0.227	5.893	25.00	25.00	25.000
8	2.027	0.272	25.00	1.972	3.803	1.160	25.00	25.00	0.160	25.00	25.00	25.00	25.00	25.00	0.0140
9	1.534	1.176	0.196	25.00	25.00	25.00	25.00	25.00	0.990	25.00	25.00	25.00	0.153	0.012	0.3060
10	2.560	2.771	0.296	0.046	0.112	25.00	5.960	0.128	25.00	0.382	0.227	0.134	0.069	0.276	0.8810
11	0.599	0.171	0.124	0.158	2.985	0.740	0.076	1.307	0.164	1.845	1.656	0.017	0.273	0.025	0.0015
12	0.587	1.903	0.416	1.926	0.565	1.922	0.266	5.067	2.998	25.00	0.612	0.112	0.194	0.178	0.1980
13	0.343	2.804	0.561	0.041	25.00	7.541	25.00	0.145	2.998	1.469	25.00	0.147	0.076	0.169	0.6560
16	0.397	1.028	0.272	0.663	0.859	2.690	25.00	25.00	1.904	25.00	0.639	0.066	0.275	0.118	0.0710
17	1.284	2.817	2.167	4.252	25.00	25.00	0.079	25.00	1.564	2.830	0.727	0.114	0.252	0.989	2.1660
18	0.523	3.165	0.154	1.175	2.208	0.593	2.402	5.959	1.063	5.349	1.295	0.095	0.032	0.091	2.1530
19	2.483	1.201	0.144	0.311	1.046	25.00	0.248	1.303	0.698	25.00	1.319	0.034	0.031	0.059	0.0670
20	1.120	0.348	25.00	0.130	1.308	25.00	0.370	25.00	25.00	3.147	25.00	0.023	0.056	0.031	0.0780
23	1.224	0.793	0.145	1.288	0.234	0.556	1.286	0.862	0.548	25.00	0.635	0.062	0.051	0.126	0.0950
24	1.169	0.555	25.00	0.045	1.658	1.164	1.359	0.879	0.041	4.944	2.902	0.041	25.00	0.062	0.0960
26	25.00	25.00	0.690	0.013	3.169	25.00	0.097	1.464	25.00	2.267	1.523	25.00	1.874	25.00	3.0650
27	1.377	1.689	0.186	1.213	3.173	2.292	0.696	25.00	2.962	25.00	0.530	0.959	0.116	0.938	2.5060
28	0.307	4.539	0.162	0.681	25.00	25.00	0.573	5.675	4.995	2.783	0.320	0.079	0.038	0.119	2.9600
29	2.720	3.050	1.499	25.00	1.063	0.531	25.00	0.109	3.043	25.00	0.269	0.033	0.617	0.135	0.2840
31	0.155	25.00	0.557	1.443	0.193	0.644	2.261	4.589	25.00	25.00	0.353	0.278	0.046	0.311	5.7240
33	1.392	0.625	0.082	0.170	0.318	1.516	0.170	25.00	1.785	5.025	1.598	0.036	0.050	0.080	0.6490
36	0.647	0.177	0.115	25.00	0.541	0.153	25.00	25.00	25.00	4.855	25.00	0.041	0.024	0.058	0.0590
38	1.365	0.526	0.097	1.314	2.222	25.00	2.091	2.251	3.003	4.694	1.046	0.018	0.017	0.021	0.0110
39	1.934	0.138	0.326	0.723	1.591	0.938	1.292	1.785	25.00	7.808	1.634	0.066	0.112	0.036	0.1480
44	5.115	2.632	0.155	0.330	1.591	5.108	0.157	25.00	25.00	2.898	1.367	0.060	0.070	0.189	0.5990
47	2.966	2.458	0.172	2.374	3.151	1.102	2.832	25.00	25.00	25.00	0.437	0.036	0.040	0.040	0.0600
49	25.00	0.673	0.235	3.173	0.151	3.132	0.506	3.195	25.00	1.248	25.00	0.054	0.062	0.043	0.2840
54	0.362	1.067	0.122	0.703	12.19	25.00	1.948	10.56	25.00	2.076	1.237	0.204	0.323	0.227	0.2710
61	0.561	25.00	0.208	1.477	2.761	0.018	3.949	2.190	3.003	25.00	0.289	0.025	0.028	0.029	2.7690
66	25.00	0.967	2.479	2.610	25.00	1.460	0.653	1.900	25.00	2.755	0.187	0.178	0.290	0.194	0.0900
72	1.431	25.00	0.083	25.00	25.00	25.00	25.00			2.408	0.565	0.791	1.430	25.00	2.2050
75	1.173	1.547	0.139	0.589	25.00		0.577			25.00		0.038	0.077	0.044	25.000
78		0.033	1.591	25.00	25.00					25.00		0.021	0.022	0.034	0.3450
87		25.00	0.282	25.00	1.411					3.427		0.063	0.172	1.449	0.0780
62		1.301	0.171	0.235								0.070	0.194	0.032	4.5620
T-39		1.491										0.032	0.042	0.037	0.1050

Table III: Minimal Inhibitory Concentration for the 28 Test Compounds Against the 54 HRV Serotypes as Measured by the TCID50 Assay

HRV	Cp31	Cp32	Cp33	Cp24	Cp35	Cp36	Cp37	Cp38	Cp39	Cp40	Cp41	Cp42	Cp43	Cp44	Cp45
1A	0.016	0.659	0.149	0.030	0.067	0.023	0.026	0.033	0.014	0.017	0.064	0.006	0.004	0.119	0.098
89	0.005	0.032	0.011	0.003	0.008	0.005	0.005	0.003	0.005	0.003	0.008	0.010	0.007	0.014	0.005
2	0.004	0.028	0.014	0.011	0.008	0.003	0.016	0.008	0.017	0.004	0.013	0.015	0.003	0.027	0.004
50	0.023	0.106	0.081	0.037	0.087	0.035	0.039	0.087	0.014	0.003	0.075	0.110	0.010	0.130	0.047
22	0.003	0.058	0.027	0.016	0.033	0.010	0.017	NA	0.029	0.006	0.022	0.048	0.015	0.010	0.008
41	0.278	1.126	0.507	0.233	0.231	0.107	0.175	0.361	0.336	NA	0.181	0.281	0.175	0.338	0.192
1B	0.011	0.143	0.035	0.010	0.034	0.009	0.016	0.031	0.004	0.009	0.020	0.016	0.003	0.054	0.035
6	0.004	0.213	0.348	1.013	0.066	0.302	0.257	0.118	0.005	0.140	0.032	0.067	0.005	0.024	0.279
14	0.004	0.598	0.219	0.759	0.038	0.072	0.079	0.054	0.007	0.039	0.023	0.085	0.005	0.022	0.089
15	0.030	0.328	0.273	0.053	0.321	0.086	0.144	0.369	0.121	0.134	0.318	0.465	0.064	0.144	0.177
21	0.002	0.018	0.013	0.003	0.006	0.004	0.005	0.013	0.005	0.004	0.005	0.009	0.003	0.015	0.010
25	1.997	0.191	0.111	0.007	0.049	0.017	0.007	0.019	0.010	0.003	0.044	0.002	0.004	0.035	0.005
30	0.003	0.072	0.040	0.026	0.049	0.013	0.036	0.063	0.040	0.016	0.044	0.051	0.036	0.047	0.057
67	0.016	0.326	0.125	0.068	0.161	0.039	0.067	0.159	0.017	0.088	0.135	0.476	0.065	0.070	0.090
86	0.015	0.470	0.299	1.127	0.086	0.364	0.168	0.159	0.018	0.080	0.049	0.335	0.013	0.029	0.335
3	0.012	1.391	0.400	1.096	0.078	0.018	0.143	0.087	0.019	0.088	0.026	NA	0.072	0.037	0.160
4	0.020	3.744	1.809	25.00	2.976	0.278	25.00	25.00	25.00	25.00	25.00	25.00	0.037	25.00	0.005
5	0.144	25.00	25.00	25.00	25.00	2.500	25.00	25.00	25.00	25.00	25.00	25.00	0.002	25.00	25.00
7	0.003	0.015	0.008	4.632	0.008	0.002	0.004	0.005	25.00	0.003	0.008	0.005	0.042	0.019	0.005
8	25.00	25.00	25.00	0.052	25.00	3.259	25.00	25.00	25.00	25.00	0.123	0.075	0.027	2.040	0.908
9	0.022	0.299	0.118	0.100	0.232	0.037	0.068	0.183	0.053	2.248	0.150	0.091	0.003	0.076	0.060
10	0.020	0.497	0.032	0.009	0.322	0.007	0.009	0.289	0.004	0.035	0.024	0.028	0.033	0.039	0.029
11	0.006	0.037	0.013	0.068	0.020	0.054	0.153	0.015	0.052	0.050	0.136	0.213	0.061	0.052	0.027
12	0.005	0.334	0.280	0.111	0.268	0.057	0.085	0.405	0.123	0.007	0.242	0.096	0.029	0.100	0.019
13	0.045	0.694	0.309	0.050	0.084	0.024	0.055	0.087	0.057	0.047	0.089	0.114	0.032	0.156	0.104
16	0.025	0.099	0.091	0.114	0.088	0.216	0.186	0.315	0.123	0.074	0.030	NA	0.047	0.147	0.102
17	0.013	1.689	0.353	2.850	0.237	0.085	0.055	0.068	0.015	0.036	0.247	0.048	0.035	0.048	0.322
18	0.055	0.451	0.279	0.025	0.019	0.027	0.040	0.184	0.059	0.095	0.052	0.079	0.002	0.089	0.039
19	0.025	0.091	0.045	0.007	0.128	0.004	0.026	0.057	0.055	0.045	0.023	0.006	0.014	0.075	0.036
20	0.003	0.072	0.022	0.043	0.076	0.034	0.009	0.011	0.059	0.026	0.084	0.061	0.008	0.020	0.010
23	0.009	0.065	0.083	0.033	0.181	0.029	0.064	0.121	0.016	0.005	0.057	0.039	NA	0.125	0.061
24	0.015	0.084	0.064	25.00	0.072	0.097	0.035	0.034	0.025	0.073	0.787	25.00	NA	0.074	0.031
26	0.003	0.416	2.503	0.057	0.209	0.733	1.346	25.00	0.060	0.027	2.567	25.00	0.011	0.341	25.00
27	0.016	0.351	1.908	4.836	0.101	0.031	2.604	0.123	0.017	2.927	0.107	0.071	0.006	0.885	1.835
28	0.005	0.245	0.133	0.037	0.031	0.018	0.074	0.044	0.043	2.215	0.078	0.115	25.00	0.056	0.026
29	0.008	0.102	0.046	0.040	0.016	0.027	0.029	0.370	0.021	0.042	0.064	0.597	0.003	0.072	NA
31	1.607	0.819	0.829	0.035	0.144	0.043	0.042	0.188	25.00	NA	0.115	0.272	0.004	0.083	0.033
32	0.026	0.127	0.119	0.020	0.060	0.010	0.054	0.030	0.060	0.020	0.039	0.035	0.029	0.091	0.087
33	0.004	0.034	0.020	0.004	0.021	0.004	0.018	0.024	0.006	0.037	0.019	0.021	0.007	0.054	0.013
36	0.008	0.025	0.014	0.011	0.087	0.010	0.007	0.083	0.039	0.010	0.036	0.063	0.007	0.038	0.008
38	0.010	0.035	0.019	0.028	0.202	0.024	0.014	0.027	0.010	0.006	0.132	0.020	0.374	0.041	0.021
39	0.009	0.417	0.112	0.014	0.062	0.009	0.048	0.083	0.022	0.030	0.048	0.072	0.015	0.105	0.026
44	0.006	0.119	0.059	0.009	0.087	0.004	0.024	0.027	0.430	0.008	0.070	0.057	0.009	0.060	0.007
49	0.352	0.036	0.066	0.035	0.520	0.022	0.048	0.044	0.018	0.004	0.070	0.032	0.007	0.005	0.005
54	0.011	0.115	0.325	0.126	0.076	0.117	0.038	0.335	0.014	0.032	0.262	0.091	NA	0.074	0.025
61	0.019	0.574	0.038	0.016	0.006	0.009	0.011	0.016	0.009	0.015	0.034	0.034	0.041	0.187	0.167
66	0.008	0.064	0.144	3.751	0.830	0.010	0.025	0.085	0.004	0.026	0.078	0.057	0.049	0.066	0.024
72	0.003	0.178	1.845	0.043	0.076	0.793	1.003	0.384	0.056	0.129	0.203	0.008	25.00	0.313	1.437
75	0.059	3.318	0.067	0.028	0.076	0.019	0.036	0.042	0.065	0.024	0.078	2.658	25.00	0.077	0.018
78	0.021	0.107	0.019	0.005	0.006	0.004	25.00	0.008	25.00	0.003	0.007	0.008	0.005	0.024	0.005
87	0.317	0.033	0.687	25.00	0.520	25.00	25.00	1.294	25.00	1.001	1.158	2.599	0.006	0.024	25.00
62	0.804	25.00	0.138	25.00	0.830	25.00	25.00	0.039	25.00	25.00	0.010	0.007	25.00	25.00	25.00
T-39	0.006	0.283	0.059	0.020	0.064	0.018	0.025	0.048	0.006	0.014	0.043	0.005	0.005	0.055	0.015

Table III continued

HRV	Cp46	Cp47	Cp48	Cp49	Cp50	Cp51	Cp52	Cp53	Cp54	Cp55	Cp56	Cp57	Cp58
1A	0.023	0.177	0.060	0.034	0.078	0.041	0.033	0.037	0.034	0.172	0.158	0.039	0.025
89	0.003	0.009	0.008	0.010	0.020	0.009	0.005	0.009	0.005	0.035	0.008	0.035	0.002
2	0.003	0.006	0.012	0.018	0.082	0.023	0.014	0.027	0.010	0.040	0.044	0.035	0.098
50	0.032	0.043	0.036	0.029	0.124	0.065	0.042	0.074	0.074	0.183	0.070	0.022	0.032
22	0.008	0.012	0.005	0.007	0.027	0.016	0.022	0.013	0.009	0.012	0.006	0.003	0.007
41	0.014	0.191	0.085	0.135	0.348	0.160	0.188	0.136	0.154	0.644	1.232	0.113	0.168
1B	0.009	0.104	0.030	0.012	0.028	0.020	0.009	0.023	0.014	0.016	0.650	0.009	0.004
6	0.136	0.044	1.083	0.071	0.087	0.042	0.243	0.073	0.073	0.043	0.043	0.054	0.147
14	0.044	0.042	0.441	0.071	0.108	0.032	0.388	0.036	0.036	0.311	0.038	0.055	0.060
15	0.148	0.200	0.081	0.059	0.252	0.257	0.150	0.094	0.117	0.098	0.090	0.063	0.072
21	0.004	0.010	0.010	0.004	0.029	0.011	0.005	0.007	0.005	0.011	0.004	0.002	NA
25	0.019	0.010	0.002	0.008	0.012	0.011	0.014	0.005	0.022	0.084	0.298	0.022	NA
30	0.015	0.034	0.062	0.008	0.057	0.157	0.027	0.032	0.016	0.158	0.013	0.008	0.010
67	0.047	0.134	0.729	0.041	0.159	0.026	0.065	0.076	0.065	0.035	0.041	0.019	0.038
86	0.078	0.026	0.712	0.110	0.071	0.077	0.109	0.058	0.034	0.034	0.039	0.033	0.075
3	0.064	0.024	1.105	0.155	NA	2.886	0.938	0.073	0.043	NA	0.034	0.063	0.097
4	1.090	0.623	25.00	1.511	NA	25.00	25.00	25.00	25.00	25.00	5.459	3.638	25.00
5	25.00	25.00	0.012	25.00	25.00	0.012	0.005	0.008	0.004	0.022	25.00	0.003	25.00
7	0.006	0.004	1.355	0.004	0.013	25.00	25.00	1.990	3.973	25.00	0.006	1.901	0.004
8	6.196	1.845	0.068	2.641	25.00	0.145	0.049	0.065	0.038	0.131	2.737	0.019	5.577
9	0.041	0.066	0.070	0.039	0.239	0.224	0.012	0.073	0.039	0.047	0.044	0.020	0.060
10	0.046	0.046	0.049	0.029	0.148	0.015	0.129	0.015	0.009	0.067	0.020	0.011	0.082
11	0.008	0.005	0.031	0.007	0.035	0.213	0.069	0.072	0.031	0.181	0.032	0.024	0.007
12	NA	NA	0.214	0.030	0.125	0.289	0.055	0.117	0.069	0.162	0.086	0.032	0.048
13	0.054	0.083	0.137	0.074	0.149	0.086	NA	0.068	0.042	0.148	0.062	0.021	0.055
16	0.073	0.005	2.502	0.032	0.039	0.111	0.070	0.088	0.048	NA	0.032	0.029	0.035
17	NA	0.035	0.023	0.124	0.141	0.581	0.040	0.058	0.120	0.048	0.395	0.097	NA
18	0.021	NA	0.061	0.067	NA	0.054	0.008	0.028	0.025	0.081	0.060	0.029	0.374
19	0.002	0.065	0.124	0.012	NA	0.022	0.028	0.011	0.009	0.033	0.008	0.004	0.012
20	0.091	0.005	0.061	0.007	0.004	0.114	0.024	0.047	0.049	0.054	0.039	0.030	0.010
23	0.026	NA	25.00	0.026	0.137	0.058	25.00	0.037	0.020	0.071	0.037	0.017	0.024
26	NA	0.035	1.842	0.025	0.059	0.436	1.741	0.866	2.424	0.151	1.623	2.596	0.019
27	4.672	1.959	0.068	1.902	25.00	0.577	0.047	0.056	1.561	NA	0.017	2.577	25.00
28	0.018	1.461	NA	1.471	0.937	0.137	0.071	0.046	0.032	0.125	0.115	0.018	25.00
29	NA	0.068	0.084	0.028	0.151	NA	0.088	0.033	0.067	0.130	0.199	0.022	0.039
31	0.171	0.055	0.081	NA	0.096	0.426	0.050	0.054	0.050	0.988	0.090	0.032	0.076
32	0.018	0.091	0.032	NA	0.127	0.121	0.051	0.025	0.014	0.200	0.198	0.016	0.044
33	0.007	0.112	0.018	0.023	0.074	0.025	0.010	0.010	0.010	0.066	0.022	0.020	0.038
36	0.012	0.028	0.038	0.007	0.021	0.014	0.036	0.010	0.010	0.735	0.021	0.011	NA
38	0.022	0.008	0.059	0.005	0.048	0.021	0.068	0.079	0.035	0.072	0.036	0.005	0.008
39	0.014	0.036	0.007	0.026	0.022	0.071	0.037	0.019	0.011	0.068	0.016	0.020	0.010
44	NA	0.061	0.013	0.004	0.067	0.011	0.006	0.006	0.007	0.026	0.024	0.004	0.052
47	0.023	0.031	0.058	0.003	0.060	0.019	0.015	0.052	0.025	0.017	0.028	0.002	0.031
49	0.346	6.000	0.091	0.014	0.011	0.058	0.061	0.105	0.283	0.039	0.714	0.010	0.012
54	0.013	0.089	0.037	0.070	0.061	0.288	0.018	0.040	0.028	0.407	0.135	0.213	0.229
61	0.039	0.036	0.034	0.008	0.599	0.037	0.043	0.036	0.054	0.054	0.202	0.005	0.009
66	0.535	0.023	25.00	0.017	0.026	0.049	1.075	0.359	0.276	0.073	0.040	0.290	0.118
72	2.037	0.286	0.030	0.686	0.110	0.379	0.049	0.048	0.040	0.193	0.019	0.024	0.263
75	0.008	0.032	0.015	0.015	0.586	0.074	0.005	0.005	0.005	0.031	5.221	0.009	0.064
78	NA	0.006	0.924	0.004	0.089	0.009	1.267	25.00	0.466	0.053	0.293	25.00	0.004
87	0.259	0.595	0.006	1.799	0.006	0.259	0.021	0.010	0.036	25.00	0.010	0.062	5.900
62	0.016	0.007	0.066	0.007	25.00	0.034	0.005	0.029	0.016	0.041	NA	0.005	0.033
T-39	NA	0.025	NA	0.016	0.020	0.052	NA	NA	NA	0.035	NA	NA	0.035

NA: Data is not available.

150   COMPUTER-AIDED MOLECULAR DESIGN

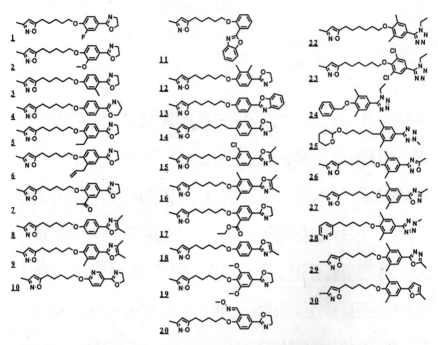

Figure 4. The structures of the 30 representative compounds selected from the collection of 769 compounds based on the atom pairs clustering results.

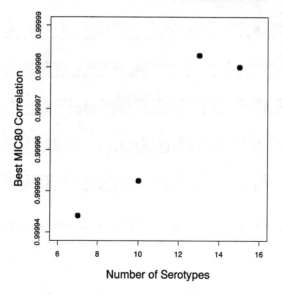

Figure 5. A plot of the best MIC80 correlation found as a function of the number of serotypes allowed in the trial set.

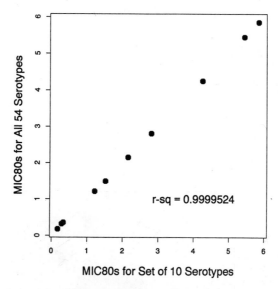

Figure 6. A plot of the MIC80 correlation between one of the solution sets found by the GASSER runs and the 54 Human Rhinovirus serotypes for the 30 representative compounds.

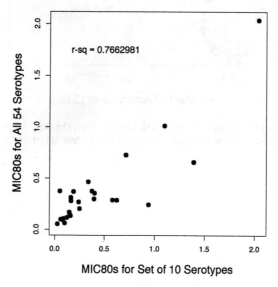

Figure 7. A plot of the MIC80 correlation between a solution set found by GASSER and the 54 HRV serotypes for the test set of 28 compounds.

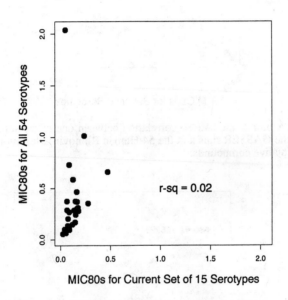

Figure 8. A plot of the MIC80 correlation between the original screening set of 15 serotypes and the 54 HRV serotypes for the test set of 28 compounds.

provides the user with a collection of solutions (when they exist) from which he may choose.

The project team applied an additional set of criteria to the 583 good solutions in order to narrow the choices. The first of these criteria was the MIC_{80} correlation for the test set of 28 compounds. Since these test compounds had already been used to assess the predictability of the solution set of viruses, it was now possible to use this correlation to rank the 583 solutions sets. The other two critera were the MIC_{50} correlation for both the training and test sets of compounds. Here the value is that concentration of the compound required to inhibit 50% of the viruses in a given screen. Applying these criteria narrowed the field to 16 candidate solutions. The final selection was based predominately on the availability or anticipated availability of crystallographic data for two viruses, HRV-16 (*19*) and HRV-3. The final chosen set of serotypes contains HRV-3, -4, -5, -9, -16, -18, -38, -66, -67, and -75; the set whose performance is demonstrated in Figures 6 and 7. This set has been established by the project team as the primary screen. Its use by the project team is the ultimate measure of the impact that this work has had.

The new screening set has been shown to have significant predictive advantages over the original set of 15 serotypes. The poor correlation of the original screen may have had the effect of steering synthetic efforts in unproductive directions by providing false indications of spectrum of activity. The new screen offers a better tool for selecting which compounds should go on for more extensive testing and which compounds should not be tested beyond the 10 serotypes.

As chemical collections becoming increasing larger and it becomes increasingly challenging to test each compound against every new screen, the sampling of compounds from a collection will become a frequent process. The use of techniques such as the ones described here have potential for the selection of a smaller set of "representative" compounds for screening purposes. It was found that the atom pairs method of classifying molecules yielded a much richer classification than the keys approach in this experiment. In this case where the resource constraints of the biological testing limited the number of compounds that could be tested, the sampling of the 769 compounds by the 30 compounds proved to be adequate.

The GA is an optimization that has been found to be useful in a number of fields (*10,21,22*). In computational chemistry, GAs have been used on a variety of problems including conformational analysis and docking (*23-28*). The GA was used here because of its ability to handle large search spaces and its ability to handle response surfaces which are not necessarily continuous. The fact that the GA was able to test 12.8 million out of a possible 10^{16} possible solutions and was able to identify 583 which met the desired criteria of good correlation of the MIC_{80} values demonstrates that this is a practical tool.

The final issue to be discussed is that of the impact of the computational methodologies on the project team. This entire process was made contingent upon project team involvement and "buy in". The experiments were begun only after the recognition by the team that the primary screen would change as a result of this work. There were several stages of impact by the project. First, was the agreement that the experiment needed to be done and that the primary screen would be replaced as a result. The second demonstration of impact was the expenditure of biology resources to test 30 compounds against the 54 serotypes. The third stage of involvement by the project teams was the comparison of the 583 results and final selection of the 10 serotypes. The fourth illustration of the impact of this experiment was the implementation of the chosen set of serotypes as the primary screen for all routine project testing. This has led to the most important impact of all: the new primary screen has provided for more accurately focussed SAR efforts in the discovery of potent HRV inhibitors.

Conclusions.

A novel means of choosing a subset of the biological tests for a primary screen has been described. The screen was selected in such a way as to have the greatest likelihood of correlating with the complete assay. This has been shown in the context of the ongoing antirhinovirus design project. The newly selected panel of serotypes was implemented by the project team as its primary screen. A subsequent analysis of the results from this screen showed that it did indeed translate to increased efficiency for the team with maintained reliability. The method is generally applicable for any system where a subset of assays would be useful as a preliminary screen if their cumulative results track well with the entire set of assays. It represents a new and unique way in which computational methods may be used to add value to the drug design process.

Acknowledgements.

The authors wish to acknowledge the support and many useful discussions with the members of the project team, who are: D. Pevear, G. Diana, TJ Nitz, F. Dutko, D. Aldous, T. Bailey, M. Eissenstat. We also acknowledge the technical assistance of W. Shave.

Experimental.

The following is a sample input file to the GASSER program.

```
    Experiments = 1
    Total Trials = 1000000
Population Size = 100
Structure Length = 60
  Crossover Rate = 0.600000
  Mutation Rate = 0.001000
 Generation Gap = 1.000000
 Scaling Window = -1
Report Interval = 10000
Structures Saved = 20
Max Gens w/o Eval = 2
   Dump Interval = 100
   Dumps Saved = 1
      Options = celL
   Random Seed = 2644486641
   Maximum Bias = 0.990000
Max Convergence = 60
  Conv Threshold = 0.900000
DPE Time Constant = 0
  Sigma Scaling = 2.000000
--
micdb30
1
10
```

References

1. Pevear, D. C.; Fancher, M. J. ; Felock, P. J.; Rossmann, M. G.; Miller, M. S.; Diana, G. D.; Treasurywala, A. M.; McKinlay, M. A.; Dutko, F. J. *J. Virol.* **1989,** 63, 2002-2007.
2. Fox, M. P.; Otto, M. J.; McKinlay, M. A. *Antimicrob. Agents Chemother.* **1986,** 30, 110-116.
3. Shepard, D. A.; Heinz, B. A.; Rueckert, R. R., *J. Virol.* **1993,** 67, 2245-2254.
4. Fields, B. N.; Knipe, D. M., eds. *Fields Virology*, Raven Press, New York, NY, 1990.
5. Smith, T. J.; Kremer, M. J.; Luo, M.; Vriend, G.; Arnold, E.; Kamer, G.; Rossmann, M. G.; McKinlay, M. A.; Diana, G. D.; Otto, M. J. *Science* **1986,** 233, 1286-1293.
6. Kim, K. H.; Willingmann, P.; Gong, Z. X.; Kremer, M. J.; Chapman, M. S.; Minor, I.; Oliveira, M. A.;' Rossmann, M. G.; Andries, K.; Diana, G. D.; Dutko, F. J.; McKinlay, M. A.; Pevear, D. C. *J. Mol. Biol.* **1993,** 230, 206-227.
7. Judson, R. S.; Jaeger, E. P.; Treasurywala, A. M.; Peterson, M. L. *J. Comput. Chem.* **1993,** 14(11), 1407-1414.
8. Goldberg, D. E. *Genetic Algorithms in Search, Optimization, and Machine Learning*, Addison-Wesley, Reading, MA, 1989.
9. Davis, L., ed. *Handbook of Genetic Algorithms*, Van Nostrand Reinhold, New York, 1991.
10. Schaffer, J .D., ed., *Proceedings of the 3rd International Conference on Genetic Algorithms*, Morgan Kaufman, San Diego, CA, 1989.
11. Belew, R. K.; Booker, L. B., eds., *Proceedings of the 4th International Conference on Genetic Algorithms,* Morgan Kaufman, San Diego, CA, 1991.
12. MACCS-II, MDL Information Systems, San Leandro, California.
13. Jarvis, R. A.; Patrick, E. A. *IEEE Trans. Computers* **1973,** C-22(11), 1025-1034,.
14. SAS, SAS Institute, Cary, North Carolina, 1985.
15. Becker, R. A.; Chambers, J. M.; Wilks, A. R. *The New S Language*, Wadsworth & Brooks, Pacific Grove, California, 1988.
16. S-PLUS, Statistical Sciences, Seattle, Washington, 1992.
17. Carhart, R. E.; Smith, D. H.; Venkataraghavan, R *J. Chem. Inf. Comput. Sci.* **1985,** 25, 64-73.
18. Jaeger, E. P.; Jurs, P. C. *Analytical Chemistry Instrumentation* (Laing, W. R., ed.), Lewis Publishers, MI, 275-283, 1986
19. Oliveira, M. A.; Zhao, R.; Lee, W.-M.; Kremer, M. J.; Minor, I.; Rueckert, R. R.; Diana, G. D.; Pevear, D. C.; Dutko, F. J.; McKinlay, M. A.; Rossmann, M. G. *Structure* **1993,** 1, 51-68.
20. Grefenstette, J. J.; Schraudolph, N.N. *GENESIS 1.2ucsd*, available via anonymous ftp from ftp.aic.nrl.navy.mil in /pub/galist/source-code/gasource/gaucsd12.tar.
21. Judson, R. S.; Rabitz, H. *Phys. Rev. Lett.* **1992,** 68, 1500.
22. Ngo, J. T.; Morris, P. G. *Magn, Res. Med.* **1987,** 5, 217.
23. Judson, R. S.; Colvin, M. E.; Meza, J. C.; Huffer, A.; Gutierrez, D. *Intl. J. Quant. Chem.* **1992,** 44, 277-290.
24. McGarrah, D. B.; Judson, R. S. *J. Comp. Chem..* **1993,** 14, 1385.
25. Judson, R. S.; Jaeger, E. P.; Treasurywala, A. M.; Peterson, M. L. *J. Comp. Chem.* **1993,** 14(11), 1407-1414.
26. Legrand, S.; Merz, K. *J. Global. Opt.* **1993,** 3, 49-66.
27. Dandekar, T.; Argos, P. Protein Engineering **1992,** 5, 637-645.
28. Judson, R. S.; Jaeger, E. P.; Treasurywala, A. M. *J. Mol. Struct. (THEOCHEM)* **1994,** 308, 191-206.

RECEIVED December 13, 1994

AGROCHEMICALS

Chapter 11

Semiempirical Quantum Chemical Probes of the Mechanism of Chorismate Mutase

Stephen B. Bowlus

Sandoz Agro, Inc., Research Division, 975 California Avenue,
Palo Alto, CA 94304

As a step in the design of novel, reaction intermediate or transition state analogs, we have studied the chorismate mutase-catalyzed rearrangement of chorismic to prephenic acid by the AM1 and PM3 methods. These methods suggest a symmetrical transition state and an exothermic reaction, but the calculated activation enthalpy is significantly greater than experiment. Chorismic acid is conformationally mobile, with the crucial, diaxial conformer calculated to be 3-4 kcal above the global minimum. Diequatorial species are stabilized by hydrogen bonds. Inclusion of cations and hydrogen bond donors simulating active site interactions gives rise to alternative salt-bridged and/or hydrogen bonded models of the TS. The reaction appears to be accelerated by neutralization of the enolpyruvyl carboxylate, and ionization of the cyclohexadienyl carboxylate. These models are discussed in light of the enzyme's recently published x-ray structure.

As the first, dedicated enzyme in aromatic amino acid biosynthesis, chorismate mutase is an attractive target for herbicide action. The enzyme catalyzes a unique reaction in the intermediary metabolism, the [3.3]-sigmatropic rearrangement of chorismic (**1**) to

0097–6156/95/0589–0158$12.00/0
© 1995 American Chemical Society

prephenic acid (2). Bacterial and plant enzymes have been intensively studied (1,2). Many mechanistic details have been elucidated or proposed, and these have formed the basis for the design of inhibitors (3). The reaction has also been studied by the extended Huckel and MINDO/3 quantum chemical methods (4,5).

We wished to develop a qualitative model of the enzyme's action sufficient to propose alternatives for potential herbicides. At the outset of our investigation, the absence of a crystal structure rendered many aspects of the enzyme's action speculative. Without this knowledge, a comprehensive model was impossible; however, extant mechanistic studies and models provided sufficient clues for direct consideration of the rigid transition state. The recent report of Chook *et al.* (6) has since provided a basis for structural binding models, yet leaves several questions unanswered. Of particular interest to us was the role of the enzyme in catalyzing the isomerization to or stabilization of the reactive, diaxial conformer. Also unclear were the roles of potentially salt-bridging or hydrogen-bonding residues in the active site: residues or interactions which stabilize the transition state are inferred to contribute to that structure's binding affinity, and presumably to the affinity of potential reaction intermediate or transition state analog inhibitors capable of similar interactions. To obtain insights useful to the design of such inhibitors, we have used the semi-empirical AM1 and PM3 methods to model these structures, interactions and their energetics.

Methods

For quantum chemical calculations, we used MOPAC ver. 6 (7). Structure building, visualization and management were accomplished with the Sybyl suite (8). All programs were implemented on an Evans and Sutherland ESV workstation.

Transition states (TS) were located using SADDLE, and fully optimized using eigenvector following. Except as discussed, gradients were reduced to meet PRECISE criteria. The TS were further characterized to have exactly one imaginary root to the Hessian matrix and could be connected to both product and starting materials by intrinsic reaction coordinate calculations.

Interesting points on the reaction coordinate were located by combined intrinsic and dynamic reaction coordinate calculations. Use of an infinite kinetic energy half-life and refinement of the structure from the first energy minimum, or use of a 1 fs half-life and refinement of the final structure gave equivalent results. For these refinements, Bartel's method (NLLSQ) was used exclusively: gradients were reduced until an apparent local minimum was approached, which was characterized by a force constant calculation. While the reported gradient norms were in several cases large, the internal coordinate derivatives and trivial mode eigenvalues were comparable to those of small structures with small gradient norms. Comparison of structures at various stages of refinement lead us to believe that further optimization would provide inconsequential geometry adjustments, and acceptable energy changes, relative to conclusions based thereon.

Results and Discussion

Structure of the Uncomplexed Transition State. We calculated the transition state using various NDDO methods which have been developed since Andrews' study (5).

As summarized in Table I, we obtained good agreement with the earlier study using MINDO/3; the geometries are nearly identical. We attribute discrepancies in the calculated energies to changes in the minimizers used in the two studies.

AM1 and PM3 give a strongly exothermic reaction, with a symmetrical TS. The symmetry of the TS is in agreement with experiment (9), which suggests C-O bond breaking is nearly complete while C-C bond formation has only started. Bond lengths of the two transition states seem to reflect a preponderance of "aromatic" TS character [cf. (10)] more typical of pericyclic reactions, as opposed to a putative, competing biradical mechanism proposed for the Claisen rearrangement. The calculated enthalpy of activation is much greater than the experimental value [20.7 kcal/mol at 37° (4)]. This relatively high value is consistent with the previously reported trend (10), where AM1 overpredicted the activation enthalpy for a series of cyano-substituted allyl vinyl ethers. While the absolute value is in substantial error, it serves as a comparative basis in further study, a point elaborated by Dewar (10,11).

Table I. Transition State Bond Lengths (Å) and Enthalpies of Activation and Reaction (kcal/mol) by NDDO Methods

Parameter	Method			
	MINDO/3	MNDO	AM1	PM3
	Andrews This work			
Diacids				
C-O Breaking	1.45 1.446	1.461	1.865	1.753
C-C Forming	1.94 1.939	1.860	2.122	2.023
Activation Enthalpy	59.1 60.6	57.3	40.2	40.7
Reaction Enthalpy	10.9 13.2	-7.4	-13.1	-18.2
Dianions				
C-O Breaking	1.42 1.429	1.458	1.823	1.774
C-C Forming	1.99 1.977	1.906	2.119	2.078
Activation Enthalpy	66.3 69.1	64.0	50.6	54.3
Reaction Enthalpy	7.6 13.7	-4.2	-12.5	-11.8

Conformational Energetics. The x-ray structure of chorismic acid has recently been solved (12). Comparison of the computed structures with this shows excellent agreement for the ring atoms and conformation. The disposition of the hydroxyl and pyruvyl side chain, however, differ significantly: in the x-ray structure **3**, the hydroxyl and pyruvyl carboxylic acid groups are involved in hydrogen bonding to water of crystallization and neighboring chorismate molecules, while the structure **4**, minimized *in vacuo* using eigenvector following, shows the two groups as internally hydrogen bonded.

3   4

Chorismate has a multitude of accessible conformations, taking into account the torsions of the enolpyruvyl moiety plus ring inversion. Table II summarizes some points of interest to our study. We infer the energy surface is relatively flat; most important, the *syn*-diequatorial conformer **5** is only marginally above the global minimum by AM1, and is the minimum by PM3. This conformer is notable in that 1) it is stabilized by hydrogen bonding between the 4-hydroxyl and enolpyruvyl carboxyl groups, and 2) it is perfectly disposed for reaction upon ring inversion. Gajewski (*9*) and others (*13*) have shown the rate retarding effect of the 4-hydroxyl group, and speculated this is due to ground-state hydrogen bonding. In fact, the transition state **6** for ring inversion appears to involve the breakage of the hydrogen bond rather than the anticipated ring flattening. The corresponding *syn*-diaxial conformer **7** is some 3 kcal/mol higher than **5**, in a shallow minimum 1 kcal/mol below the transition state **6**.

Table II. Enthalpy of Formation for Significant Chorismate Conformers

Conformer	ΔH_f (kcal/mol)	
	AM1	PM3
syn-diequatorial (**5**)	-210.45	-204.88
anti-diequatorial	-210.84	-204.62
TS - *syn*-ee to *syn*-aa (**6**)	-206.59	-200.79
TS - *anti*-ee to *syn*-ee (rotation)		-204.45
syn-diaxial (**7**)	-207.71	-201.43
anti-diaxial	-206.04	-201.64

While the energy difference between conformers **5** and **7** seems small, this is consistent with the 0.9-1.4 kcal/mol estimated from NMR in methanol-d_4 (*14*). In polar media, the stabilizing effect of the internal hydrogen bond would be attenuated; our *in vacuo* calculation reasonably gives the higher value. Qualitatively, the gas phase calculation is consistent with the solution experiment in showing that the

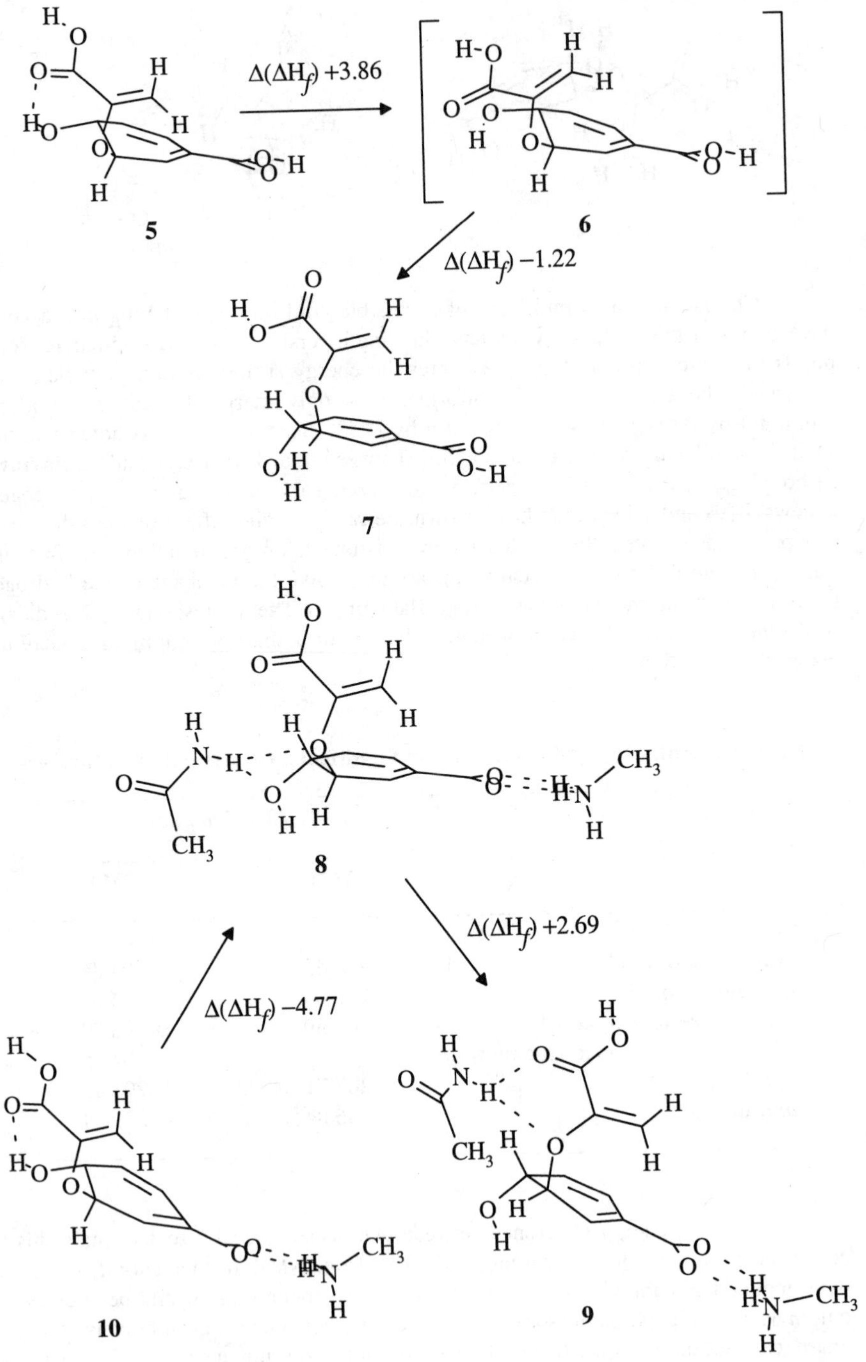

Δ(ΔH_f) +3.86
5
6
Δ(ΔH_f) −1.22
7
8
Δ(ΔH_f) +2.69
9
Δ(ΔH_f) −4.77
10

pseudodiaxial conformer is energetically accessible even with a stronger hydrogen bond.

Introduction of groups anticipated to be involved in binding chorismate (*cf. 13,15*; illustrated with acetamide as a glutamine surrogate) gives rise to no significant change in the enthalpy of ring inversion (**8** → **9**), compared to the uncomplexed case. The overall process, however, is rendered favorable by the enthalpy gained during initial binding through disruption of internal hydrogen bonds and formation of new bonds with the binding residue (**10** → **8**). The reorganization of hydrogen bonds seen in **8** → **9** does not appear to have any substantial affect on the process. A tyrosine surrogate gives the same pattern, albeit with less advantageous energetics (Table III).

The diaxial conformers for these comparisons were generated by dynamic reaction coordinate calculations, and their refinement was rendered difficult by a tendency to invert to the diequatorial conformer. As in the case of non-hydrogen bonded species, the potential well containing the diaxial local minimum may be quite shallow. Hydrogen-bonding residues in the enzyme's active site may stabilize such a conformation; however, the stabilization would be due to constraints imparted by the enzyme's rigidity (*6,16*).

Thus, as anticipated by Hoare (*13*), one potential aspect of CM's mode of action is facilitation of ring inversion through destabilization of the hydrogen-bonding network of the diequatorial conformer. This may be reinforced by further coordination of hydrogen bonding centers, which stabilizes the reactive, diaxial conformer.

Table III. AM1 Enthalpies of Ring Inversion

Isomerization	Enthalpy (kcal/mol)			
	Syn-EE	"Bound" Syn-EE	"Bound" Syn-AA	$\Delta(\Delta H_f)$
Uncomplexed (**1**)[a]	-210.45		-207.71	+2.64
"Glutamine"-Stabilized (**16**)[a,b]	-208.54	-264.00	-261.31	-2.08
"Tyrosine"-Stabilized (**17**)[a,b]	-208.54	-234.01	-231.79	-1.00

[a] Energies taken from the corresponding entry of Table IV. The *syn*-diequatorial (Syn-EE) conformer for the enzyme-like systems is **15**.

[b] Acetamide and phenol ("glutamine" and "tyrosine") have calculated ΔH_f -50.69 and -22.25 kcal/mol, respectively.

Structure and Energetics of Pseudo-enzymatic Transition States. We have attempted to mimic the enzymatic transition state, drawing on the models of Andrews (*17,18*) and Knowles (*15*): we assume both carboxylic acids are coordinated to neighboring lysines (*19*), and the ethereal oxygen is an hydrogen bond acceptor from an active site *e.g.* glutamine or tyrosine. We have examined several cases to gain some sense of the interactions causing significant changes in TS geometry or activation enthalpy, as summarized in Tables IV and V.

The disposition of the hydrogen-bonding prosthesis, adjacent to the pyruvyl carboxylate (*e.g.* **8**), was arrived at by trial and error. In principle, coordination with the ethereal oxygen lone pair neighboring the cyclohexadienyl carboxylate is possible. In our hands, however, stationary transition states of this geometry could not be obtained.

11, 12   X = O⁻, Y = OH
13, 14   X = OH, Y = O⁻

Of the cases examined, only those involving a fully developed enolpyruvate anion (**11**) show large changes in the geometry of the TS, compared to chorismic acid. The structures are reminiscent of "biradicaloid" TS discussed by Dewar (*10*), an observation which was not further pursued. The geometry moves toward that of the expected "aromatic" mechanism if the enolic oxygen is hydrogen bonded to an active-site "glutamine" (**12**), or if the carboxylate is bound to a basic residue, *e.g.* lysine (**18**, Table V, structure not shown).

Discussion of the calculated enthalpies of activation for the reaction must be preceded by the caveat that, in model systems for which relative rates have been measured (*9*), no meaningful correlation was found between the reported rates and calculated energy changes, simple electronic or geometric descriptors (*20*). As previously noted, Dewar (*10*) has shown good correlation between computed and experimental activation energies for closely analogous structures; the correlation with more diverse structures is not addressed. The choice of reference structure for estimation of the enthalpy of activation is also not clear. While a fully relaxed structure, with appropriate groups to mimic enzymatic interactions may be more representative of the overall energetics of the reaction, we have elected to use the pseudodiaxial conformer obtained by gradient minimization of the final structure from the intrinsic reaction coordinate calculation. Relative to chorismic acid, enthalpy changes involving this structure are believed to reflect structural and electronic effects involved in bond reorganization. For these structures, gradients are still substantial (Table IV), and although other criteria give some reassurance (see Methods), the structures may not strictly represent a diaxial stationary state.

Focusing upon actual bond breakage and formation, the data of Table V suggest three major electronic effects:

1) Ionization of the cyclohexadienoic acid accelerates the reaction compared to the carboxylic acid (compare **1**, **13**). Coordination of the cyclohexadienyl carboxylate reduces the effect of charge on this locus (**13**, **15**), which is restored by

Table IV. AM1 Enthalpies of Formation and Gradients for Pseudo-enzymatic Transition State Species

Compound	ΔH_f (kcal/mol) (Gnorm)		
	Ground State	Transition State	Diaxial Isomer
Chorismic Acid (1)	-210.45 (0.100)	-170.61 (0.008)	-207.71 (2.40)
Pyruvyl Half-anion (11)	-243.46 (0.008)	-197.70 (0.003)	-233.70 (0.93)
"Glutamine"-Stabilized (12)	-298.30 (0.009)	-261.68 (0.045)	-297.65 (2.49)
Cyclohexadienyl Half-anion (13)	-236.62 (0.010)	-203.61 (0.004)	-236.52 (0.79)
"Glutamine"-Stabilized (14)	-296.03 (0.009)	-263.88 (0.016)	-294.62 (2.92)
Cyclohexadienyl Half-Salt (15)	-208.54 (0.009)	-171.27 (0.007)	-206.04 (4.70)
"Glutamine"-Stabilized (16)	-264.00 (0.010)	-227.79 (0.009)	-261.31 (6.29)
"Tyrosine"-Stabilized (17)	-234.01 (0.006)	-197.77 (0.006)	-231.79 (4.40)
Pyruvyl Half-Salt (18)	-207.04 (0.009)	-169.16 (0.006)	-205.99 (4.26)
Bis-salt (19)		-164.98 (0.760)	-200.14 (6.42)
Pyruvate protonated (20)	-226.18 (0.007)	-190.45 (0.014)	-223.50 (6.05)
Tri-coordinate (21)	-272.67 (0.008)	-232.63 (0.009)	-267.88 (6.99)
Pyruvate protonated (22)	-281.92 (0.318)	-246.64 (0.040)	-280.42 (5.13)

Table V. AM1 Enthalpies of Activation and Bond Lengths for Pseudo-enzymatic Transition States

System	$\Delta(\Delta H_f)$ (kcal/mol)	Bond Length (Å)	
		C-C	C-O
Chorismic Acid (1)	37.10	2.122	1.865
Pyruvyl Half-anion (11)	36.00	1.908	1.593
"Glutamine"-Stabilized (12)	35.97	2.080	1.776
Cyclohexadienyl Half-anion (13)	32.91	2.138	1.900
"Glutamine"-Stabilized (14)	30.74	2.165	1.925
Cyclohexadienyl Half-Salt (15)	34.77	2.132	1.885
"Glutamine"-Stabilized (16)	33.52	2.143	1.905
"Tyrosine"-Stabilized (17)	34.02	2.165	1.912
Pyruvyl Half-Salt (18)	37.83	2.118	1.833
Bis-salt (19)	35.16	2.129	1.852
Pyruvate protonated (20)	33.05	2.125	1.879
Tri-coordinate (21)	35.25	2.141	1.863
Pyruvate protonated (22)	33.78	2.150	1.888

15

16 L = NHC(=O)CH₃
17 L = C₆H₅O

19, 21 X = O⁻, Y = NH₃⁺
20, 22 X = OH, Y = NH₂

polarization of the ethereal oxygen bond (**13**, **14** and **15**, **16**, **17**). Considering the relative electron withdrawing character of the carboxyl and carboxylate groups, as indicated by substituent constants, one would anticipate this trend from comparisons of terminally substituted cyano- and methoxyallyl vinyl ethers (*10*). The effect is here greater than the earlier computation would suggest, although it is more in line with the reported experimental values (*10*).

2) Ionization of the enolpyruvate moiety has no substantial effect (**11**, **12**, **18**), except to mask the effects of the hydrogen-bonding prosthesis. Protonation of this group, particularly in the presence of other, hydrogen-bonding or salt-bridging groups allows the full effect of the prosthesis to be seen. The relative effects of the carboxylic acid and carboxylate anion are just reversed from expectations from the 2-cyano- and 2-methoxyvinyl allyl ethers (*10*), an anomaly we are at a loss to rationalize. Given the enzyme's wide pH profile, it seems unlikely that such a proton transfer takes place; on the other hand, the charge status of this group may not be crucial to the enzyme's action, a point to be included in the design of potential inhibitors (*3*).

3) When the enolpyruvyl moiety is protonated, the prosthesis adds 1 - 2.5 kcal of stabilization to the TS (compare **15** to **16**, **17**, and **21**, **22**). Previous AM1-SM2 studies (*21*) have attributed the aqueous acceleration of the Claisen rearrangement to polarization and hydration effects, arguably principally involving the ethereal oxygen. Simulations (*22*) have demonstrated explicitly the effect of hydrogen bonding to the ethereal oxygen. Thus, the observed stabilization through hydrogen bonds in this locus

is completely consistent with the known kinetics as the reaction is transferred from gas to solution phase.

Comparison of the computed models *vs*. crystal structures. The recently published structures of chorismate mutase (*6*) bound to the transition state analog shown below (*3*), as well as a monoclonal antibody (*16*) raised against the same analog, permit us to rationalize our findings and methods against definitive criteria.

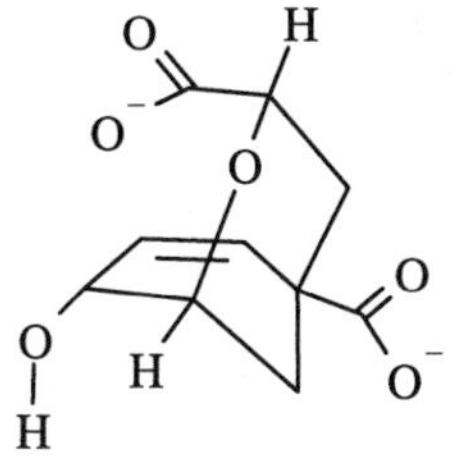

In both structures, the binding cavity is shaped to contain only the pseudo-diaxial conformer. Thus, any apparent contribution to rate enhancement through catalysis of the isomerization must lie in organization of the system during binding of the substrate (*16*). A cartoon of this entire process is difficult to envision. However, in the enzyme (*6*), the potential contributions of Arg-7 and -90 to destabilization of an internally hydrogen-bonded ground state should be clear, as we demonstrate for other hydrogen-bonding residues (Table III).

We have recalculated portions of our model, using elements of the experimental structures which better mimic the environment of the reaction (Table VI). Introduction of water (*24*) and N-methylacetamide to simulate the polar contacts of the cyclohexadienyl carboxylate with the backbone of Lys-60 (**23**) shows that the previously found stabilization of the TS by the carboxylate is maintained. However, if methylammonium is used to model the ionic contact with Arg-116 (*19*) in the presence of water (**24**), the more weakly bound group is displaced, a strong salt bridge is formed, and the stabilizing effect is attenuated as before. Thus the apparent effect of the charged side chain on **15** (Table V) is an artifact of our method: unconstrained minimizations lead to much closer contacts than observed in the enzyme [2.7 *vs*. 3.1 - 4.1 Å (*6*)]. This suggests that a more effective contact with Arg-116 might be a desirable feature of an inhibitor. One speculates whether the inhibitory properties of adamantane-1-phosphonic acid (*24*) may not be due in part to such an improved contact.

Replacement of the glutamine or tyrosine surrogates with methylammonium to mimic Arg-90 (*19*) gives models which demonstrate the effect of ionization of the cyclohexadienyl carboxylate (**26**).   Stabilization contributed by the Arg-90 surrogate through coordination to the ethereal oxygen is obscured: as the TS is relaxed to the diaxial conformer, the conformational energetics become dominated by formation of an hydrogen bond between the 4-hydroxyl and the prosthesis (**25, 27**). Thus again, use of unconstrained minimization of these systems proves misleading. Inclusion of a mimic for Tyr-108 (**28, 29**) together with the Arg-90 mimic provides

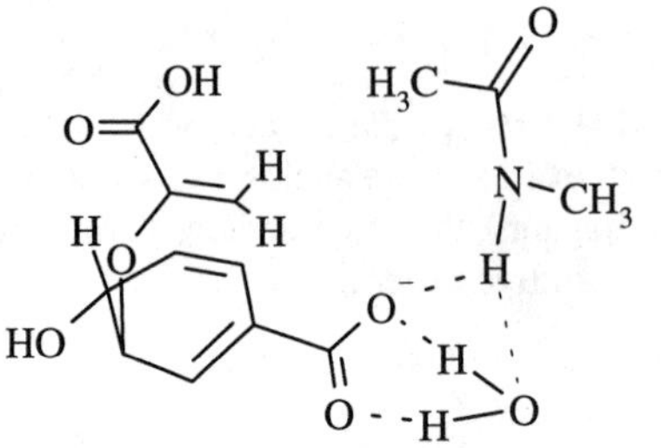

23

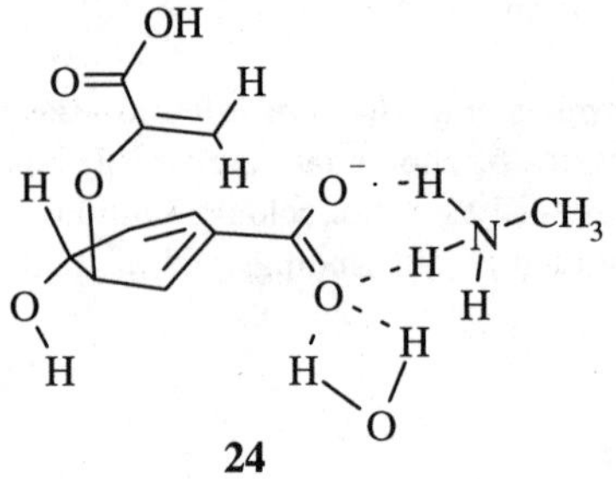

24

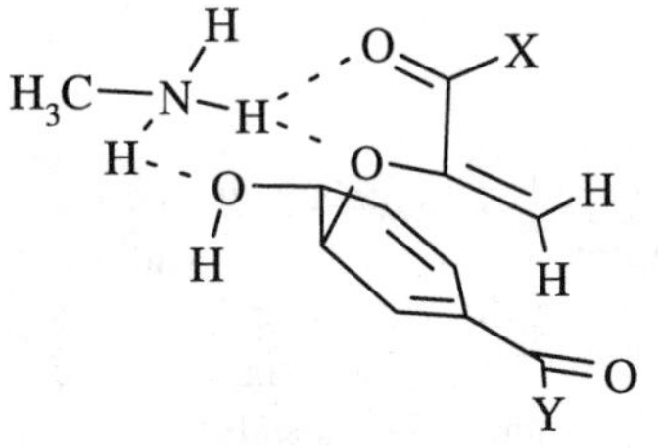

25 X = Y = OH
26 X = OH, Y = O⁻
27 X = O⁻, Y = OH

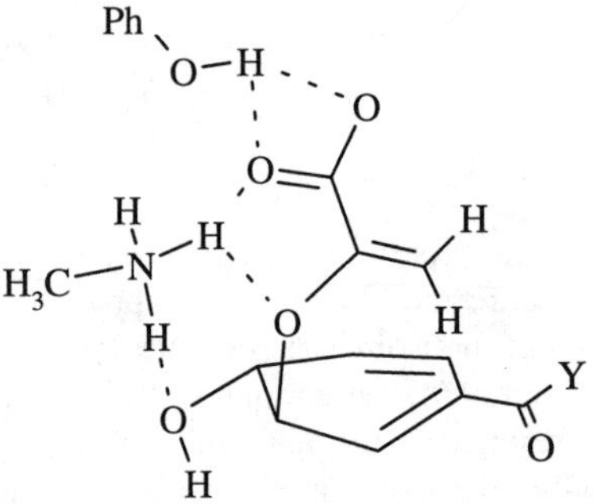

28 Y = OH
29 Y = O⁻

Table VI. AM1 Energies, Gradients and TS Geometries for Species Modelled
on the X-ray Crystal Structure

System	ΔH_f (kcal/mol) (Gnorm)		$\Delta(\Delta H_f)$ (kcal/mol)	Bond Length (Å)	
	Transition State	Diaxial Isomer		C-C	C-O
1			37.10	2.122	1.865
23	-336.41 (0.050)	-366.21 (4.99)	29.80	2.134	1.897
24	-235.12 (0.029)	-270.53 (2.84)	35.41	2.131	1.884
25	-47.33 (0.009)	-90.47 (1.23)	43.14	2.193	1.925
26	-136.91 (0.14)	-168.73 (1.69)	31.82	2.299	2.006
27	-152.28 (1.09)	-193.1 (4.0)	40.5	2.173	1.834
28	-184.49 (0.094)	-221.0 (10.0)	36.5	2.173	1.845
29	-212.87 (2.67)	-250.2 (9.8)	37.3	2.195	1.882

significant stabilization in comparison to **25** and **27**. Thus, while proton transfer to the carboxylate is unlikely, the combination of several hydrogen bonds from Tyr-108, Arg-7 and Arg-90 may suffice to diffuse the locally high negative charge. Unresolved in this instance is whether the the unusually close contact between Tyr-108 and the carboxylate (2.85 *vs.* 3.1 Å) and contacts between the Arg-90 mimic and the ethereal and carboxylate oxygens [2.9 and 2.6 *vs.* 3.1 and 3.1 Å, respectively (*6*)] are sufficient to create this anomaly, as discussed above.

Conclusions

The semi-empirical AM1 and PM3 methods provide a qualitative picture of major effects in the chorismate mutase-catalyzed conversion of chorismic to prephenic acid. Conformational analysis reveals the potential for rate acceleration through destabilization of the internally hydrogen-bonded diequatorial conformer, and possibly through stabilization of the crucial, reactive diaxial conformer.

   Calculations of various pseudo-active sites have identified two effects which contribute to the enzyme's acceleration of the reaction: ionization of the cyclohexadienyl carboxylate, and hydrogen bonding to the ethereal oxygen. These effects are qualitatively consistent with experimental and theoretical studies of the Claisen reaction. A third, potential effect, protonation (or polarization) of the enolpyruvyl carboxylate, may be an artifact of the method.

   There remains to be tested the central assumption of our approach: incorporation of structural features promoting TS stabilization into analogous, potential inhibitors will contribute significantly to their binding affinity, and this will translate into the desired, biological effect. These results bear directly on the design of new inhibitors in two ways:

   1) The different energies calculated for different permutations of charge and binding groups in the transition state suggest *alternative* structures for inhibitors which may be proposed based on more conventional tools of the medicinal chemist, *e.g.* bioisosterism. These may serve as probes better to refine the model or elucidate optional hypotheses, besides being target molecules in their own right;

   2) The pseudo-bound transition states establish probable spatial and vector relationships among putative catalytic and/or prosthetic groups and the complementary functional groups of the substrate and potential inhibitors. Thus, even in the absence of detailed knowledge of the active site, pseudoreceptors may be suitable for exploitation by recently developed *de novo* design tools.

   The major deficiency of our approach is the use of unconstrained geometry optimization, which leads to unrealistic, imputed motions of the binding residues, especially as the TS models are relaxed to the pseudodiaxial conformer of the ligand. These motions result in closer contacts than observed experimentally, with corresponding potential error in the calculated, relative energies. With the availability of the x-ray structures of bound transition state inhibitors (*6,16*) comes the opportunity better to model these interactions with appropriate constraints.

Literature Cited

1.   Haslam, E. *The Shikimate Pathway*; Halstead Press, Wiley, New York, 1974.
2.   Poulsen, C.; Verpoorte, R. *Phytochemistry* **1991**, *30*, 377-386.

3. Bartlett, P. A.; Johnson, C. R. *J. Amer. Chem. Soc.* **1985**, *107*, 7792-7793.
4. Andrews, P. R.; Smith, G. D.; Young, I. G. *Biochemistry* **1973**, *12*, 3492-3498.
5. Andrews, P. R.; Haddon, R. C. *Aust. J. Chem.* **1979**, *32*, 1921-1929.
6. Chook, Y. M.; Ke, H.; Lipscomb, W. N. *Proc. Natl. Acad. Sci. USA* **1993**, *90*, 8600-8603.
7. Stewart, J. J. P. QCPE 455 (6.0), 1990.
8. Sybyl 6.03, Tripos Associates, Inc., St. Louis, MO.
9. Gajewski, J. J.; Jurayj, J.; Kimbrough, D. R.; Gande, M. E.; Ganem, B.; Carpenter, B. K. *J. Amer. Chem. Soc.* **1987**, *109*, 1170-1186.
10. Dewar, M. J. S.; Jie, C. *J. Amer. Chem. Soc.* **1989**, *111*, 511-519.
11. A referee has questioned the appropriateness of using NDDO methods in our studies, citing the methods' shortcomings in accurately describing the Claisen rearrangement and hydrogen bonds. While we agree that our results considered in isolation may be misleading, by adopting Dewar's strategy of examining a series of related calculations, and relating these results to trends in experimental observations, we may derive a sound, if qualitative, model.
12. Afshar, C.; Jaffe, E. K.; Carrell, H. L.; Markham, G. D.; Rajagopalan, J. S.; Rossi, M.; Glusker, J. P. *Bioorg. Chem.* **1992**, *20*, 323-333.
13. Hoare, J. H.; Policastro, P. P.; Berchtold, G. A. *J. Amer. Chem. Soc.* **1983**, *105*, 6264-6267.
14. Copley, S. D.; Knowles, J. R. *J. Amer. Chem. Soc.* **1987**, *109*, 5008-5013.
15. Guilford, W. J.; Copley, S. D.; Knowles, J. R. *J. Amer. Chem. Soc.* **1987**, *109*, 5013-5019.
16. Haynes, M. R.; Stura, E. A.; Hilvert, D.; Wilson, I. A. *Science* **1994**, *263*, 646-652.
17. Andrews, P. R.; Heyde, E. *J. Theor. Biol.* **1979**, *78*, 393-403.
18. Gething, M.-J. H.; Davidson, B. E. *Eur. J. Biochem.* **1977**, *78*, 111-117.
19. While there is evidence of at least one lysine in the CM active site in some species (*18*), our use of two was determined by difficulties modeling arginine: in our hands, AM1 gave a dish-shaped guanidinium function, with complete loss of hydrogen bond directionality.
20. S. B. Bowlus, unpublished. Calculations as detailed in this report were done for the compounds discussed in Table II of (*9*).
21. Cramer, C. J.; Truhlar, D. G. *J. Amer. Chem. Soc.* **1992**, *114*, 8794-8799.
22. Severance, D. L.; Jorgensen, W. L. *J. Amer. Chem. Soc.* **1992**, *114*, 10966-10968.
23. Wiest, O.; Houk, K. N. *Chemtracts--Org. Chem.* **1994**, *7*, 45-48. This feature is not apparent or discussed in (*6*).
24. Chao, H. S.-I.; Berchtold, G. A. *Biochemistry* **1982**, *21*, 2778-2781.

RECEIVED October 13, 1994

Chapter 12

Rational Design of Novel Ergosterol Biosynthesis Inhibitor Fungicides

Charles H. Reynolds and Steven H. Shaber

Rohm and Haas Company, 727 Norristown Road, Spring House, PA 19477

New ergosterol biosynthesis inhibitor based fungicides have been developed using computer aided molecular design. This work used a receptor model inferred from active analogs and cytochrome P-450 oxidase crystal structures. The objective was to design new inhibitors which would bind more strongly because of stronger interaction with the heme iron. In order to accomplish this semi-empirical MO calculations, qualitative MO theory, and classical molecular mechanics calculations were all exploited. The MO calculations were used to predict bond strengths between prospective ligands and iron. It was hypothesized that ligands which bound more strongly to the heme would be better inhibitors. Comparison of the computed bond strengths with efficacy for a set of known inhibitors indicated that this hypothesis was correct. This approach was then used to screen compounds for synthesis, and some of the most promising compounds identified by the modeling study were subsequently made. As predicted these compounds were found to have excellent fungicidal activity.

Many human (*1*) and plant (*2*) fungal diseases are effectively controlled by compounds which inhibit biosynthesis of ergosterol. These ergosterol biosynthesis inhibitors (EBIs) act by shutting down cytochrome P_{450} catalyzed oxidation of the C_{14} methyl in lanosterol (Figure 1). This mode of action was first demonstrated by Ragsdale and Sisler (*3*) in 1972. In the years since many labs have aggressively pursued EBIs (*2*) for both agricultural and pharmaceutical applications. Rohm and Haas developed two triazole based commercial EBI fungicides (*4,5*) which are sold under the trade names Systhane (**1**) and Indar (**2**). One of our goals for this program was to directly apply the SAR (*6,7*) for **1** and **2** to a new class of chemistry where triazole is replaced with another heterocycle. In order to accomplish this we relied heavily on molecular modeling.

EBI Binding Site

Qualitative models for the EBI active site can be constructed (*8-11*) from the crystal structure for analogous P_{450} oxidases with bound camphor (*12*). At the simplest

0097–6156/95/0589–0171$12.00/0
© 1995 American Chemical Society

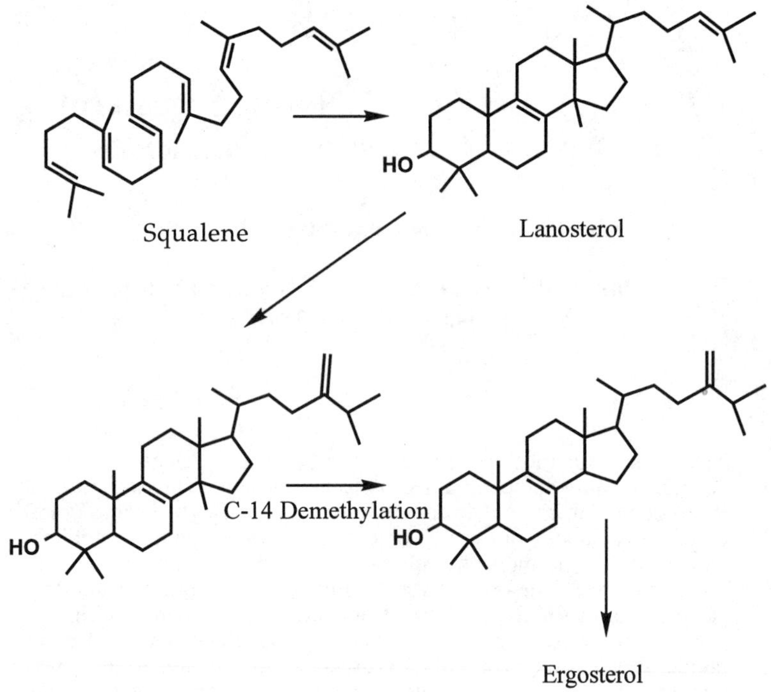

Figure 1. Ergosterol biosynthesis pathway.

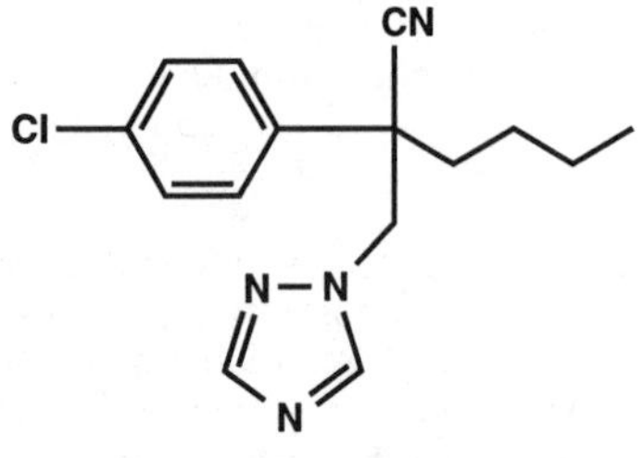

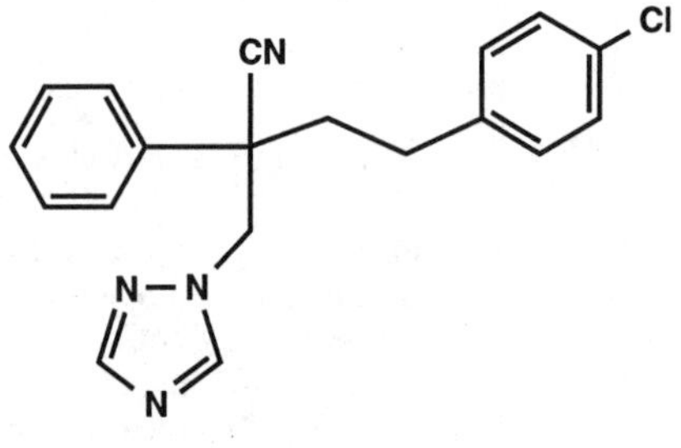

level this crystal structure tells us that the binding site consists of a large mostly lipophilic pocket with a bound Fe-porphyrin complex (heme) which is responsible for catalyzing oxidation of lanosterol. Three factors contribute to binding and inhibition of this active site: (1) hydrophobic interactions between the inhibitor and the active site binding pocket, (2) hydrogen bonding and electrostatic interactions between inhibitor and binding site, and (3) the heterocycle-Fe bond. The hydrophobic and Fe-heterocycle bond strength are thought to be dominant. Since we had considerable knowledge of the SAR and QSAR (6,7) for the cyano-triazoles where R_1 and R_2 have been varied (Figure 2), we wanted to preserve this knowledge as much as possible in the pursuit of new heterocycles to replace triazole. The most obvious role of the triazole is to act as a ligand for Fe^{+3}, therefore, we hypothesized that a heterocycle which binds more strongly to Fe would be a good starting point for more potent EBIs. The task was to determine what these heterocycles might be.

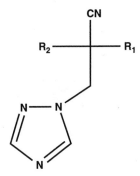

Figure 2. Generic Rohm and Haas triazoles.

Modeling Iron Binding

Computing bond strengths for organo-Fe complexes is a daunting task. In general, classical approaches such as molecular mechanics are not applicable because they do not allow for bond breaking/forming processes. In addition parameters for metals are virtually nonexistent. The most straightforward approach would be to compute relative binding enthalpies using molecular orbital methods. Unfortunately, this approach is extremely expensive computationally given the size of the systems involved and the extra difficulties incurred for transition metals. Therefore, we wanted to develop a method for modeling the relative ligand binding ability of heterocycles without including the metal explicitly. To do this we turned to simple concepts from perturbation theory (*13*). When two AOs or MOs interact they are split into a lower lying bonding and higher lying antibonding pair. The magnitude of the split is determined by the overlap and energy difference between AOs or MOs (see equation 1).

$$\Delta E = \beta^2/(E_L - E_M) \tag{1}$$

β = overlap, E_L = Ligand MO energy, E_M = Metal MO energy

If the energies are close the denominator in equation 1 becomes smaller, mixing is greater and the resulting bond is stronger. If the energies are far apart the denominator in equation 1 becomes larger, mixing is poorer and the resulting bond is weaker. This is illustrated in Figure 3 for a filled vacant interaction such as one might propose for interaction of the N lone pair in triazole with an empty MO on Fe. Thus all we need to estimate relative bond strengths are the overlap and difference in energies between the Fe and N MOs. Since the energy of the metal MO is invariant and the overlap between N lone pairs and Fe should be fairly constant, the only quantity left is the energy of the filled heterocycle N lone pair. This quantity can be computed using any of a variety of molecular orbital methods. Our first choice was MNDO (*14*) because of its speed and success reproducing experimental ionization potentials.

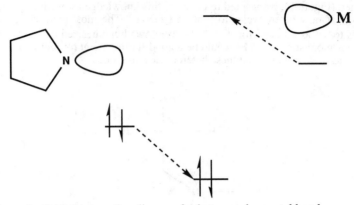

Figure 3. Orbital interaction diagram for heterocycle - metal bond.

Testing the Model

According to our theoretical model, heterocycles which possess high lying N lone pairs should be better ligands for Fe than heterocycles with low lying N lone pairs. It follows that heterocycles which are better ligands should lead to more negative free energies of binding and more potent EBIs. In order to evaluate this hypothesis we needed to compare computed orbital energies for the N lone pair with measured activity for a variety of heterocycles. This tests two vital links in our logic. First, that the Fe-N bond strength can be estimated by computing lone pair energies; and second, that heterocycles which bind more strongly to heme are inherently more active. Brown et al. (*15*) at Lilly published biological data (Table I) for a series of EBIs where the heterocycle was varied systematically while keeping the rest of the molecule constant (**4-9**).

4　　　　　　　**5**　　　　　　　**6**

7　　　　　　　**8**　　　　　　　**9**

For the Lilly compounds R=

Het

This series provided a perfect test case. We computed N lone pair energies for the 3-methyl analogs of heterocycles **4-9** using MNDO (Table I). The computed lone pair energies are plotted against the Lilly data in Figure 4. The correlation is very good. Similar results are obtained using energies derived at the ab initio (*16*) HF/6-31G* level (Figure 5).

Table I. Calculated N Lone Pair Energies and Observed Activities (*13*)

heterocycle	activity[a]	MNDO eV	HF/6-31G* eV
4	10	-10.68	-10.79
5	10	-10.78	-11.02
6	8	-10.90	-11.02
7	6	-11.29	-11.50
8	6	-11.65	-11.99
9	4	-12.32	-12.76

(a) Activities were reported on a 0-10 scale with 10 being most active.

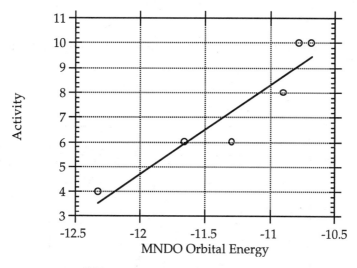

Figure 4. Correlation of MNDO lone pair orbital energies with Lilly activities (see Table I).

These results show that the simple MO model is predictive, and relatively independent of level of theory. It is important to emphasize that this approach was derived using basic principles of molecular orbital theory, and was *not* developed by regressing molecular properties against activity until a correlation was found. The correlation found in Figures 4 and 5 simply confirmed our hypothesis regarding ligand binding in the active site. Comparison of in-house EC_{75} data with MNDO computed lone pair energies also showed the trend of highest energy lone-pairs leading to highest activities (Figure 6). The correlation is poor relative to the Lilly data, but many compounds in this series have structures which vary in addition to containing different heterocycles. Therefore, one would expect significantly more scatter.

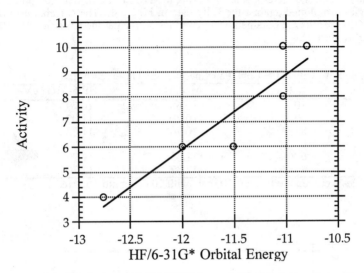

Figure 5. Correlation of HF/6-31G* lone pair orbital energies with Lilly activities.

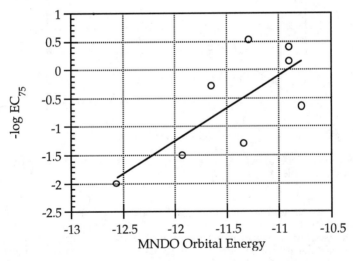

Figure 6. Correlation of MNDO lone pair energies with Rohm and Haas *in vitro* EC75 data.

Another test of this approach is to apply it to compounds which have puzzling SAR. For example compounds **1** and **3** vary only in the structure of the triazole, but have very different activities. The 1H-1,2,4-triazole (**1**) is extremely active. The 4H-1,2,4-triazole (**3**) is almost totally inactive. This is surprising given the small change in structure, but can be explained by a MO based analysis of ligand-Fe binding. In **3** two adjacent lone pair MOs are in a position to interact with the metal. Since these MOs are in close proximity and degenerate in energy they interact with each other and split into a symmetric low lying MO and an antisymmetric high lying MO (Figure 7). Interaction of the metal with the antisymmetric MO leads to poor overlap due to the node between adjacent lone pair p-orbitals. Interaction with the symmetric MO provides good overlap, but since the symmetric MO is much lower in energy the interaction energy is smaller. In addition, N-bridged structures significantly alter the geometric relationship between the heterocycle-Fe bond and the ring substituents with potentially significant impact on steric fit within the receptor pocket. This later problem might be addressed by structural changes elsewhere in the molecule, but is beyond the scope of the present discussion.

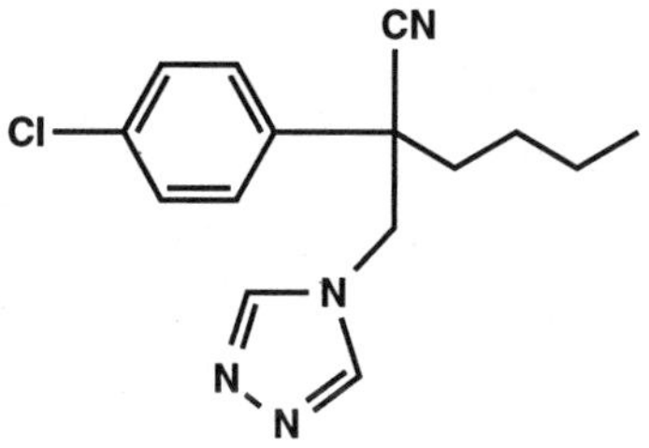

3

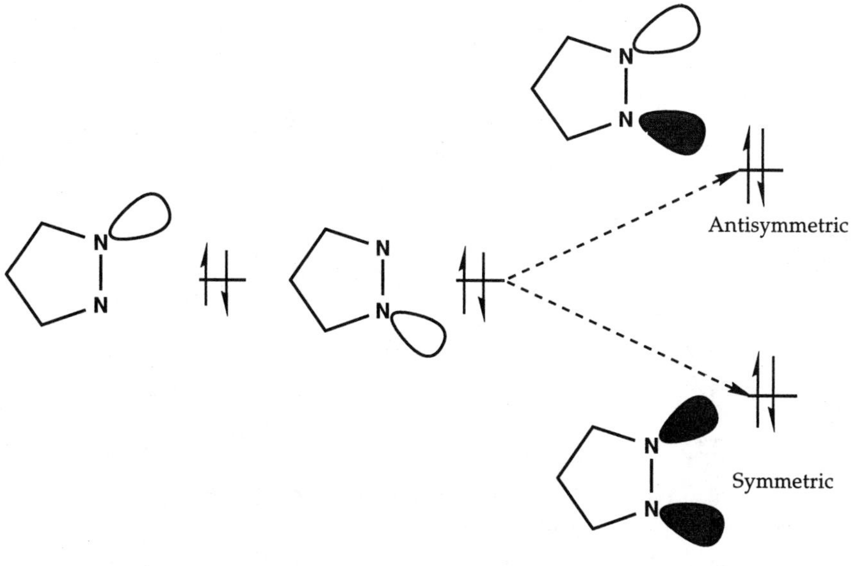

Figure 7. Orbital splitting of adjacent lone pairs in heterocycles such as 4H-1,2,4-triazole.

Whether due to poor geometric fit of bridged ligands or the less accessible nitrogen lone pair, it is not surprising that compounds such as **3** would have significantly different activities. We have concluded that adjacent nitrogens in the metal binding portion of the heterocycle are generally detrimental to activity. This has been observed in both five and six membered heterocycles.

Designing New Inhibitors

The challenge is to use the modeling results to design new ergosterol biosynthesis inhibitors. To do this we calculated lone pair orbital energies for 35-40 prospective 3-methyl substituted heterocycles. This list was then ordered from best predicted ligand to worst. Some of the heterocycles at the top of the list were pyridines, pyrimidines and pyrazine (Table II, Figure 8). Other promising heterocycles include 5-substituted imidazole (**14**) and the corresponding thiazole which have both been shown to be active 5 membered ring heterocycles (*17,18*). Pyridazine (**9**) was predicted to be a very poor ligand for the same reason the 4H-1,2,4-triazoles are poor ligands. This is consistent with the low activity of **9** in the Lilly screen (*15*).

Table II. MNDO Computed N Lone Pair Energies for 3-Methyl Substituted Heterocycles

heterocycle	MO eV	rank	heterocycle	MO eV	rank
4	-10.68	2	10	-9.79	1
5	-10.78	3	11	-10.86	5
6	-10.90	6	12	-10.80	4
7	-11.29	7	13	-11.59	9
8	-11.65	10	14	-11.34	8
9	-12.32	12	15	-11.94	11

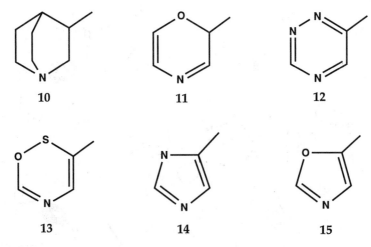

Figure 8. Heterocycles with favorable computed lone pair energies (predicted to be good ligands for Fe).

Based on this list one would expect direct substitution of pyridine for triazole in either **1** or **2** to lead to an increase in activity. Actually the opposite occurs (Table III). The pyridine analogs of Systhane and Indar are much less active. This means that either our ligand binding approach has failed, or other factors such as fit in the receptor or transport have changed in a way which overcomes the stronger N-Fe bond. One obvious difference is changing from a five membered heterocyclic ring to a six membered heterocyclic ring.

Table III. EC_{90} (ppm) as a Function of Heterocycle for Systhane and Indar Analogs

	Systhane		Indar	
disease	triazole	3-pyridyl	triazole	3-pyridyl
WPM	1.6	30	1	>200
WLR			6	>200
SNW			25	150

WPM= wheat powdery mildew; WLR= wheat leaf rust; SNW=septoria nodorum.

In order to understand this change it is necessary to examine how going from a five to six membered heterocycle affects the structure of these compounds. Building on previous work in this area *(7,8,19)* we constructed a model using the structural relationship between the heme active site and bound camphor to infer the structural relationship between the heme active site and bound lanosterol. The large size of heme and lanosterol greatly reduce the degrees of freedom for rotating around the C_{14} methyl and make it possible to construct a reasonable structure for the active site heme lanosterol complex. If a solvent accessible surface is added to lanosterol the result is the binding site model shown in Color Plate 1. This inferred binding site can be used to rationalize the fit of EBI fungicides in the active site. Molecules which can bind to the metal and fill the hydrophobic pocket defined by the solvent accessible surface should be good inhibitors. Systhane (**1**) fits these criteria (Color Plate 2) and is an excellent inhibitor. The pyridine analog while a better ligand for Fe does not fit the binding site pocket, as can be seen in Color Plate 3. Going from a five to six membered heterocycle causes too much structural change. It is impossible for the 3-pyridyl nitrogen to adopt the proper orientation for good binding with Fe without forcing the hydrophobic regions well out of the binding site. This means good binding to the metal or good fit within the binding pocket must be sacrificed. Most likely the former is lost.

We can improve this situation tremendously by simply removing the methylene (spacer methylene; Figure 9) between pyridine and the quaternary carbon. This small structural change has a large effect on fit within the putative receptor. This 3-pyridyl analog (**16**) without the spacer methylene is shown bound to our binding site model in Color Plate 4. Removal of this carbon leads to a 3-pyridyl compound which fits our binding site model just as well as Systhane. It is important to note that these conclusions are insensitive to the specifics of our qualitative binding site model. Indeed, one could make the same arguments based only on the ligand structures. For example, if one flexibly superimposes **1** and its 3-pyridyl derivative the resulting fit is poor. If, however, **1** and **16** are superimposed the heterocyclic, aryl, cyano and butyl regions all match very well.

The observation that removal of the spacer methylene in 6-membered heterocycles should improve activity led to synthesis of compounds such as **16** and **17**. These compounds are found to have activities comparable to Systhane (Table IV) against many fungal diseases, and they serve as valuable leads for additional synthetic optimization. Thus using molecular modeling we were able to take our triazole chemistry and move directly to new classes of compounds which contain other heterocycles with a minimum of synthetic effort. This has enabled us to patent new

NOTE: Color plates appear in color section.

classes of compounds (20,21) and propose other novel chemistries (for example compounds based on quinuclidine (**10**)) which might prove to be potent EBIs.

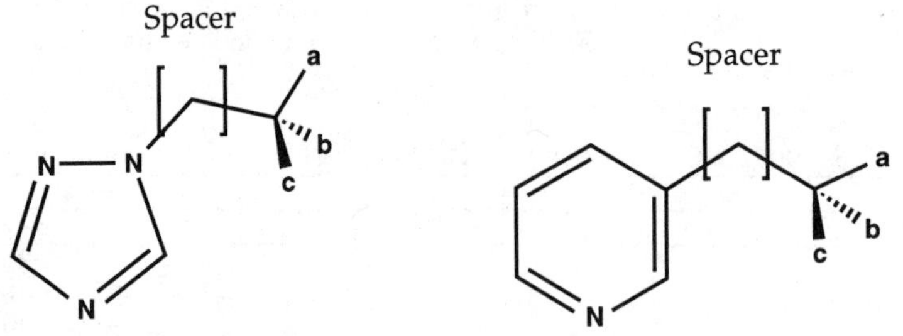

Figure 9. Structural comparison of five and six membered heterocycles ("spacer" methylene is shown in brackets).

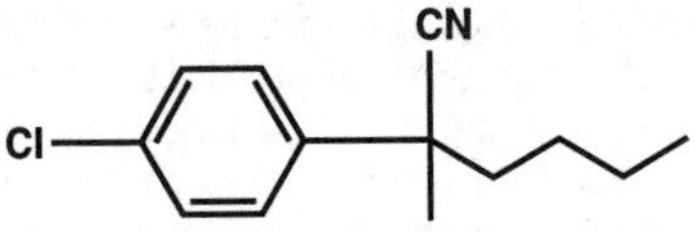

16 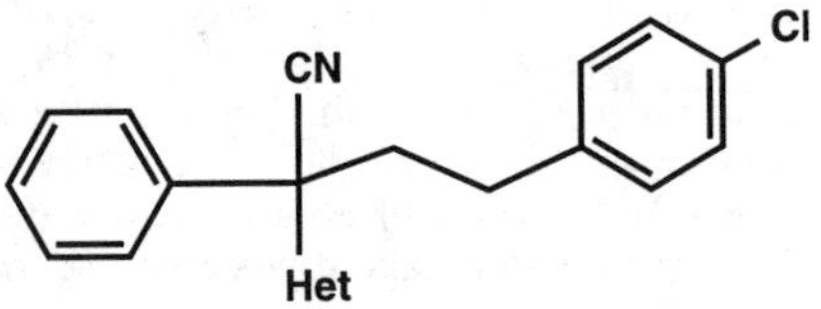   **17**

Het = 3-pyridyl, 3-pyrimidyl

Table IV. EC$_{90}$ (ppm) as a Function of Heterocycle and Spacer for Systhane and Indar Analogs

	Systhane analogs			Indar analogs		
disease	triazole	3-pyridyl	3-pyridyl w/o spacer	triazole	3-pyridyl	3-pyridyl w/o spacer
WPM	1.6	30	1	1	>200	1.5
WLR				6	>200	50
SNW				25	150	25

WPM= wheat powdery mildew; WLR= wheat leaf rust; SNW=septoria nodorum.

Conclusion

We have used insights gained from molecular modeling to systematically design new EBIs. By computing relative ligand binding strengths we were able to quickly focus on a small number of target heterocycles for investigation. This resulted in a tremendous saving of experimental resources. In addition, our binding site model allowed us to understand and affect structural changes brought about by switching between five and six membered heterocycles. Without this insight we might not have ever made compounds such as **16** and **17** where the "spacer" methylene has been deleted, and a very active series would have been overlooked.

We can also draw a number of general conclusions about using molecular models in designing bioactive molecules. First, it is often possible to derive considerable insight into a problem using relatively simple computational methods. This is illustrated by our use of MNDO to model the ligand-Fe bond strength and an inferred binding site model to rationalize the structural effects of different heterocycles on binding. Although these models did not give us quantitative predicted binding constants, they were crucial in pointing us in the right direction. This is usually much more important than quantitative predictions. Our EBI project also benefited from close collaboration between the modeling and synthesis efforts. The cross functional team approach provides for maximum synergy between synthetic and modeling efforts. Finally, this work clearly illustrates the value of modeling for rationalizing SAR and guiding synthesis of new bioactive molecules.

Literature Cited

1   Georgopapadakou, N. H.; Walsh, T. J. *Science*, **1994**, *264*, 371.
2   Koller, W. In *Target Sites of Fungicidal Action*; Koller, W., Ed.; CRC Press: Boca Raton, FL, 1992, 119-206
3. Ragsdale, N. N.; Sisler, H. D. *Biochem. Biophys. Res. Commun.* **1972**, *46*, 2048.
4. (a) Orpin, C.; Bauer, A.; Bieri, R.; Faugeron, J. M.; Siddi, G. *Proceedings - British Crop Protection Conf. - Pests. Dis.* **1986**, *1*, 55; (b) Fujimoto, T. T. US 4920139, 1990.
5. (a) Driant, D.; Hede-Hauy, L.; Perrot, A.; Quinn, J. A.; Shaber, S. H. *Brighton Crop Prot. Conf. - Pest. Dis.* **1988**, *1*, 33; (b) Shaber, S. H. US 5087635, 1992
6. Fujimoto, T. T.; Shaber, S. H.; Chan, H. F.; Quinn, J. A.; Carlson, G. R.   In *Synthesis and Chemistry of Agrochemicals*; Baker, D. R.; Fenyes, J. G.; Moberg, W. K.; Cross, B., Eds.; ACS Symposium Series 355; American Chemical Society: Washington, DC, 1987, 318-327.
7. (a) Fujimoto, T. T.; Quinn, J. A.; Egan, A. R.; Shaber, S. H.; Ross, R. R. *Pesticide Biochem. Physiol.* **1988**, *30*, 199; (b) Shaber, S. H.; Quinn, J. A.; Fujimoto, T. T. In *Synthesis and Chemistry of Agrochemicals IV*, in press.
8. Marchington, A. F. In *Pesticide Synthesis Through Rational Approaches;* Magee, P.; Kohn, G. K.; Menn, J. J., Eds.; ACS Symposium Series 255; American Chemical Society: Washington, DC, 1984, 173-183.
9. Worthington, P. A. In *Synthesis and Chemistry of Agrochemicals*; Baker, D. R.; Fenyes, J. G.; Moberg, W. K.; Cross, B., Eds.; ACS Symposium Series 355; American Chemical Society: Washington, DC, 1987, 302-317.
10. Koymans, L. M. H.; Vermeulen, N. P. E.; Baarslag, A.; den Kelder, G. M. D.-O. *J. Comput.- Aided Mol. Design*, **1993**, *7*, 281.
11. Odell, B. *J. Comput.-Aided Mol. Design*, **1988**, *2*, 191.
12. (a) Poulos, T. L.; Finzel, B. C.; Howard, A. J. *J. Mol. Biol.* **1987**, *195*, 687; (b) Poulos, T. L.; Perez, M.; Wagner, G. C. *J. Biol. Chem.* **1982**, *257*, 10427.
13. Dewar, M. J. S.; Dougherty, R. C. *The PMO Theory of Organic Chemistry*; Plenum Press: New York, 1975.
14. Dewar, M. J. S.; Thiele, W. *J. Am. Chem. Soc.* **1977**, *99*, 4899, 4907.
15. Brown, Jr., I. F.; Taylor, H. M.; Hackler, R. E. In *Pesticide Synthesis Through Rational Approaches;* Magee, P.; Kohn, G. K.; Menn, J. J., Eds.; ACS Symposium Series 255; American Chemical Society: Washington, DC, 1984, 65-82.
16. Frisch, M.J.; Head-Gordon, M.; Trucks, G.W.; Foresman, J.B.; Schlegel, H.B.; Raghavachari, K.; Robb, M.A.; Binkley, J.S. ; Gonzalez, C.; Defrees, D.J.; Fox, D.J.; Whiteside, R.A.; Seeger,R.; Melius,C.F. ; Baker,J. ; Martin,R.; Kahn, L.R.; Stewart, J.J.P.; Topiol,S. ; Pople,J.A, Gaussian90, Gaussian Inc., Pittsburgh, PA 1990.

17. Wilson, J. H.; Sawney, I. US 5057529, 1990.
18. Merritt, L.; Pioch, R. US 4281141, 1981.
19. Fujimoto; T. T.; Reynolds, C. H. *Chemical Design Automation News*, **1988**, *3*, Number 7, 1.
20. Shaber, S. H.; Sharma, A. K.; Reynolds, C. H. EP 276920, 1988.
21. Shaber, S. H.; Szapacs, E. M.; Reynolds, C. H. EP 451948, 1991.

RECEIVED January 19, 1995

Chapter 13

Design and Synthesis of 5,6-Dihydro-4H-1,3,4-oxadiazines as Potential Octopaminergic Pesticides

Mark A. Dekeyser[1], W. Ashley Harrison[1], Paul T. McDonald[2], G. W. Angle, Jr.[2], Saad M. M. Ismail[3], and Roger G. H. Downer[3]

[1]Research Laboratories, Uniroyal Chemical Ltd., Guelph, Ontario N1H 6N3, Canada
[2]New Product Research, Uniroyal Chemical Inc., Bethany, CT 06525
[3]Biology Department, University of Waterloo, Waterloo, Ontario N2L 3G1, Canada

The application of Computer Aided Molecular Design (CAMD) tools to the process of insecticide discovery is examined as a key component in a biorational design of octopaminergic pesticides. A study of novel, conformationally restricted analogs of octopamine resulted in the selection of dihydrooxadiazines as candidates for octopaminergic pesticides. A series of dihydrooxadiazines containing N-H, N-methyl and N-fluoroethyl substituents were synthesized and tested for useful pesticidal properties. Certain dihydrooxadiazines showed activity against economically important agricultural pests and interacted with their octopaminergic system.

One biorational approach in the development of safer and more selective pesticides is to select targets which are vital and specific to pest species, thereby minimizing toxicity to non-target organisms (1). Much attention has been directed at the biogenic amine, octopamine (p-hydroxyphenylethanolamine), as a valid target in the search for novel pesticides (2,3). Octopamine is one of the most abundant biogenic amines in the insect and mite nervous system (4,5). Compounds that stimulate octopaminergic systems in insects and mites have the potential to cause physiological, behavioral and lethal effects that often are highly compatible with integrated pest management (IPM) systems. The octopamine receptor has been

0097–6156/95/0589–0183$12.00/0
© 1995 American Chemical Society

recognized as a useful target since the discovery of the insecticidal and miticidal properties of formamidines (6-9), which function as octopamine agonists by stimulating cyclic adenosine monophosphate (cyclic AMP) (10-14). Two formamidines, chlordimeform and amitraz, are commercial pesticides (15).

Known octopaminergic pesticides were not discovered through structural modifications of octopamine, but were largely the result of random and biochemical screening efforts. We reasoned that it should be possible to design agonists which would avoid the problems of low bioavailability and rapid metabolic degradation associated with octopamine (16), thereby becoming suitable pesticide candidates. It was hypothesized that these problems were largely due to an unoptimized lipophilic profile of the molecule. A parabolic relationship often exists between pesticidal activity of a compound and its lipid solubility. When the lipophilicity of the compound is low, this implies that it is poorly absorbed by a biological membrane and when its lipophilicity is high, it becomes trapped in the membrane and does not pass on to the site of action. Our strategy for avoiding these problems involved modulation of the physical properties of octopamine by modifying the ethanolamine portion of the molecule and introducing suitable substituents in the phenyl portion. These modifications were designed to enhance penetration of the cuticle and CNS of pest species as well as to afford resistance to oxidative degradation. Towards this end, we used Computer Aided Molecular Design (CAMD) tools to identify novel octopamine mimics as potential pesticide candidates. The candidates were then synthesized and evaluated for pesticidal and octopaminergic activities. We recently reported selected results from this work (17-19).

Molecular Modeling

Molecules were constructed using the SYBYL software package (Tripos Associates, St. Louis, MO). The minimum energy conformations were determined from the MAXIMIN program. An X-ray crystallographic study of octopamine hydrochloride (20) showed that the molecule adopts a conformation whereby the ethanolamine portion is extended away from the phenyl ring. A minimum energy conformation of octopamine which closely resembled the X-ray conformation was used for this study.

It is reasonable that low-energy conformers of potential octopamine mimics should be capable of superimposition on a low-energy conformer of octopamine, providing a match between corresponding phenyl rings and nitrogen atoms. The superimposed energy-minimized structures of octopamine and chlordimeform are shown in Figure 1. A close match between these two structures was

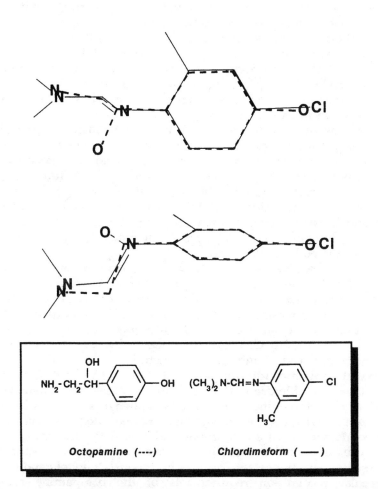

Figure 1. Superimposition of computer-generated low-energy conformers of octopamine and the octopaminergic pesticide chlordimeform (orthogonal views). H atoms have been omitted for clarity.

observed. Minimum energy conformers of additional proposed octopaminergic pesticides, imidazoline (NC-5) and oxazoline (AC-6), overlayed well with octopamine (not shown).

It has been shown that the hydroxy group in the side-chain portion is not critical to octopamine agonist activity in phenylethanolamine analogs (21). Therefore, it appears that the dimethylformamidine side-chain [N=CH-N(CH$_3$)$_2$] of chlordimeform mimics the ethanolamine side-chain [CH(OH)CH$_2$NH$_2$] of octopamine, while the 4-chloro-2-methylphenyl group of chlordimeform mimics the 4-hydroxyphenyl group of octopamine. Obviously, these modifications to the octopamine molecule must result in dramatically altered physicochemical properties.

A comparison of calculated nitrogen-to-phenyl distances in low-energy conformers and relative lipophilic properties (calculated Log P values) of octopamine and some known octopaminergic pesticides are shown in Table I. On the basis of these values, a minimum pharmacophoric binding model was proposed for octopaminergic pesticide candidates. The model included a phenyl ring and a basic nitrogen atom separated by a distance of 3.4-3.7 angstroms, an angle of 75-100° between the phenyl ring and side-chain and a calculated log P value >1.

The objective was to design an isostere with increased lipophilicity while retaining the functionality essential for agonist action. To investigate the potential of nitrogen-containing heterocycles to act as novel octopamine mimics related to NC-5 and AC-6 (Table I), we evaluated the structural similarities between octopamine and several 5,6-dihydro-4H-1,3,4-oxadiazines. Figure 2 shows the superimposed energy-minimized structures of octopamine and a proposed dihydrooxadiazine analog (adapted from ref. 17). In this compound, a dihydrooxadiazine ring replaces the ethanolamine portion of octopamine while a bromophenyl group replaces the hydroxyphenyl group of octopamine. We reasoned that this type of compound would be a conformationally restricted lipophilic analog. The use of conformationally restricted analogs of octopamine represents an important approach towards understanding the molecular recognition requirements of the octopamine receptor. Additional analogs that are substituted on the dihydrooxadiazine nitrogen by a methyl and small fluorine-containing groups were expected to further enhance lipophilic properties without affecting the nitrogen-to-phenyl distance. Since chlordimeform possesses N-methyl groups and fluorine-containing compounds are noted for their high level of biological activity (22), these substituents are likely to increase the potential pesticidal activity of dihydrooxadiazines. The number of polar groups in dihydrooxadiazine analogs is less than in octopamine. Therefore, there are fewer sites for potential enzymatic degradation in these compounds compared to octopamine.

**Table I. Nitrogen-to-Phenyl Distances and LogP Values for
Octopamine and Octopaminergic Pesticides.**

Compound	N-Phenyl (Angstroms)	LogP*
Octopamine	3.7	-0.1
Chlordimeform	3.5	3.5
NC-5	3.5	3.2
AC-6	3.4	2.2

SOURCE: Adapted from ref. 17.
*Estimated Using PrologP (CompuDrug, Rochester, NY)

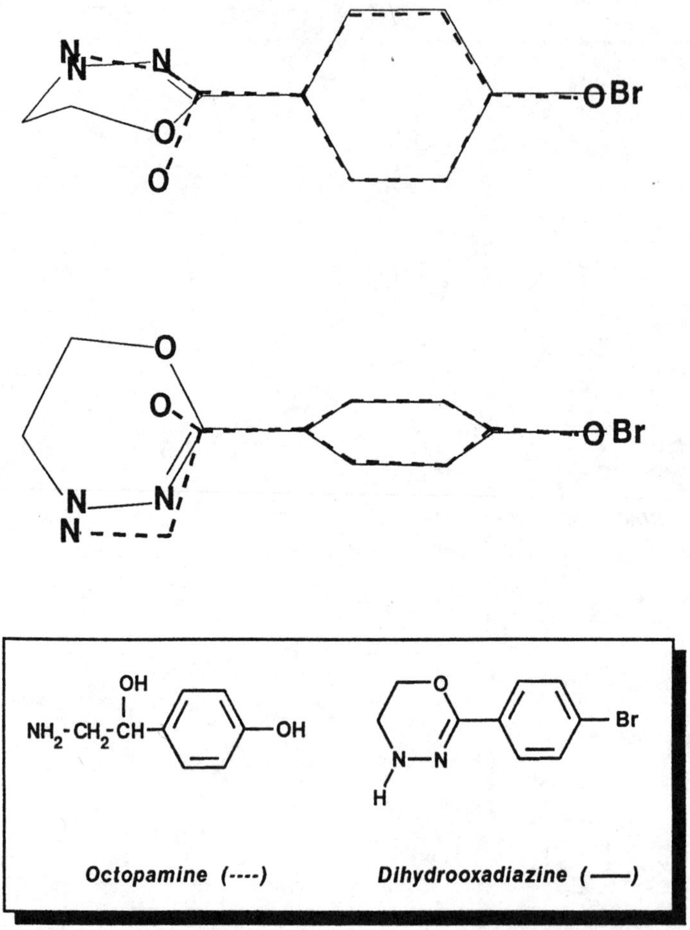

Figure 2. Superimposition of computer-generated low-energy conformers of octopamine and a proposed dihydrooxadiazine analog (orthogonal views). H atoms have been omitted for clarity. (Adapted from ref. 17.)

A comparison of calculated nitrogen-to-phenyl distances in low-energy conformers and lipophilic properties (calculated Log P) of octopamine and several dihydrooxadiazines are shown in Table II. The proposed octopaminergic pesticide candidates satisfied the criteria for the binding model, possessing a nitrogen-to-phenyl distance of 3.6 angstroms, a dihedral angle between the phenyl ring and dihydrooxadiazine ring of 94°, a good overlay between one of its low-energy conformers and that of octopamine and also a calculated LogP >1.

Synthesis

Dihydrooxadiazines are known in the literature (23). Using a modification of the methods of (23), a series of N-H , N-methyl and N-fluoroethyl dihydrooxadiazines, having various substituents in the phenyl ring, were prepared as shown in Figure 3, in 10-60% overall yields (17, 18). The N-H and N-fluoroethyl derivatives were prepared from the appropriate substituted benzoic acids which were converted, via benzoyl halides, to the substituted benzhydrazides then reacted with bromofluoroethane. The N-methyl derivatives were prepared from the appropriate substituted benzoic acids which were first converted to the substituted benzoates, then to the substituted 2-methylbenzhydrazides and finally reacted with bromofluoroethane. The types of substituents on the phenyl ring were lipophilic groups, such as halogen and alkyl, rather than the polar hydroxy group found in octopamine.

Pesticidal Properties

Several dihydrooxadiazines were shown to possess potent ovicidal toxicity to two-spotted spider mites, Tetranychus urticae Koch, and tobacco budworms, Helicoverpa virescens (F.) in greenhouse screening following contact treatment (see Table III). Adult female mites were allowed to deposit eggs on cowpea leaves for one day before treatment when they were removed, leaving the eggs. Plants were sprayed to run-off with three rates of the test compound in acetone and distilled water containing a wetting agent. Nine days following treatment, the number of hatched eggs were counted with an estimated percent mortality based on the number of eggs hatched on the check plants. Budworm eggs were immersed on cheesecloth for one minute with three rates of the test compound in acetone and distilled water containing a wetting agent. The cheesecloth samples were set on moist filter paper for five days when the numbers of hatched eggs were counted and an adjusted percent control determined based on the number of eggs hatched in the checks.

Table II. Nitrogen-to-Phenyl Distances and LogP Values for Octopamine and Dihydrooxadiazine Analogs.

Compound	N-Phenyl (Angstroms)	LogP *
Octopamine	3.7	-0.1
N-H Analog	3.6	1.6
N-Methyl Analog	3.6	2.3
N-Fluoroethyl Analog	3.6	2.5

* Estimated Using PrologP (CompuDrug, Rochester, NY)

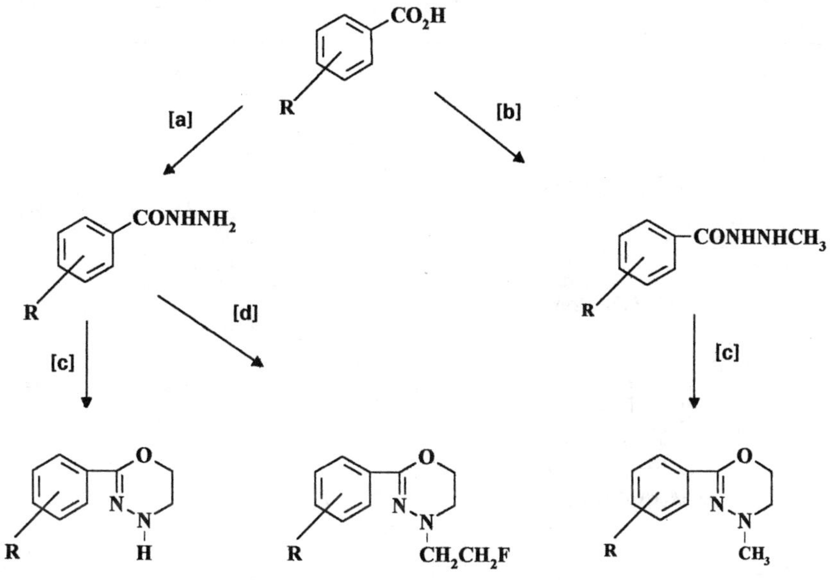

Reagents:

[a] SOCl$_2$; NH$_2$NH$_2$

[b] CH$_3$OH, H$_2$SO$_4$; NH$_2$NHCH$_3$

[c] FCH$_2$CH$_2$Br, NaOH

[d] xs FCH$_2$CH$_2$Br, NaOH

Figure 3. Synthesis of N-H (I), N-Methyl (II) and N-Fluoroethyl (III) Dihydrooxadiazines. (Adapted from refs. 17, 18.)

Table III. Pesticidal Activity of Dihydrooxadiazines on Mite (MIOVO) and Budworm (TBOVO) Eggs.

Compound	% Control @ 100 ppm	
	MIOVO	TBOVO

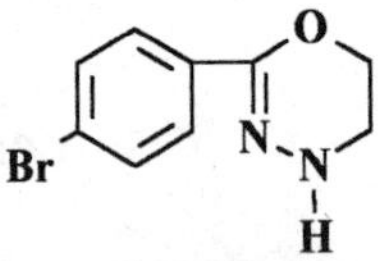

N-H Analog — 88, 97

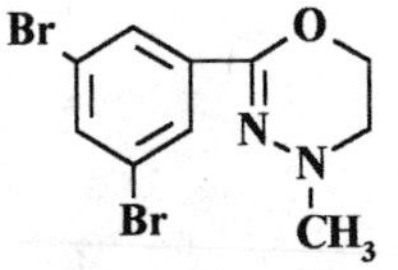

N-Methyl Analog — 80, 14

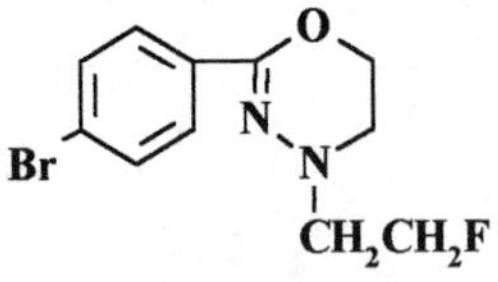

N-Fluoroethyl Analog — 99, 100

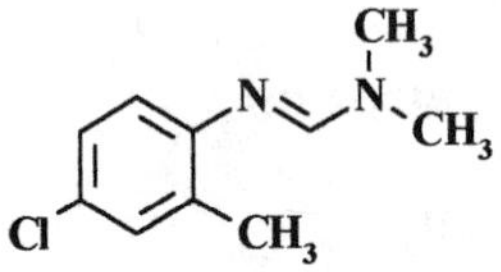

Chlordimeform — 50, 83

SOURCE: Adapted from refs. 18,19.

Pesticidal activity appeared to be most enhanced by dihydrooxadiazines substituted in the   phenyl ring by halogens, especially bromine, and on the dihydrooxadiazine nitrogen by a fluoroethyl   group.   The   4-bromophenyl   N-fluoroethyl dihydrooxadiazine showed the greatest ovicidal activity on both mites and budworms and was more potent than the formamidine, chlordimeform. A similar spectrum of biological activity to the dihydrooxadiazines was reported for the formamidines (24).

Octopaminergic Action

Using homogenates of the two-spotted spider mite and the American cockroach nervous system, selected pesticidal dihydrooxadiazines were shown to interact with octopamine receptors by stimulating adenylate cyclase activity (Table IV). All compounds tested caused elevation of cyclic AMP levels at 1×10^{-5} M. This effect was concentration-dependent (not shown). The 4-bromophenyl N-H dihydrooxadiazine (I) showed the greatest octopamine agonist activity.

The possibility that dihydrooxadiazine pesticides may interact with the same binding site as octopamine in mite homogenates and cockroach nervous system was investigated by examining the additive effects of (I) and octopamine in elevating cyclic AMP production at maximally effective concentrations. The results shown in Table IV indicate that the level of cyclic AMP production due to a combination of octopamine and (I) was not significantly different from that caused by octopamine alone in either preparation. Thus, it is possible that dihydrooxadiazines and octopamine affect the same receptor and elevate cyclic AMP levels through interaction with an octopamine-sensitive receptor coupled to adenylate cyclase. However, other sites associated with the complex may also be important to the pesticidal action of dihydrooxadiazines, for example, inhibition of N-acetyltransferase activity, which is the main pathway for biogenic amine degradation (25).

Conclusions

Using CAMD, we identified dihydrooxadiazines as a potential new class of octopaminergic pesticides by evaluating the common structural features, lipophilic profiles and biochemical potentcies of established octopaminergic pesticides, including formamidines (26), imidazolines (27) and aminooxazolines (28). Octopamine and several octopaminergic pesticides have been shown to have certain structural similarities, including the superimposability of essential functionality

Table IV. Octopaminergic Activity of Dihydrooxadiazines in Mite Homogenates and the Cockroach Nervous System.

Compound	Cyclic AMP Production @ 10 μM *	
	Mites	Cockroach
N-H Analog (I)	287	1111
N-Methyl Analog	184	768
N-Fluoroethyl Analog	206	805
Octopamine	371	2366
Octopamine + (I)	386	2397
Control	159	440

*pmol/min/mg protein

(phenyl group and a basic nitrogen atom) which were separated by a distance range of 3.4-3.7 angstroms and a dihedral angle between the phenyl ring and the side-chain of 75-100°. Additionally, octopaminergic pesticides possessed enhanced lipophilic values (Log P>1) compared to octopamine. 5,6-Dihydro-4H-1,3,4-oxadiazines were shown to satisfy the criteria for proposed octopamine mimics and, upon biological evaluation, to possess pesticidal and octopaminergic properties.

Acknowledgements

The contributions to the preparation of this manuscript by Dr. Derek McPhee and helpful discussions with Dr. Barry Friedlander are acknowledged and greatly appreciated.

Literature Cited

1. Goosey, M. W. *Pestic. Sci.* **1992,** *34,* 313-320.
2. Jennings, K. R.; Kuhn, D. G.; Trotto, S. H.; Whitney, W. K. In *Trace Amines: Their Comparative Neurobiology and Clinical Significance*; Boulton, A. R.; Juorio, A. V.; Downer, R. G. H., Eds.; Humana Press: New Jersey, 1987; pp 53-63.
3. Downer, R. G. H. In *Cockroaches as Models for Neurobiology: Applications in Biomedical Research*; Huber, I.; Masler, E.P.; Rao, B.R., Eds.; Vol. 2; CRC Press: Boca Raton, USA, 1990; pp. 103-124.
4. Dekeyser, M. A.; Downer, R. G. H. *Pestic. Sci.* **1994,** *40,* 85-101 .
5. Evans, P. D. In *Comprehensive Insect Physiology, Biochemistry and Pharmacology*; Kerkut, G. A.; Gilbert, L. I., Eds.; Pergamon Press: Oxford, 1985; Vol. 11, pp 499-530.
6. Dittrich, V. *J. Econ. Entomol.* **1966,** *59,* 889-893.
7. Beeman, R. W.; Matsumura, F. In *Pesticide and Venom Neurotoxicity*; Shankland, D. L.; Hollingworth, R. M.; Smyth, T., Jr., Eds.; Plenum Press: New York, 1978; pp. 189-206.
8. Knowles, C. O. In *Pesticide Chemistry: Human Welfare and the Environment*; Miyamoto, J.; Kearney, P. C., Eds.; Pergamon Press: Oxford, Vol 1., 1983; pp. 265-270.
9. Drabek, J.; Neumann, R. *Progr. Pestic. Biochem. Toxicol.* **1985,** *5,* 1-7.
10. Hollingworth, R. M.; Murdock, L. L. *Science* **1980,** *208,* 74-76.
11. Evans, P. D.; Gee, J. D. *Nature* **1980,** *287,* 60-62.
12. Nathanson, J. A.; Hunnicutt, E. J. *Mol. Pharmacol.* **1981,** *20,* 68-75.
13. Gole, J. W. D.; Orr, G. L.; Downer, R. G. H. *Life Sci.* **1983,** *32,* 2939-2947.

14.   Ismail, S. M. M.; Matsumura, F. *Insect Biochem.* **1989,** *19,* 715-722.
15.   Ware, G.W. *The Pesticide Book.* Thomson Publications: Fresno, CA, 1989, 336p.
16.   Hirashima, A.; Yoshii, Y.; Eto, M. *Comp. Biochem. Physiol.* **1992,** *103C,* 321-325.
17.   Dekeyser, M. A.; Harrison, W. A.; McDonald, P. T.; Downer, R. G. H. *Pestic. Sci.* **1993,** *38,* 309-314.
18.   Dekeyser, M. A.; McDonald, P. T.; Angle, G. W., Jr.; Downer, R. G. H. *J. Ag. Food Chem.* **1993,** *41,* 1329-1331.
19.   Dekeyser, M. A.; McDonald, P. T.; Angle, G. W., Jr.; Borth, D. M.; Downer, R. G. H. *J. Econ. Entomol.* **1993,** *86,* 1339-1343.
20.   Paxton, K.; Hamor, T. A. *Acta Crystallogr.* **1977,** *B33,* 2143-2146.
21.   Hollingworth, R. M.; Johnstone, E. M.; Wright, N. In *Pesticide Synthesis Through Rational Approaches*; Magee, P. S.; Kohn, G. K.; Menn, J. J., Eds.; ACS Symposium Series 255, American Chemical Society: Washington, DC, 1984; pp 103-125.
22.   Haas, A. *Actual. Chim.* **1987,** *5,* 183-188.
23.   Trepanier, D. L.; Spracmanis, V.; Eble, J. N. *J. Med. Chem.* **1966,** *9,* 753-758.
24.   Gemrich, E. G.; Lee, B. L.; Tripp, M. L.; Vandestreek, E. *J. Econ. Entomol.* **1976,** *69,* 301-306.
25.   Downer, R. G. H.; Martin, R. J. In *Sites of Action for Neurotoxic Pesticides*; Hollingworth, R. M.; Green, M. B., Eds.; ACS Symposium Series 356; American Chemical Society, Washington, DC, 1987; pp 202-210.
26.   Hollingworth, R. M.; Murdock, L. L. *Science* **1980,** *208,* 74-76.
27.   Nathanson, J. A. *Proc. Natn. Acad. Sci. U. S. A.* **1985,** *82,* 599-603.
28.   Jennings, K. R.; Kuhn, D. G.; Kukel, C. F.; Trotto, S. H.; Whitney, W. K. *Pestic. Biochem. Physiol.* **1988,** *30,* 190-197.

RECEIVED November 18, 1994

Chapter 14

Insect Aggregation Pheromone Response Synergized by "Host-Type" Volatiles

Molecular Modeling Evidence for Close Proximity Binding of Pheromone and Coattractant in *Carpophilushemipterus (L.)* (Coleoptera: Nitidulidae)

Richard J. Petroski[1] and Roy Vaz[2]

[1]Bioactive Constituents Research, National Center for Agricultural Utilization Research, Agricultural Research Service, U. S. Department of Agriculture, Peoria, IL 61604
[2]Marion Merrell Dow, Cincinnati, OH 45242

The driedfruit beetle, *Carpophilus hemipterus* (L.) is a worldwide pest of avariety of fruits and grains, both before and after harvest. Attractiveness of the male-produced aggregation pheromone is enhanced by the presence of a "host-type" volatile coattractant. A set of 26 compounds was used to explore relationships between pheromone structure and activity by 3D-QSAR/CoMFA methods. Significant differences in aggregation pheromone CoMFA-coefficient contour maps were observed in the presence and absence of the "host-type" volatile coattractant.

The driedfruit beetle, *Carpophilus hemipterus* (L.) (Coleoptera: Nitidulidae), attacks a large number of agricultural commodities in the field, during storage after harvest or in transport (*1*). It is also able to vector microorganisms responsible for the souring of figs (*1*) and mycotoxin production in corn (*2*).

Both sexes of *C. hemipterus* respond to a male-produced aggregation pheromone (*3*). A wind tunnel bioassay guided the isolation of eleven all-*E* tetraene hydrocarbons, two *Z*-isomer tetraene hydrocarbons and one all-*E* triene hydrocarbon (*3,4*). The pheromone components were tentatively identified by spectroscopic methods then the assigned structures were proven by synthesis (*3-5*). Structures of the synthesized compounds are shown in Figure 1. Compounds A to N have been identified in the *C. hemipterus* pheromone blend (*3,4*); the additional compounds were prepared to explore structure activity relationships (*4*).

Previous studies have shown that aggregation pheromone activity may be enhanced when the pheromone is used in combination with attractive chemicals produced by the host plant or associated microorganisms, termed host-type volatiles or host-type coattractants (*6-9*). In order to investigate relationships between the structure of the pheromone molecule and biological activity, as well as explore possible additional relationships between the coattractant and pheromone structure-activity relationships, all compounds (A to Z) were tested for activity both with and

0097-6156/95/0589-0197$12.00/0
© 1995 American Chemical Society

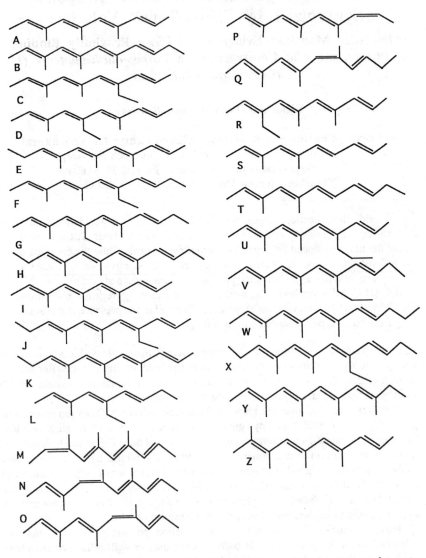

Figure 1. Hydrocarbon structures used in the *Carpophilus hemipterus* data set.

without adding a host-type coattractant (propyl acetate) to the bioassay treatments (*4*). The results of this previous work are summarized in Table 1.

Individual compounds are capable of eliciting the pheromonal response, as opposed to an obligate requirement for a blend of compounds (*4*). This observation is consistent with a hypothesis that all the structures interact with a single recognition site or a family of component recognition sites having conservation of the required bioactive conformation of the ligands at the recognition sites.   An obligate requirement for a blend of compounds would indicate the action of distinctly different recognition sites, each with its own structural and conformational requirements.

Recent advances in computational chemistry enabled us to probe quantitative structure-activity relationships (QSAR) in three-dimensional space. We report 3D QSAR studies with the aid of Comparative Molecular Field Analysis (CoMFA) methodology (*10-13*). With CoMFA, a suitable sampling of the steric and electrostatic fields surrounding a set of ligand molecules might provide all the information necessary for understanding their observed biological properties (*13*).

Materials and Methods

Data. Chemical structures (Figure 1) and the corresponding bioassay data with and without coattractant (Table 1) were taken from Bartelt et al. (*4*).   The data corresponded to a counting of the number of beetles alighting on pieces of filter paper in a wind tunnel bioassay; two treatment preparations to be compared (pheromone versus control or pheromone plus coattractant versus control) were applied to pieces of filter paper, and those were hung side by side in the upwind end of the wind tunnel. The coattractant (propyl acetate) alone vs a blank filter paper control was only minimally attractive to *C. hemipterus*; relative bioassay activity was less than 5 percent (*4*). Two CoMFA analyses were done, one for the data without a coattractant (propyl acetate) and one for the data with the coattractant.

Establishing the Conformation of Each Molecule. A computation using the MOPAC (*14*) program and the AM1 Hamiltonian was done on the sequence of model structures shown in Figure 2 which shows the optimal geometries as well as the bond orders. The doubly substituted structure is twisted more than the singly substituted structure. The amount of delocalization decreases as substituent methyl groups are introduced in the progression. Some conformational searching is required to find the low energy conformations. Hence, the single bonds in structure A were assumed to be rotatable with a reasonable energy barrier in terms of all states being populated at room temperature.

The 3D structures represented in Figure 1 were constructed using structure A from the figure as a template.   Structure A was subjected to conformational searching about the rotatable bonds using the Tripos 5.2 Molecular Mechanics Force Field (*10*).

The minima encountered in the conformational search of compound A were optimized with the AM1 Hamiltonian and the minimum conformation was used as the template. If there were any extensions made, in terms of adding rotatable bonds to structure A such as in structure B, a molecular mechanics force field conforma-

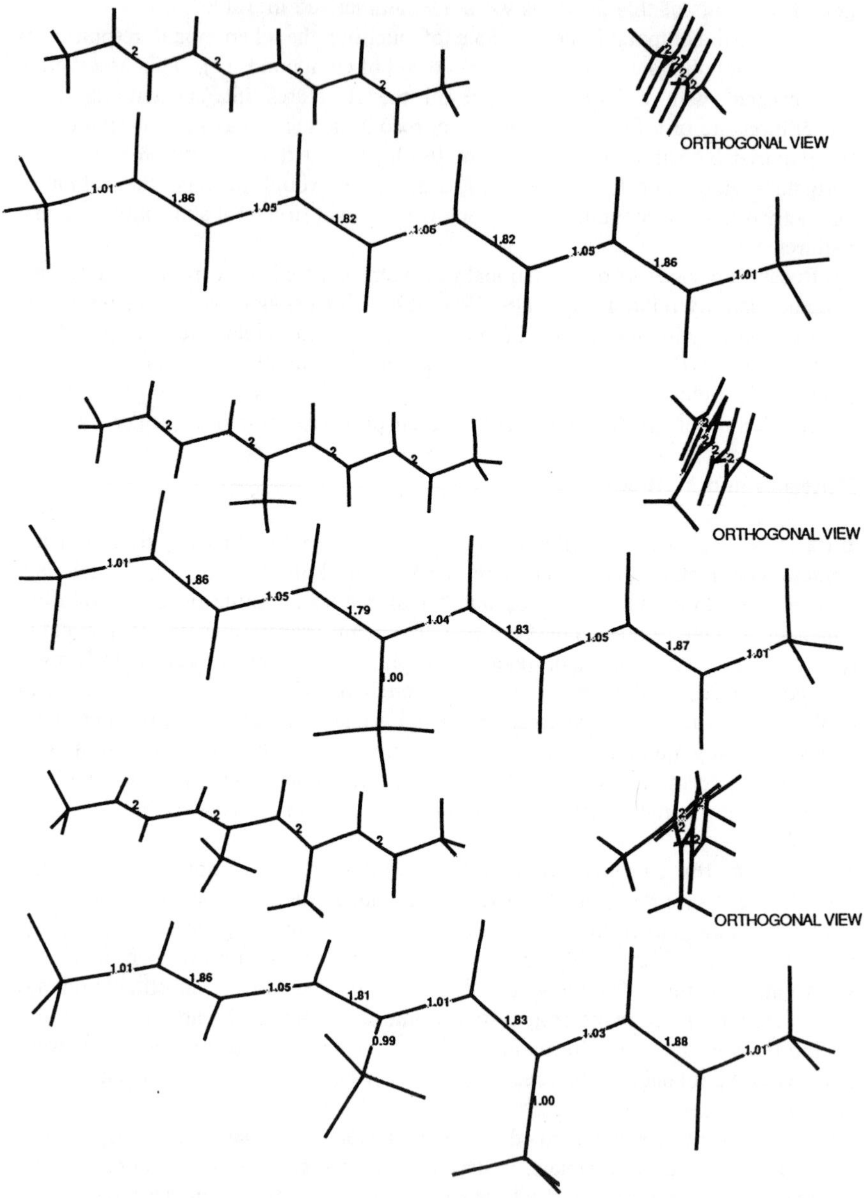

Figure 2. A sequence of model structures (2E,4E,6E, 8E,-tetradecenes having 0, 1, or 2 methyl substituents on carbons 5 and 7) showing lower delocalization and thus lowering the rotational barrier for the single bond between carbons 5 and 6 with lesser substitution.

tional search was again done on the additional rotatable bond and the minima obtained, optimized using the AM1 Hamiltonian and the energies compared, choosing the minimum energy conformation again. Conformations described here are vapor phase. Since all the compounds examined in this study are only unsaturated hydrocarbons, "solvent" effect on conformations at the putative pheromone recognition site located on an insect antenna should be minimal.

Superimposing the Molecules Within a Region. Once an optimal conformation was obtained for all the structures, the structures were then overlapped via an RMS fit using the atoms labeled with an asterisk in Figure 3. The overlapped structures are depicted in Figure 4. A region, as shown in Figure 4, was then constructed such that all structures fell at least 2 A° away from the region extents. The region only had carbon atoms used as probes and the lowest and highest points had the coordinates of 9.6170, 7.2766, -5.0496) and (10.4965, 4.2695, 6.5532) respectively. The points were separated at intervals of 2 A° along each axis.

Comparative Molecular Field Analysis. This region containing the superimposed structures was utilized in a Comparative Molecular Field Analysis (CoMFA) experiment (Figure 5). Normally in a CoMFA, two probes are used. One probe is a carbon atom with no charge and the other probe is a positive charge (with no mass or van der Waals radius). The energy of the probe at each point of the region is calculated using the Tripos 5.2 force field (*10*). The two terms of interest in the force field are the 6-12 van der Waals terms which account for the London dispersion forces, and the coulombic terms representing coulombic forces arising from point charges. The positive charge probe was not used in this study because the π electron clouds of the unsaturated hydrocarbons in the analysis are not reflected by the charge on the carbon atoms. Points in the region where the energy of the carbon probe exceeded 30 kcal/mol were dropped from the analysis. The biological activities used were those listed in Table 1. CoMFA columns whose standard deviation was less than 2.0 kcal/mol were ignored in the calculation. This reduced the number of columns involved in the Partial Least Squares (PLS) statistical analysis (*15*) substantially. Also, changing the dropped columns to those having a standard deviation of less than 1.0 kcal/mol did not have any significant impact on the statistics.

The predictive ability of both models (3D QSAR with and without coattractant) were evaluated using cross-validation in which the cross-validation was done using as many groups as there were rows except as noted. Cross-validation involves pretending that one of the rows does not have experimental data. The resulting equation is used to predict the experimental measurement for the omitted compound. The cross-validation cycle is repeated, leaving out one different compound until each compound has been excluded and predicted exactly once. The resulting individual squared errors of prediction are accumulated. The result of the cross-validation is the sum of squared prediction errors, sometimes termed the PRESS (Predictive Residual Sum of Squares). In PLS, the iterations are continued until the PRESS no longer decreases significantly. Substituting PLS, which operates on all independent variables simultaneously, for regression, which operates on one independent variable at a time, reduces the probability of accepting a chance correlation (*13*).

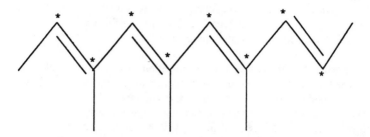

Figure 3. The carbons marked by an asterisk are used to match the other structures after they are optimized.

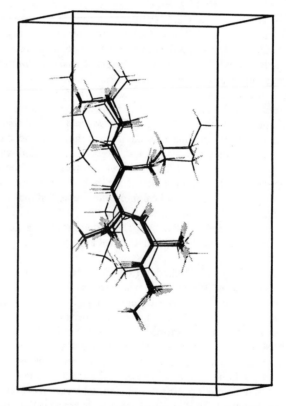

Figure 4. All optimized structures aligned using the atoms marked in Figure 3.

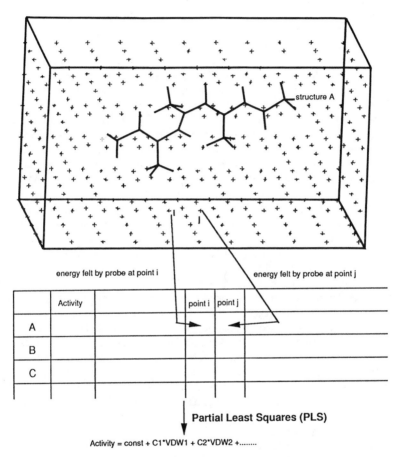

Figure 5. Schematic of the CoMFA steric field for structure A.

Conventional R-squared values (regression range from 0 to 1; however, the cross-validated R-squared values (PLS) range from negative infinity to 1.0.

Results and Discussion

Some synthetic analogs, which were never detected from the beetles (compounds O to Z in Table 1), showed activity in the bioassay (e.g. compounds O, P, V and W). This observation led us to the conclusion, shared by others (*16*), that insect pheromone communication systems are not as rigid as once thought.

Some generalizations can be made about structure-activity relationships (*4*): (1) The left-hand terminal alkyl group (as drawn in structure A) should be methyl; substitution of ethyl for methyl renders the compound inactive or nearly so. Compounds E, H, J, K, and X have low or no activity. (2) The left-hand alkyl branch should be methyl, but the one example with an ethyl group at that position (compound R) did have slight activity. (3) An ethyl group as the middle alkyl branch (e.g., the 5-ethyl group in compound D) also renders the compound inactive. (4) The right-hand alkyl branch (e.g., the 7-position of compound B) can be methyl, ethyl (as in compound F), or propyl (as in compound V) and still have activity; however, only a hydrogen in that position (compounds S and T) renders the compound inactive. (5) The right-hand terminal alkyl group can be methyl (compound A), ethyl (compound B) or propyl (compound W) and still have activity, but the ethyl group seems most consistent with high activity. (6) Alkyl groups in the 9-position (compound Y) or in the 2-position (compound Z) greatly reduce activity. (7) The presence of *cis* double bonds at any position reduces activity (compounds M, N, O, P and Q).

Another important general feature was evident from the results shown in Table 1. Relative activity was often enhanced when each unsaturated hydrocarbon was separately tested in the presence of the coattractant but the proportion of enhancement varied from hydrocarbon to hydrocarbon tested. In some cases (e.g. compounds A, N, and O), activity decreased in the presence of the coattractant. These observations revealed a relationship between the structure of the hydrocarbon tested and the role of the coattractant; maximal activity in the presence of the coattractant was observed when the right-hand terminal alkyl group was ethyl. Compounds A, N, and O all have methyl as the terminal alkyl group. Beyond this observation, it is hard to imagine a more precise role for the coattractant without use of modern computational tools.

Although insights can be acquired by looking at two-dimensional representations of structures as are shown in Figure 1, the compounds are actually three-dimensional. A more refined examination is gained by using modern 3D QSAR methods.

The predicted versus actual plots for the 3D-QSAR analyses with and without the coattractant show that both CoMFA models are workable predictors of biological activity (Figure 6). The R-squared values and other relevant statistics for both analyses are reasonable (Table 2). Thus, the CoMFA results also support the hypothesis of either a single pheromone recognition site or (less likely) a family of

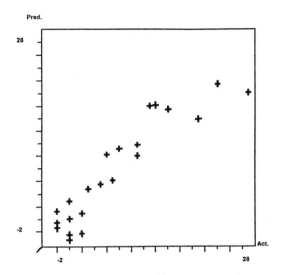

Predicted vs actual for structures without coattractant

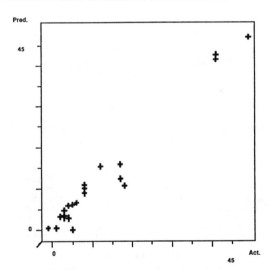

Predicted vs actual for structures with coattractant

Figure 6. Plots of predicted versus actual biological activity values for structures with and without coattractant.

Table 1
Bioassay Activity for Individual Hydrocarbons[a]

| | Relative Activity (%) | |
Hydrocarbon	Without Coattractant	With Coattractant
A	24	18
B	29	60
C	21	41
D	0	2
E	2	3
F	16	49
G	2	8
H	0	17
I	0	1
J	0	0
K	4	5
L	5	5
M	5	3
N	11	3
O	11	8
P	13	17
Q	7	4
R	6	6
S	0	0
T	0	1
U	8	1
V	35	41
W	14	12
X	0	11
Y	0	8
Z	3	4

[a] Data from Bartelt et al., *J. Chem. Ecol.* 18(3) 379-402 (1992).
The coattractant was propyl acetate (20 µl, 10% in mineral oil).

Table 2
Partial Least Squares (PLS) Analysis of CoMFA Data
(only steric field included)

Statistics	Without Coattractant	With Coattractant
Number of components	4	3
Standard error of estimate	3.727	2.957
R-squared	0.851	0.964
R-squared (crossvalidated[a])	0.490	0.811
Standard error of prediction	6.891	6.730
Compounds dropped from analysis	V, K, W	U, J, X, I

[a] The R-squared is related to the "PRESS" via the equation:
(S.D. - PRESS)/S.D. where S.D. is the sum over all molecules
of squared deviations of each biological parameter from the
mean and PRESS (Predictive Sum of Squares) is the sum over
all molecules of the squared differences between the actual
and predicted biological parameters (range is neg infinity to 1).

pheromone component recognition sites having bioactive conformation conservation in *C. hemipterus*.

The main analysis tools, in terms of computer aided molecular design, are the coefficient plots as shown in figures 7 and 8. These plots are actually contours of the standard deviation times PLS coefficient [(std dev)*(coefficient)] at each point in the region that fall in a particular range. The field is created as the point by point product of the PLS coefficient and the standard deviation of energies at the point among all compounds in the study. The view of this field is preferred to the view of only the PLS coefficients field because it reduces the visual cluster of moderately large coefficients that arise by chance association with larger scale trends.

The contours are centered at -0.7 (light gray, both Figures) and 0.14 (black, Figure 7) or 0.19 (black, Figure 8) with structure A embedded in the contour plots. These contours have the same meaning as the plots in reference 3 viz. if the contour is for a region corresponding to a negative value, in that case, that region in space would need lower van der Waals interaction energies if a carbon probe atom were placed in that region or at the very most, no change would be made in that region for increased activity. Similarly, a positive region would prefer increased van dar Waals interaction energies for a carbon probe for increased activity. This can be derived from the equation in Figure 5.

The positive and negative coefficient regions show that extending the structure A by a methylene in the direction as in structure B puts the methylene in a positive coefficient region and similarly extending structure A by a methylene in the other direction, as in structure E, puts this latter methylene in a negative coefficient region. Also, extending molecule A by a methylene such as in structure B or differently such as in structure C, even though both regions have positive coefficients, their relative values are different and thus the structural extensions have different consequences on the activity.

The contour plot from the analysis of the structures with a coattractant is quite interesting. A new, sharp, and very well-defined most-negative region seems to have been created which could possibly be attributed to the coattractant occupying this region on the putative receptor. The region corresponding to the most negative coefficient region for the analysis without the coattractant is still present in the analysis with the coattractant.

The field times coefficient field where the field value represents the product of the molecule's field energy and the PLS coefficient from the appropriate analysis from which they were dropped between the two analyses did not lead to any activity in the new negative region for the appropriate dropped molecules, thereby eliminating this new region as arising from the outliers not used in the analysis. The (field)*(coefficient) plot represents the contribution of this field for this molecule to its predicted activity.

The significant change in the CoMFA contour plot with coattractant versus the corresponding CoMFA contour plot without coattractant suggests a close proximity in binding sites for the pheromone and the coattractant near the 10-position of the pheromone (e.g., compound A in Figure 8) If compound B were pictured in the figure, the new most negative area would still reside at the 10-position, which would be over the methylene portion of the right-hand terminal ethyl group. An

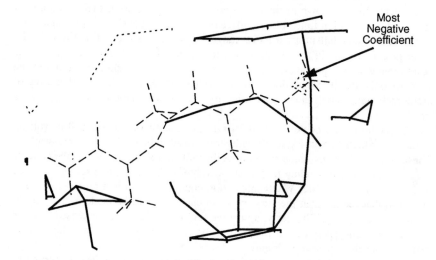

Figure 7. The std. dev. * coefficient Co MFA contour plot for the structures without coattractant showing only structure A.

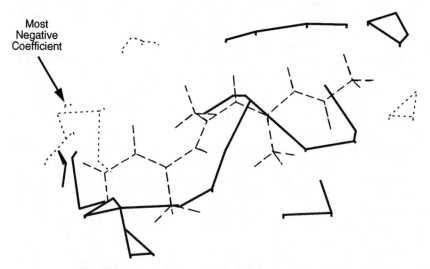

Figure 8. The std. dev. * coefficient CoMFA contour plot for the structures with coattractant showing only structure A.

alternative, but less likely, interpretation of our CoMFA data would be binding of the coattractant at a separate binding site but affecting the binding of the pheromone (allosterism). This placement of the coattractant would not have been possible without a 3D analysis.

If *C. hemipterus* pheromone recognition sites have enough fluidity then it is theoretically possible to develop pheromone analogs that serve as species-specific insect control agents. Based on our CoMFA results, it might be possible that an oxygen atom from the coattractant (propyl acetate) resides at this new most negative site. It would be interesting to place an oxygen atom in a pheromone analog at this position in 3D space, but this has yet to be tested experimentally.

Carefully designed pheromone analogs, or blends thereof, might surpass the natural pheromone in biological activity, ease of preparation, or stability under field conditions. Such analogs would improve our ability to monitor pest populations, lower pest populations by mass trapping, or lower pest populations by use of combinations of pheromone and either insecticides or biological control agents. It is also theoretically possible that pheromone perception inhibitors (antagonists) could be developed against *C. hemipterus*. Pheromone perception inhibitors could be used for the protection of commodities during storage or transport.

Literature Cited

1. Hinton, H. E. *A Monograph of the Beetles Associated with Stored Products*; Jarrold and Sons: Norwich, U.K., 1945, 443 pp.
2. Wicklow, D. T. In *Phytochemical Ecology: Allelochemicals, Mycotoxins and Insect Pheromones and Allomones*; Chou, C. H.; Waller, G. R., Eds.; Institute of Botany, Academia Sinica Monograph Series No. 9: Taipei, ROC., 1989, p263.
3. Bartelt, R. J.; Dowd, P. F.; Plattner, R. D.; Weisleder, D. *J. Chem. Ecol.* 1990, *16*, 1015.
4. Bartelt, R. J.; Weisleder, D.; Dowd, P. F.; Plattner, R. D. *J. Chem. Ecol.* 1992, *18*, 379.
5. Bartelt, R. J.; Weisleder, D.; Plattner, R. D. *J. Agric. Food Chem.* 1990, *18*, 2192.
6. Walgenbach, C. A.; Burkholder, W. E.; Curtis, M. J.; Khan, Z. A. *J Econ. Entomol.* 1987, *80*, 763.
7. Oehlschlager, A. C.; Pierce, A. M.; Pierce H. D. Jr.; Bprden, J. H. *J. Chem. Ecol.* 1988, *14*, 2071.
8. Birch, M. C. In *Chemical Ecology of Insects*; Bell, W. J.; Carde, R. T., Eds.; Sinauer Assoc.: Sunderland, Massachusetts, 1984; Chapter 12.
9. Bartelt, R. J.; Schaner, A. M.; Jackson, L. L. *Physiol. Entomol.* 1986, *11*, 367.
10. Clark, M.; Cramer III, R. D.; Van Opdenbosh, N. *J. Comp. Chem.* 1989, *10*, 982.
11. Cramer III, R. D.; Patterson, D. E.; Bunce, J. D. *J. Am. Chem. Soc.*, 1988, *110*, 5959.
12. Cramer III, R. D., DePriest, S. A., Patterson, D. E., Hecht, P. in "*3D QSAR in Drug Design: Theory and Applications*"; Kabinyi, H, Ed; ESCOM, The Netherlands, 1993, p 443.

13. Cramer III, R. D., Simeroth, P., Patterson, D. E. in "*QSAR: Rational Approaches to the Design of Bioactive Compounds*"; Silipo, C. and Vittoria, A., Eds; Elsevier Science Publishers B. V., Amsterdam

14. MOPAC 5.0 is available from QCPE, Indiana University, Bloomington, IN.

15. Cramer III, R. D., Bunce, J. D., Patterson, D. E., Frank, I. E. *Quant. Struct.-Act. Relat. Pharmacol., Chem. Biol.* 1988, *7*, 18.

16. Carlson, D. A.; McLaughlin, J. R. *Experientia* 1982, *38*, 309.

RECEIVED January 31, 1995

Chapter 15

Predicting Activity of Protoporphyrinogen Oxidase Inhibitors by Computer-Aided Molecular Modeling

Krishna N. Reddy, Ujjana B. Nandihalli[1], Hee Jae Lee, Mary V. Duke, and Stephen O. Duke[2]

Southern Weed Science Laboratory, Agricultural Research Service, U. S. Department of Agriculture, P.O. Box 350, Stoneville, MS 38776

Protoporphyrinogen oxidase (Protox) is the primary site of action of herbicides belonging to at least ten chemical classes. Structure-activity relationships studies of diphenyl ether and phenopylate herbicides have shown the bicyclic structure of Protox inhibitors to approximate one half of the enzyme substrate, protoporphyrinogen IX. We determined the effect of a member of each of ten different chemical classes of Protox inhibitors on Protox activity and ability to induce protoporphyrin IX accumulation and to cause light-induced cellular leakage. Molecular properties determined with MOPAC of these compounds were then correlated with these biological activities by regression analysis. The regression equations based on the bulk, electronic, energy, and lipophilicity properties accounted for most of the variation in the three biological activities, and appear to be specific to the particular biological activity. Predictive equations generated by this method were then tested on enantiomers with chirality in the *meta*-substitution of the phenyl ring. The equations were somewhat predictive for active compounds, but were not predictive for the inactive compounds.

The molecular target site of photobleaching herbicides such as diphenyl ethers (DPE) is the enzyme protoporphyrinogen oxidase (Protox) of the porphyrin pathway, the last enzyme common to both chlorophyll and heme biosynthesis (*1-7*). Several chemical classes of herbicides cause

[1]Current address: Hazleton Laboratories, Madison, WI 53707
[2]Corresponding author

0097–6156/95/0589–0211$12.00/0
© 1995 American Chemical Society

rapid peroxidative bleaching and/or desiccation of plant tissues (3, 5, 7-14). They do this by causing accumulation of massive amounts of protoporphyrin IX (Proto IX) (4, 6), the product of Protox, rather than the substrate, protoporphyrinogen IX (Protogen IX). A model that describes the mechanism of the accumulation of Proto IX in plant cells in the presence of the Protox-inhibiting herbicides has been proposed (6, 15-17). By inhibiting plastid Protox, these herbicides induce Protogen IX accumulation which is exported out of the plastid envelope into the cytoplasm. The exported Protogen IX is oxidized by herbicide-resistant extraorganellar oxidases associated with the plasma membrane and microsomes. High levels of Proto IX build up in extraplastidic sites. Some of this may re-enter the porphyrin pathway, but most accumulates and appears to participate in a type II photoperoxidation process which damages plants.

Just as the discovery of the D-1 protein binding site of photosystem II (PS II) inhibitors spurred a better understanding of the QSAR of PS II-inhibiting compounds (18), discovery of Protox as the binding site of a large number of peroxidizing herbicides has important implications for those interested in QSAR studies of this category of herbicides. However, unlike PS II inhibitors, Protox inhibitors are competitive inhibitors (19) that apparently mimic a portion of the substrate Protogen IX or transition state structure (7, 14, 20, 21). Protox inhibitors are perhaps the largest group of effective herbicides that are competitive inhibitors of an enzyme target site. Whether all classes of Protox inhibitors mimic the same portions of the Protogen IX molecule is unknown.

These studies have demonstrated that at least three classes of Protox inhibitors (DPEs, O-phenyl pyrrolidinocarbamates, and O-phenyl piperidinocarbamates) are structurally and electronically similar to one half of the Protogen IX molecule, mimicking two of the porphyrin rings (7, 14, 20, 21). The most active analogues generally most closely approximate the molecular properties of one-half of Protogen IX (14, 20). Molecular geometry information from X-ray crystallography data of four Protox inhibitors (22) supports the view that they mimic one half of the Protogen IX molecule. Slight steric changes, as studied in chiral pairs with very similar molecular properties, can have overriding influences on Protox-inhibiting action (23).

Several previous QSAR studies of Protox inhibitors have been published (reviewed in 14). Some of these studies have relied heavily on computational chemistry to relate structure to activity at the molecular or physiological level (e.g., 20, 21, 23), whereas others have related structure only to whole plant activity (e.g., 24-29). Whole plant QSAR studies can be confounded by different molecular requirements for uptake, metabolic degradation, and movement to the site of action than for activity at the molecular level. Furthermore, the DPE and probably other classes of Protox inhibitors can have several molecular sites of action (30), each with its own set of molecular requirements.

In this chapter we provide a computational analysis of ten Protox inhibitors of different chemical classes, in an attempt to predict the biological activities of compounds of similar structure. Molecular properties of these compounds are shown to be good predictors of Protox inhibition, accumulation of Proto IX, and herbicidal activity.

Materials and Methods

The Compounds. Herbicidally active members of ten classes of Protox inhibitors were used (Figure 1). Technical grade compounds with 95% or greater purity were used. The source of the compounds was as follows: AH 2.430 (Monsanto Co.); UCC-C4243 (Uniroyal Co.); PPG-1013 (PPG Corp.); oxadiazon and RH-0978 (Rohm and Haas Co.); FMC-14 and F-6285 (FMC); LS 82-556 and M&B 39279 (Rhône Poulenc); and TNPP-ethyl (Otsuka Chemical Co.).

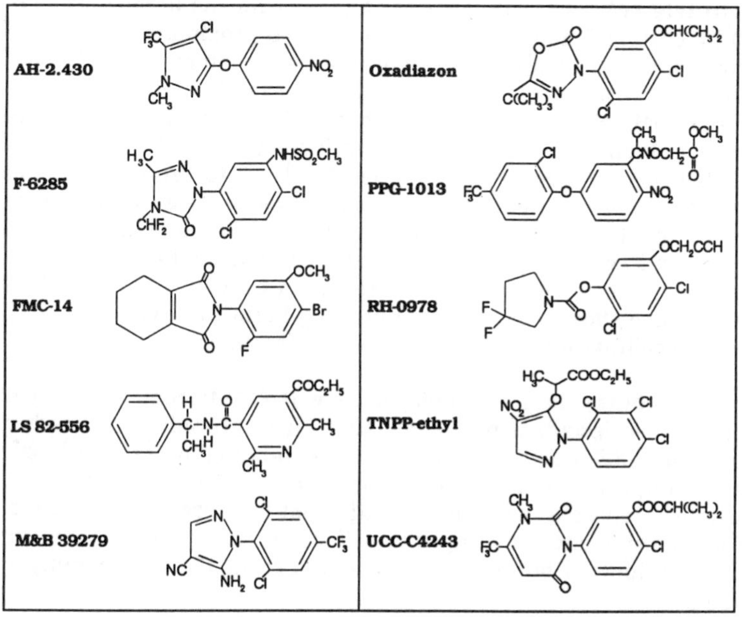

Figure 1. Protox-inhibiting compounds used in these studies.

Plant Material. For Protox assays and tissue bioassays, barley (*Hordeum vulgare* L. [cv. Morex]) was grown in flats in a commercial greenhouse substrate (Jiffy Mix; JPA, West Chicago, IL) and watered with distilled water. Etiolated barley plants were grown in the dark at 25°C for 6 days and green barley plants were grown at 25°C under continuous white light of 500 μmol m^{-2} s^{-1} photosynthetically active radiation (PAR) and >90% relative humidity for 8 days.

Protox Assay. Etioplast Protox preparations were made from 6-day-old, dark-grown barley leaves as before (*13*). Prior to assay, protein content of plastid preparations was determined by the method of Bradford (31) with bovine serum albumin as a standard. Extracts were diluted to 4 mg protein per ml in resuspension buffer and Protox activity determined as before (13).

Porphyrin Determinations. All extractions for HPLC were made under a dim, green light source. Samples (0.22 g of barley leaf sections) were homogenized and extracted as before (13). Determinations of Proto IX, Mg-Proto IX, Mg-Proto IX monomethyl ester, and protochlorophyll-ide (PChlide) were made as before (23) by HPLC with spectrofluorometric detection. All porphyrin compound levels are expressed on a molar basis per gram of fresh weight. All treatments for porphyrin samples were triplicated.

Electrolyte Leakage. Tissues were treated with the herbicides as before (13) by cutting 5-mm barley leaf sections (approximately 0.22 g) with a razor blade, and then placing them in a 6-cm-diameter polystyrene Petri dish in 5 ml of sucrose (10 g L^{-1}), 2-[N-morpholino]ethanesulfonic acid (MES; 1 mM, pH 6.5) medium with or without herbicide dissolved in absolute ethanol. The discs were then incubated at 25°C in darkness for 20 h before exposure to 500 μmol cm^{-2} s^{-1} PAR for up to 24 h. Cellular damage was measured by detection of electrolyte leakage into the bathing medium with a conductivity meter. Because of differences in background conductivity of different treatment solutions, results were expressed as changes in conductivity upon exposure to light. Previous studies have shown that Protox-inhibiting herbicides have no significant effect on cellular leakage in darkness. All treatments were triplicated.

Binding Studies. Barley etioplasts were isolated from etiolated barley seedlings as described previously (21). The procedure of Tischer and Strotmann (32) was used to determine binding of ^{14}C-acifluorfen to barley etioplasts in the presence or absence of the herbicides. Using previously reported methods (21), etioplast membranes were mixed with ^{14}C-acifluorfen and centrifuged to pellet the membranes and the amount of ^{14}C-acifluorfen bound was calculated from the radioactivity in the pellets.

Estimation of Partition Coefficients. Reversed-phase high-performance liquid chromatography (RP-HPLC) was used for the determination of octanol/water partition coefficients (P) of herbicides as before (20, 21, 23, 33).

Molecular Properties and Statistical Analysis. The molecular properties of herbicides were calculated using the computer software and procedures as described in our previous work (20, 21, 23). Three dimensional chemical structures were built from standard atoms and fragments stored in the library file of Chem-X software (Chemical Design Limited, Oxford, England). The structures were optimized by a molecular orbital program, MOPAC (Quantum Chemistry Program

Exchange, No. 560, Dept. Chemistry, Indiana University, Bloomington, IN, USA; Version 6.0) using AM1 (Austin Model) hamiltonian. Molecular electrostatic potential maps were generated as before (*21*).

Multiple regression analysis was performed to establish relationships between biological activities and molecular properties. Molecular properties of bulk, electronic, energy, and lipophilicity were considered independent variables and three biological activities (Protox I_{50}, Proto IX accumulation, and electrolyte leakage) were considered as dependent variables. The RSQUARE method of SAS (SAS Institute, Inc., Cary, NC) was used to choose the variables for building various regression models.

Results

Herbicidal Activity. All ten compounds caused herbicidal damage as measured by electrolyte leakage from barley leaf sections (Figure 2). The activity varied about two-fold from the most active compound (UCC-C4243) to the least active (LS 82-556). The leakage rate was approximately linear from 2 to 8 h after the beginning of light exposure.

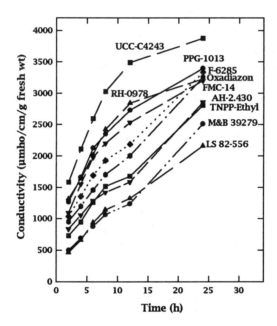

Figure 2. Herbicidal activity of ten compounds, each at 100 µM, used in these studies as measured by electrolyte leakage of green barley leaf sections.

***In Vitro* Activity and Binding Studies.** The barley etioplast Protox I_{50} values ranged from 21 to 33,200 nM (Table I). The compounds (each at 100 µM) caused Proto IX to accumulate from about 20- to 60-fold over the control levels in dark-incubated (20 h) green barley leaf sections

(Table I). PChilde levels were generally also increased in this assay, although the maximal effect was only a three-fold increase. PChlide and Proto IX accumulation levels were positively correlated with an r value of 0.87.

Table I. Effects of Ten Protox Inhibitors on Protox Inhibition and Porphyrin Accumulation

Herbicide	Protox I_{50} (μM)	Proto IX	PChlide
		- (nmoles/g fresh weight) -	
Control		0.27	5.04
PPG-1013	0.021	10.08	10.66
UCC-C4243	0.030	13.36	15.89
RH-0978	0.042	9.32	7.84
FMC-14	0.110	7.56	5.54
AH-2.430	0.200	9.37	10.23
Oxadiazon	0.400	11.61	10.64
F-6285	1.100	15.95	17.88
TNPP-Ethyl	3.800	7.01	10.53
M&B 39279	4.200	5.01	6.85
LS 82-556	33.200	5.44	8.02

There was not a good correlation between Protox I_{50} levels and Proto IX accumulation. However, there was a positive correlation, although not quite significant at the 95% confidence level, between Proto IX levels and cellular damage as measured by electrolyte leakage (Figure 3).

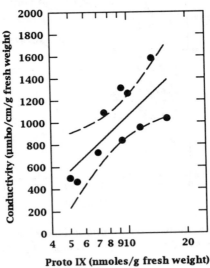

Figure 3. Relationship between Proto IX accumulated (Table I) and electrolyte leakage from the tissues at 2 h after exposure to light (Figure 2). Dotted lines are the 95% confidence intervals.

All ten compounds reduced the binding of ^{14}C-acifluorfen to Protox-containing etioplast membranes (Figure 4 provides examples), suggesting that they share common binding sites on Protox. Reliable binding constants could not be determined from these data, due to variability.

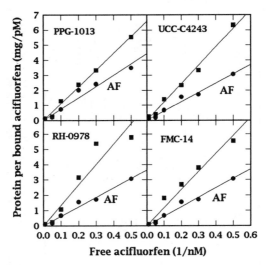

Figure 4. Binding of radiolabeled acifluorfen (AF) in the presence of 100 nM unlabeled compound for four representative compounds of the ten compounds in this study.

Molecular Properties. Molecular properties varied considerably between representatives of the different Protox inhibitor classes (Table II). There were large differences in molecular properties such as log P, various superdelocalizabilities, and dipole moments among these herbicidal inhibitors. The best Protox inhibitor (PPG-1013, Table I) was one of the most lipophilic and had the highest molecular volume and total electrostatic interaction.

The molecular electrostatic potential (MEP) distributions of the ten compounds are presented in Plate 5. The MEP is a physical property that can be used as a tool for understanding and predicting biological recognition processes and molecular reactivity. Because of the highly concentrated nature of the nuclear charge, the positive MEP provides no clear guideline as to the interaction of nucleophiles with these regions.

In general, all ten compounds have a strongly negative MEP concentrated around one of the ring structures. These negative potentials appear to be the major regions through which the herbicide molecule interacts with the Protox binding site (*20, 21*). Furthermore, these regions carry at least one substituent group that is capable of hydrogen bonding with a receptor site.

NOTE: Color plates appear in color section.

Table II. Molecular Properties of Protox Inhibitors of Figure 1 that Are Used in Equations

Herbicide	Log P	VDW$_V$ $Å^3$	TEI (eV)	PMEPV $Å^3$	PMEPA $Å^2$	NMEPA $Å^2$	ϵ_{HOMO}	S_E (eV)	S_N
PPG-1013	4.29	297	291	231	233	273	-10.28	-8.91	41.81
UCC-C4243	2.82	266	250	161	171	186	-9.59	-7.85	-15.03
RH-0978	3.40	233	205	132	154	202	-9.26	-6.76	21.64
FMC-14	3.80	220	207	119	152	219	-8.83	-6.46	84.20
AH-2.430	4.02	207	199	242	205	188	-9.71	-6.12	107.53
Oxadiazon	4.78	240	226	56	96	170	-9.37	-7.10	29.56
F-6285	1.04	232	163	208	196	241	-9.22	-7.12	17.68
TNPP-Ethyl	4.37	281	240	177	193	258	-10.44	-7.56	109.22
M&B 39279	1.46	203	177	144	146	133	-9.32	-5.76	-9535
LS 82-556	2.38	240	248	110	144	245	-9.49	-7.31	28.83

Note: S_E = electrophilic superdelocalizability; S_N = nucleophilic superdelocalizability; NMEPA = negative molecular electrostatic potential area; PMEPA = positive molecular electrostatic potential area; PMEPV = positive molecular electrostatic potential volume; VDW$_V$ = van der Waals volume; ϵ_{HOMO} = energy of the highest unoccupied molecular orbital; TEI = total electrostatic interaction.

Quantitative Structure-Activity Relationship Analysis. The molecular properties used in various equations are presented in Table II. Simple regression analysis revealed that no single molecular property gave equations with r^2 values greater than 0.29 for any biological activity. Multiple regression analysis of these ten inhibitors indicated that combinations of the bulk, electronic, energy, and lipophilicity descriptors together were responsible for variation in each of the three biological activities. The most predictive equations using four independent variables are shown below:

$$\log \text{Protox } I_{50} = 0.1797 - 0.00033(S_N) + 0.0655 \text{ (PMEPV)} - 0.1280(\text{PMEPA}) + 0.0493(\text{NMEPA})$$
$$F_{(4,5)} = 4.4, \; s = 0.67, \; r^2 = 0.78 \qquad [1]$$

$$\log \text{Proto IX} = 0.7766 - 0.286 \,(S_E) + 0.00004 \,(S_N) - 0.0052(\text{TEI}) - 0.0031(\text{NMEPA})$$
$$F_{(4,5)} = 17.0, \; s = 0.06, \; r^2 = 0.93 \qquad [2]$$

$$\log \text{conductivity (12 h)} = 5.779 + 0.0479 \,(\log P) + 0.456 \,(\epsilon_{\text{HOMO}}) + 0.0062(\text{VDW}_V) + 0.0015 \text{ (PMEPV)}$$
$$F_{(4,5)} = 5.2, \; s = 0.09, \; r^2 = 0.81 \qquad [3]$$

The relationship between predicted values with these equations and measured values are illustrated in Figure 5.

These equations were tested by predicting previously determined activities of three enantiomeric pairs of Protox inhibitors (*23*). The (R) enantiomers of these pairs were more active than the (S) enantiomers. The predicted values differed from experimental values by a factor of only 1 to 21 for active (R) enantiomers, compared to a factor of 2 to about 80,000 for inactive (S) enantiomers for these three biological activities (Table III).

Table III. Actual and Predicted Biological Activities of Three Enantiomeric Pairs of Protox Inhibitors

Herbicide[a]	Protox I_{50}		Proto IX		Conductivity	
	Meas.[b]	Pred.[b]	Meas.	Pred.	Meas.	Pred.
	(nM)		(nmol/g fresh wt)		(μmho/cm/g fresh wt)	
RH-4639 (S)	650	0	2.7	9.56	215	3326
RH-4638 (R)	55	3	10.5	7.98	2604	3587
AH 2.439 (S)	300	153	2.0	8.51	235	10188
AH 2.440 (R)	30	41	11.4	8.20	1180	8173
AH 2.442 (S)	520	27	0.4	9.11	373	10497
AH 2.441 (R)	12	16	14.8	7.94	2195	8612

[a]from ref. *23*.

[b]Meas. = measured; Pred. = predicted from respective equations.

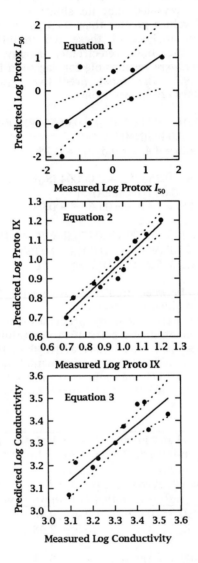

Figure 5. Relationships between measured biological activities and those predicted by equations 1, 2, and 3. Dotted lines are the 95% confidence intervals.

Discussion

A DPE analogue (PPG-1013) and an analogue of pyrimidinedione (UCC-C4243) were the most active Protox inhibitors. LS 82-556, a pyridine derivative, was the least active as a Protox inhibitor. Oxadiazon and AH-2.430, a pyrazole phenyl ether, were intermediate in their Protox-inhibiting capacity. A similar pattern was observed with respect to accumulation of Proto IX and herbicidal damage. However, only the correlation between Proto IX accumulated and cellular leakage of electrolytes was good. The poorer correlation beween *in vitro* Protox inhibition and *in vivo* Proto IX accumulation could be due to any of several factors, including differential movement of the Protox inhibitor to the site of action, metabolic inactivation, bioactivation, or other sites of herbicide action. At least one class of herbicides, the thiadiazolidines, have been demonstrated to be bioactivated to become Protox inhibitors (*34*). Also, members of the DPE Protox inhibitors have alternative mechanisms of herbicide action (*30*).

Few QSAR studies of Protox inhibitors have utilized compounds from different chemical classes. Akagi and Sakashita (*35*) showed that the four compounds, oxyfluorfen, chloro-phthalim, M&B-39279, and pyrimidinedione, which belong to different classes, exhibited some common electronic features. All showed a low value of energy of the lowest unoccupied molecular orbital (ϵ_{LUMO}). The phenyl rings on which the LUMO was located had non-planar groups at the *para* position and functional groups containing a lone electron pair at the *ortho* position. From QSAR analyses of various classes of peroxidizing herbicides, it appears that Protox activity responds primarily to the substitutional modifications on the phenyl ring (*p*-nitrophenyl in DPEs). Furthermore, a substitution at the *meta* position on the *p*-chloro (phenopylate) or *p*-nitro (DPE) ring is essential for greater herbicidal activity. Hayashi (*36*) made similar conclusions regarding the *meta* position of the *p*-nitro ring in a QSAR study of 22 analogues of a DPE herbicide.

In the present study, no set of three or fewer molecular parameters appear to be good predictors of biological activity. However, biological activities can be largely explained by a combination of four bulk (shape/size), electronic, energy, and lipophilicity parameters. In previous studies with one (*20*) or two (*21*) chemical families, highly predictive equations were possible by utilizing only two or three independent variables. It is interesting that log P was important in prediction of biological activity at the physiological levels, but was not important in prediction of activity at the molecular level. However, Niclolaus *et al.* (*37*) also found lipophilicity to be crucial in prediction of Protox-inhibiting capacity of 11 phenyltetrahydrophthalimides. Lipophilicity should be especially important in the activity of a membrane-bound enzyme such as Protox. Even when Protox is removed from the membrane, the relative inhibitory activities of Protox inhibitors with a wide range of log P values remains about the same (*38*)

In our two previous studies (*20, 21*), the most active Protox inhibitors had negative MEP distributions as two distinct fields, at opposite ends of the molecule. This distribution is similar to that of one half of the Protogen IX molecule (20). In the present study, electrostatic potential maps derived from atomic point charges as described in our earlier studies (20, 21) the negative charge distributions were not always similar to that of one half of the Protogen IX molecule (Plate 5). For example, oxadiazon had only one negative MEP distribution (Plate 5F). However, all ten compounds have a negative MEP distribution around a ring structure with at least one substituent that is capable of hydrogen bonding to a receptor site.

The three equations reported in this study were obtained by evaluating a relatively large number of independent variables for a small number of compounds. As Topliss and Edwards (39) have pointed out, fortuitous correlations are not uncommon when the number of variables screened for correlations is large compared to the number of observations. Furthermore, the equations generated in the present studies do not provide the predictive power of models (r^2 greater than 0.93) based on a group of very similar phenopylate analogue Protox inhibitors (21). However, the major limitation of these previous models for only only one chemical class is that they are specific to structurally similar compounds and, at best, have limited applicability.

Nevertheless, the predicted (calculated from respective equations) values had a mean absolute error of log 0.424 in Protox I_{50}, log 0.031 in Proto IX, and log 0.049 in electrolyte leakage. The low mean absolute error terms indicate that molecular properties can be used as crude predictors of herbicidal activity. This was verified by the level of prediction of the biological activities of the most active members of enantiomeric pairs (Table III). Chirality in the *meta*-substitution of the *para*-nitrophenyl ring of Protox-inhibiting herbicides strongly influences activity at both the molecular and whole plant levels (20, 23, 40, 41). The poor predictive performance of the equations on the less active members of the pairs is not unexpected, as the molecular properties used in these analyses do not include steric properties.

Ultimately, the determination of the primary and tertiary structures of Protox and the Protogen IX binding niche will facilitate QSAR approaches to herbicide design as it has for PS II inhibitors. Cloning of the Protox-encoding gene is the most likely means of determining primary structure of Protox, and Protox mutants that are resistant to Protox-inhibiting herbicides have the potential to reveal the Protogen IX binding site. The gene for Protox has been cloned in *E. coli* (42) and *Bacillus subtilis* (43). However, bacterial Protox appears to be more like the herbicide-insensitive Protogen IX oxidizing activity of plant plasma membranes (16, 44), and is thus probably irrelevant to herbicide studies. At least two laboratories are making progress toward characterizing eucaryotic, organellar Protox (herbicide sensitive) (38, 45). Positive results from their work should provide the opportunity for more fundamental QSAR studies of Protox inhibitors.

Acknowledgments

We thank Monsanto, Uniroyal, PPG, Rohm and Haas, FMC, Otsuka Chemical Co., and Rhône Poulenc for generously providing technical grade chemicals. Dr. Alfred French of USDA, ARS, SRRC in New Orleans generously provided molecular modeling facilities.

References

1. Matringe, M.; Camadro, J. -M.; Labbe, P.; Scalla, R. *FEBS Lett.* 1989, *245*, 35-8.
2. Witkowski, D. A.; Halling, B. P. *Plant Physiol.* 1989, *90*, 1239-42.
3. Duke, S. O.; Becerril, J. M.; Sherman, T. D.; Lydon, J.; Matsumoto, H. *Pestic. Sci.* 1990, *30*, 367-78.
4. Duke, S. O.; Lydon, J.; Becerril, J. M; Sherman, T. D.; Lehnen, L. P.; Matsumoto, H. *Weed Sci.* 1991, *39*, 465-73.
5. Scalla, R.; Matringe, M.; Camadro, J.-M.; Labbe, P. *Z. Naturforsch.* 1990, *45c*, 503-11.
6. Duke, S. O.; Nandihalli, U. B.; Lee, H. J.; Duke, M. V. *Amer. Chem. Soc. Symp. Ser.* 1994, *559*, 191-204.
7. Nandihalli, U. B.; Duke, S. O. *Amer. Chem. Soc. Symp. Ser.* 1993, *524*, 62-78.
8. Matringe, M.; Scalla, R. *Pestic. Biochem. Physiol.* 1988, *32*, 164-72.
9. Lyga, J. W.; Patera, R. M.; Theodoridis, G.; Halling, B. P.; Hotzman, F. W.; Plummer, M. J. *J. Agric. Food Chem.* 1991, *39*, 1667-73.
10. Lyga, J. W.; Halling, B. P.; Witkowski, D. A.; Patera, R. M.; Seeley, J. A.; Plummer, M. J.; Hotzman, F. W. *Amer. Chem. Soc. Symp. Ser.* 1991, *443*, 170-81.
11. Mito, N.; Sato, R.; Miyakado, M.; Oshio, H.; Tanada, S. *Pestic. Biochem. Physiol.* 1991, *40*, 128-35.
12. Morishima, Y.; Osabe, H.; Goto, Y. *J. Pestic. Sci.* 1990, *15*, 553-9.
13. Nandihalli, U. B.; Sherman, T. D.; Duke, M. V.; Fischer, J. D.; Musco, V. A.; Becerril, J. M.; Duke, S. O. *Pestic. Sci.* 1992, 227-35.
14. Nandihalli, U. B.; Duke, S. O. *Amer. Chem. Soc. Symp. Ser.* 1994, *559*, 133-46.
15. Jacobs, J. M.; Jacobs, N. J.; Sherman, T. D.; Duke S. O. *Plant Physiol.* 1991, *97*, 197-203.
16. Lee, H. J.; Duke, M. V.; Duke, S. O. *Plant Physiol.* 1993, *102*, 881-9.
17. Jacobs, J. M.; Jacobs, N. J. *Plant Physiol.* 1993, *101*, 1181-8
18. Draber, W. In *Rational Approaches to Structure, Activity, and Ecotoxicology of Agrochemicals*; Draber, W.; Fujita, T., Eds.; CRC Press, Boca Raton, FL, 1992, pp. 277-313.
19. Camadro, J. -M.; Matringe, M.; Scalla, R.; Labbe, P. *Biochem. J.* 1991, *227*, 17-21.
20. Nandihalli, U. B.; Duke, M. V.; Duke, S. O. *Pestic. Biochem. Physiol.* 1992, *43*, 193-211.
21. Nandihalli, U. B.; Duke, M. V.; Duke, S. O. *J. Agric. Food Chem.* 1992, *40*, 1993-2000.

22. Kohno, H.; Hirai, K.; Hori, M.; Sato, Y.; Böger, P.; Wakabayashi, K. *Z. Naturforsch.* **1993**, *48c*, 334-8.
23. Nandihalli, U. B.; Duke, M. V.; Ashmore, J. W.; Musco, V. A.; Clark, R. D.; Duke, S. O. *Pestic. Sci.* **1994**, *40*, 265-77.
24. Lambert, L.; Sandmann, G.; Böger, P. *Pestic. Biochem. Physiol.* **1983**, *19*, 309-20.
25. Ohta, H.; Suzuki, S.; Watanabe, H.; Jikihara, T.; Matsuya, K.; Wakabayashi, K. *Agric. Biol. Chem.* **1976**, *40*, 745-51.
26. Sato, Y.; Yokawa, H.; Katsuyama, N.; Oomikawa, R.; Wakabayashi, K. *Bull. Fac. Agric., Tamagawa Univ.* **1991**, *31*, 45-57.
27. Ohnishi, J.; Yukitake, K.; Eto, M. *J. Fac. Agric., Kyushu Univ.* **1993**, *37*, 239-46.
28. Osabe, H.; Morishima, Y.; Goto, Y.; Masamoto, K.; Nakagawa, Y.; Fujita, T. *Pestic. Sci.* **1992**, *34*, 17-25.
29. Osabe, H.; Morishima, Y.; Goto, Y.; Masamoto, K.; Nakagawa, Y.; Fujita, T. *Pestic. Sci.* **1992**, *34*, 27-36.
30. Kunert, K.-J.; Sandmann, G., Böger, P. *Rev. Weed Sci.* **1987**, *3*, 35-55.
31. Bradford, M. M. *Anal. Chem.* **1976**, *72*, 248-54.
32. Tischer, W.; Strotmann, H. *Biochim. Biophys. Acta* **1977**, *460*, 113-25.
33. Nandihalli, U. B.; Duke, M. V.; Duke, S. O. *J. Agric. Food Chem.* **1993**, *41*, 582-7.
34. Sato, Y.; Hohsi, T.; Iida, T.; Ogino, C.; Nicolaus, B.; Wakabayashi, K.; Böger, P. *Z. Naturforsch.* **1994**, *49c*, 49-56.
35. Akagi, T.; Sakashita, N. *Z. Naturforsch.* **1993**, *48c*, 345-9.
36. Hayashi, Y. *J. Agric. Food Chem.* **1990**, *38*, 839-44.
37. Nicolaus, B.; Sandmann, G.; Böger, P. *Z. Naturforsch.* **1993**, *48c*, 326-33.
38. Camadro, J.-M.; Matringe, M.; Brouillet, N.; Thome, F.; Labbe, P. *Amer. Chem. Soc. Symp. Ser.* **1994**, *559*, 91-105.
39. Topliss, J. G.; Edwards, R. P. *J. Med. Chem.* **1979**, *22*, 1238-44.
40. Camilleri, P.; Gray, A.; Weaver, K.; Bowyer, J. R. *J. Agric. Food Chem.* **1988**, *37*, 519-23.
41. Hallahan, B. J.; Camilleri, P.; Smith, A.; Bowyer, J. R. *Plant Physiol.* **1992**, *100*, 1211-6.
42. Sasarman, A.; Letowski, J.; Czaika, G.; Ramirez, V.; Nead, M. A.; Jacobs, J. M.; Morais, R. *Can. J. Microbiol.* **1993**, *39*, 1155-61.
43. Dailey, T. A.; Meissner, P.; Dailey, H. A. *J. Biol. Chem.* **1994**, *269*, 390-5.
44. Lee, H. J.; Duke, S. O. *J. Agric. Food Chem.* **1995**, In press.
45. Sato, R.; Yamamoto, M.; Shibata, H.; Oshio, H.; Harris, E. H.; Gillham, N. W.; Boynton, J. E. *Amer. Chem. Soc. Symp. Ser.* **1994**, *559*, 81-90.

RECEIVED October 13, 1994

Chapter 16

Experimental Design in Organic Synthesis

Lawrence H. Brannigan, Mark V. Grieshaber, and Dora M. Schnur

Monsanto Company, 800 North Lindbergh Boulevard,
St. Louis, MO 63167

We have previously demonstrated that the correlation of biological activity within a general class of compounds with the chemical properties of substituents can be discovered with fewer compounds than traditionally prepared in such a study. This smaller number of compounds can be chosen systematically; applying the principle of multiple variation so that the "training set" is representative of structural space. Through the use of the SYBYL programming language, SPL (Tripos, Inc), we have written a program that allows a chemist to use Experimental Design to create a training set for lead follow-up. This training set of molecules is created using a Plackett-Burman design as a template for multi-dimensional space based on variations in steric, electronic and lipophilic properties of molecular fragments. These fragments are substituents on the parent molecule representing the lead chemistry. The set of molecules should provide the optimal information for an efficient SAR study.

In the early 1980's it became apparent that the cost in resources for the development of lead biologically active compounds was becoming very high. Examination of several areas in which we were working revealed that often several thousand compounds were prepared and tested to evaluate a lead chemistry. We sought to determine whether or not it was feasible to choose synthetic candidates more effectively and limit the number of compounds needed to evaluate the commercial potential of lead activity for a class of compounds. We sought to apply QSAR early in the development process to achieve this end. The result of this study was the development of the application of experimental design to the choice of candidates. In order to bring this application to the synthetic chemist and to provide the most general method we prepared the computer program called MOD (Multivariation in Organic Design) described herein.

QSAR, Quantitative Structure Activity Relationships, has been applied to many retrospective studies of the biological activity of families of chemical compounds. The objective of QSAR is to map the variation of a biological response (biological activity) in structural space. The QSAR principle is extra-thermodynamic in character because it is the intuitive proposition that chemical structure is related to biological activity. QSAR arises from the Hammet[1] proposition that the reactivity of organic

0097–6156/95/0589–0225$12.00/0
© 1995 American Chemical Society

molecules is related to chemical structure. These relationships are often referred to as Linear Free Energy Relationships, LFER. The biological extension is usually expressed in the Hansch Equation[2], equation 1. The parameter P is the octanol/water partition coefficient for the molecule, σ is the electronic constant[3] for the relevant

$$A = a + b\text{LogP} + c[\text{LogP}]^2 + c\sigma + dE \qquad \text{equation 1}$$

substituent, and E is the steric constant[4] for that substituent. There are a large number of parameters that have been developed since the 1930's that describe the chemical and physical nature of hundreds of molecular fragments. These are contained in the Pamona College Data Base[5]. For our purposes LogP values have been calculated using a computerized version of the method of Leo, Nys and Rekker[6,7].

The most ambitious expectation of QSAR studies was that the most active molecules of a family might be predicted before many hundreds of compounds had been prepared. This predictive expectation has not been realized. QSAR has been applied retroactively in many cases to describe parametric dependence of biological activity. Much insight into the mechanism of action of many biologcially active chemical families has been gained through QSAR studies. We proposed the application of Statistical Experimental Design methods to utilize QSAR early in the lead development process.

Monovariation.

In the early 1980's we began the examination of the relationship of structure of a series of α–chloroacetamidochloroacetanilides[8] (Figure 1) to biological activity[9]. The goal of this QSAR study was to determine whether or not the optimum activity had been attained with the 300 compounds which had already been prepared. These relatives of the α-chloroacetanilide herbicides have the remarkable property of stunting the growth of turf grasses without killing the grasses while enriching the green texture of the turf. These compounds are referred to as "Turf Retardants". For this study the biological activity is measured by the decrease in vertical growth of the turf grass tall fescue. The activity was measured as an average of several different applications at a rate of 2lb/acre. This activity is designated as A. Among the 300 compounds prepared there were 110 (table 1) for which there was reliable activity data and 92 of these had only ortho substituents on the aromatic ring and can be described parametrically. All compounds with any substitution other than ortho were completely inactive. The set of 92 was chosen for this study.

Figure 1, Structure of the α–chloroacetamidochloroacetanilides

In the development of lead biological activity of a family of chemistry, the synthetic chemist tends to hold all but one of the potential subtituent positions constant at some set of structural features and vary only one substituent. This strategy is often referred to as the One Variable at a Time strategy (OVAT). The many compounds prepared in the acetamidomethyl-α-chloroacetanilide series were prepared according to this strategy. Most of the compounds prepared, in fact, were those in which R2 and R3 were held constant as ethyl groups while R1 was extensively varied. In searching for QSAR models in the diethyl series we discovered

Table 1

No.	A, Turf	LogP calc'	R_1	R_2	R_3	No.	A, Turf	LogP calc'	R_1	R_2	R_3
1	62	3.29	H	Et	Et	56	72	1.97	Et	Me	OMe
2*	0	4.88	Ph	Et	Et	57*	66	3.02	Me	Me	OBu
3	70	3.22	Me	Et	Et	58	66	2.49	Me	Me	OPr
4	0	5.65	CCl3	Et	Et	59	41	0.79	Me	H	OMe
5	53	4.28	n-Pr	Et	Et	60	63	3.02	Me	Et	OPr
6	10	6.05	2-Naph	Et	Et	61	75	1.48	CH2OMe	Me	OMe
7	56	4.05	(CH2)2OMe	Et	Et	62	65	2.49	Me	Et	OEt
8	30	3.84	CH2Cl	Et	Et	63	70	1.95	CH2CHCH2	Me	OMe
9	49	4.06	CHMe2	Et	Et	64	61	1.79	Cyclopropyl	Me	OMe
10	0	6.40	Cyclohexyl	Et	Et	65	53	2.36	Me	CHMe2	OMe
11	46	3.82	(CH2)2Cl	Et	Et	66	58	2.89	Me	Me	O-i-bu
12	65	3.75	Et	Et	Et	67	60	2.89	Me	CHMe2	OEt
13	0	4.93	CH2Ph	Et	Et	68	41	3.42	Me	Et	O-i-bu
14	35	4.13	CHMeCH2Cl	Et	Et	69*	12	0.87	Me	H	H
15	65	3.52	CHCH2	Et	Et	70*	28	0.88	Me	OMe	OMe
16*	42	1.52	Me	H	Me	71	74	3.21	CH2CHCH2	Me	Et
17*	33	3.56	H	Me	CMe3	72	67	2.99	CH2CH2OMe	Me	Et
18	71	2.70	Me	Me	Et	73	65	3.22	Et	Me	Et
19	49	3.82	CMeCH2	Et	Et	74	70	2.84	CH2COOH	Me	Et
20	60	3.49	Me	Me	CMe3	75	70	2.73	CH2OMe	Me	Et
21	63	3.09	Me	Me	CHMe2	76	13	4.55	CHCl2	Et	Et
22	67	2.17	Me	Me	Me	77	65	2.70	Et	Me	Me
23	67	3.85	Cyclopropyl	Et	Et	78	64	2.68	CH2CHCH2	Me	Me
24	72	3.05	Cyclopropyl	Me	Et	79	61	2.46	CHCH2	Me	Me
25	44	2.76	Me	CMe3	OMe	80	60	2.52	Cyclopropyl	Me	Me
26	72	3.52	(CH2)2OMe	Et	Et	81	68	2.46	(CH2)2OMe	Me	Me
27*	42	4.02	Me	Et	CMe3	82*	0	5.34	(CH2)4Me	Et	Et
28	40	4.00	(CH2)3Cl	Et	Et	83	3	7.46	(CH2)6Me	Et	Et
29	40	4.02	Me	CHMe2	CHMe2	84	42	4.81	(CH2)3Me	Et	Et
30	74	3.74	CH2CHCH2	Et	Et	85	36	2.91	CH2SMe	Me	Me
31	21	5.68	CHCHPh	Et	Et	86	34	3.96	CH2SMe	Et	Et
32	16	6.93	(CH2)7Me	Et	Et	87	0	5.87	CH2S(CH2)5H	Et	Et
33*	37	3.69	CMeCH2	Me	CHMe2	88	47	3.39	Me	O-n-pr	O-i-bu
34*	69	3.26	CH2OMe	Et	Et	89	21	2.29	Me	Cl	Cl
35	46	3.96	CH2CH2Br	Et	Et	90	36	2.23	Me	Me	Cl
36	53	4.04	CHCHMe	Et	Et	91	3	5.52	CEt3	Me	Et
37	26	4.14	(CH2)3Br	Et	Et	92*	15	4.73	Cyclohexyl	Me	Et
38	5	5.32	CH2CH2Ph	Et	Et	93	54	3.08	Cyclobutyl	Me	Me
39	14	7.11	CHCPh2	Et	Et	94	38	3.61	Cyclobutyl	Me	Et
40	28	4.66	(CH2)4Br	Et	Et	95	8	4.99	CEt3	Me	Me
41	19	7.99	(CH2)8Me	Et	Et	96	43	3.17	(CH2)3Cl	Me	CF3
42*	16	4.46	CMe3	Et	Et	97*	28	3.73	CHCl2	Me	CF3
43	7	6.28	CHPh2	Et	Et	98	0	6.05	CEt3	Et	Et
44	53	4.15	Me	Et	Et	99	44	3.67	CH2cycloprop	Me	Et
45	3	4.62	CMeCH2	CHMe2	CHMe2	100*	50	2.56	H	Me	O-n-pr
46	0	4.69	CHMe2	Et	Et	101	5	2.63	Me	CF3	CF3
47	0	1.99	Me	CH2Cl	CH2Cl	102*	53	2.91	CH2CHCH2	Me	CF3
48*	49	3.56	Me	CMe3	Cl	103	34	1.15	Me	F	F
49	49	1.73	Me	H	CF3	104	52	2.93	Et	Me	CF3
50*	62	2.40	Me	Me	CF3	105	61	2.75	Cyclopropyl	Me	CF3
51	60	1.44	Me	Me	OMe	106	21	2.59	Me	Br	Br
52	62	1.97	Me	Me	OEt	107*	45	1.5	H	Me	OMe
53*	57	1.97	Me	Et	OMe	108	65	2.76	H	Me	Et
54	52	3.79	CH2OEt	Et	Et	109	50	2.47	H	Me	CF3
55*	52	3.26	CH2OEt	Me	Et	110	58	2.03	H	Me	OEt

three major relationships. In the first case we isolated the list of compounds in which R_1 had been varied through the lower alkyl groups, methyl through iso-butyl. The activity of this list was modeled very nicely by equation 2, in which E_s is the Taft Steric constant for the alkyl substituent at R_1.

$$A = a + bE_s \qquad \text{equation 2}$$

A second list from the diethyl series in which R_1 was varied through alkyl groups substituted with electron withdrawing hetero-atoms, O,N,Cl,Br and S,was isolated. The activity of this list was modeled very nicely by equation 3, in which σ^* is the alkyl electronic constant for the substituent at R_2.

$$A = c + d\sigma^* \qquad \text{equation 3}$$

The third correlation initially discovered was the classical correlation with the partition coefficient, LogP, equation 4. This correlation is quadratic indicating a maxima in the biological response curve and an optimum LogP. Optima in LogP are very common in QSAR studies.

$$A = d + e\text{LogP} + f[\text{LogP}]^2 \qquad \text{equation 4}$$

Each of these equations were statistically valid having R^2 greater than 0.9 with at least 8 examples for each equation. These relationships suggested that the application of the QSAR methodology was valid for these compounds.

General Quadratic Response Models.

While all of the above correlations could be found in the first thirty compounds by the isolation of specified groups of compounds prepared in the acetamido-α-chloroacetanilide series a general model could not be detected with these thirty compounds. When the activity and fragment parameters of all 92 compounds in the set was subjected to stepwise multiple linear regression analysis the correlation equation 5 was obtained. This correlation had R^2 of 0.8, a reasonably valid equation. Note that this more general analysis picked up a parametric dependence which was not apparent in the first thirty compounds. The new dependence of activity is on the σ^+ electronic constant for the ortho substituents and on the steric constant, E, for the ortho substituents. This is a reasonable expectation since there was little variation of the ortho groups (R_2 and R_3) in the first thirty compounds. They were held constant at the ethyl group throughout the first thirty candidates. Note also that this general response model does not contain any parameter for R_1. The coefficients for these parameters were not statistically significant in the regression analysis.

$$A = g + h\text{LogP} + i[\text{LogP}]^2 + j\sigma^+_{R23} + kE_{R23} + l[\sigma^+_{R23}]^2 + m[E_{R23}]^2 \qquad \text{equation 5}$$

Because of this loss of significance of parameters of R_1 we performed an extensive bootstrapping operation on the data set. In this operation we selected 10 different random sets of 29 candidates from the 92 in the set. Each random set was subjected to regression analysis. Most of the sets produced correlation equations very similar to equation 5, all with R^2 between 0.7 and 0.8. Parameters for R_1 were significant in several of the equations.

The reason for the loss of significance of the R_1 parameters in equation 5 was in the structure of the variation in the set of 92 compounds. For most of the variation in R_1 the other two positions, R_2 and R_3 were held constant at the ethyl group. For most of the variation of R_2 and R_3 the position R_1 was held constant at the methyl group. The parametric sampling of structural space was not symmetrical. It was possible that a symmetrical sampling of structural space would reveal which parameters are significant.

Multivariate Sampling of Structural Space

Equations 2-4 represent limited cross sections of the biological response surface in structural space. Equation 5 represents a section which is overpopulated by methyl groups at R_1 and ethyl groups at R_2 and R_3. These correlation equations are valid for their appropriate cross sections, but do not give a reliable model for the whole response surface.

We reasoned that a correlation equation which would model the response surface more accurately might be obtained by choosing a set of compounds from the 92 by application of an Experimental Design which would give widest variation in the three substituted positions. This "Training Set" was chosen by inspection of table 1 and is indicated in table 1 by the twenty examples with the asterisk. Regression analysis of these twenty candidates produced correlation equation 6. Equation 6 has R^2 0.91.

$$A = a' + b'LogP + c'[LogP]^2 + d'\sigma^*_{R1} + e'E_{R1} + f'[E_{R1}]^2 + g'\sigma^+_{R23} \qquad \text{equation 6}$$

This correlation contains the original parameters for R_1 with only the σ^+, aromatic electronic parameter, for R_2 and R_3. More importantly, the observed activity of all 92 compounds in the full set of compounds fit equation 6 with R^2 0.78. This observation suggests that the 72 compounds not in the Training Set have actually provided no information concerning the response surface beyond that provided by those within the set. Therefore, it is clear that the biological response surface can be accurately modeled with less than 25% of the number of compounds actually prepared. A general application of this approach to the development of lead areas of chemistry could provide a strategy significantly more efficient than traditional strategies. This strategy also is an application of the QSAR method early in the lead development process.

Experimental Design

A systematic means of choosing a set of compounds rationally with the assurance of multivariation of substituents is to apply one of formal Statistical Experimental Design methods to the lead structure. Discussions of the application of experimental design strategies to the optimizaion of organic reactions and biological activity can be found in references 11 and 12. The first step of this method is the placement of the parent structure at the origin of a structural coordinate system (a Structural Matrix) in which the axes are the physical and chemical parameters, steric constant, electronic constant, and partition function. There is one set of such axes for each substitutable position on the parent structure. A test set, then, is chosen from the Structural Matrix according to one of the specific Experimental Designs.

Consider the development of a multivariate training set for the pyrazole shown in Figure 2. For the purpose of demonstration assume that variation of R_1, R_2 and R_3 will provide an appropriate training set for development of the lead. The first priority is to determine the main structural effects on herbicide acitivity. Since variation of structural fragments at each R position carries with it a variation in three parameters (partition, sterics and electronics) there is a total of nine variables associated with the three substitutable positions.   An effective design for the determination of main effects with large numbers of variables is the Plackett-Burman Design. Table 2 shows a Plackett-Burman[13] design matrix for two levels with nine variables. The + signs designate that the parameter is at a predetermined high level, the - signs designate that the parameter is at a designated low level. These designs are constructed at the high and low levels. Other designs which could be used as templates might contain center points, other designs are constructed at more than two levels.

Figure 2, Structure of a Pyrazole Training Set

The parameter set most familiar to the organic chemist is E, σ and π, the Taft steric constant, electronic constant, and partition parameter respectively. The design is applied by using the first three columns of the matrix to represent the parameters for R_1, the center three to represent parameters for R_2 and the last three to represent R_3 according to the scheme suggested by Clementi[14]. The design applied in this way is shown in Table 3.

The next step in building the design is to assign specific structural fragments to the level sequences for each substitutable position, R. Table 4 shows the Taft constant, E, the electronic constant, σ_p (electronic constant for substituents on an aromatic ring), and the fragment partition function, π, for a series of common fragments. The range for E is 0 to 1.45. The parameter value of 0 is for a very small group. The value 1.45 is for a very large group. Values of 0.74 to 1.45 are assigned the + level. Values of E of 0.59 to 0.7 the 0 level and 0 to 0.58 the - level. Similarly, the range for σ is 0.78 (electron withdrawing groups with strong resonance, i.e. NO_2) to -0.9 (electron donating, i.e. NHEt). Values of σ of 0.06 to 0.78 are +, -0.09 to 0.06 are 0 level and -0.09 to 0.90 are -. Finally π values range from 3.79 (strongly lipophilic groups) to -1.80 (strongly hydrophilic groups). Values of π 3.79 to 0.54 are assigned level +, 0.50 through -0.03 are 0 level and -0.08 to -1.80 are -. The levels for the fragments are shown in table 4.

The final design is prepared by substituting the fragment from table 4 for its matching sign sequence in table 3. The resulting structural design is shown in table 5. The compounds suggested by the matrix are prepared and evaluated with respect to the response variable of interest. The combined design and response variable matrix is then evaluated by regression analysis (or other appropriate method).

This particular design is rather robust in that substitutions are made with highly varied fragments. It is a first level design. After this design has identified the main variables a more finely tuned design can be generated by the above method. Most likely only 3 or 4 of the variables would emerge from the evaluation as main effects. A second level design would include only those main variables. This second level design would most likely be done with one of the factorial designs which can detect

These color plates are for Chapter 12.

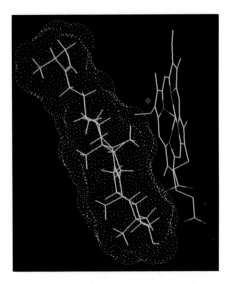

Color Plate 1. EBI binding site model. The white solvent accessible surface defines the binding site for lanosterol.

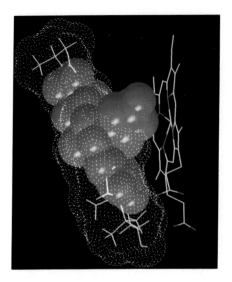

Color Plate 2. Systhane docked in EBI binding site model. The ligand (red) is rendered as van der Waals spheres and fits easily within the proposed binding site.

These color plates are for Chapter 12.

Color Plate 3. 3-Pyridyl analog of Systhane (yellow) docked in binding site (fit is poor).

Color Plate 4. 3-Pyridyl analog of Systhane without the "spacer" methylene (blue) docked in binding site (fit is excellent).

This color plate is for Chapter 15.

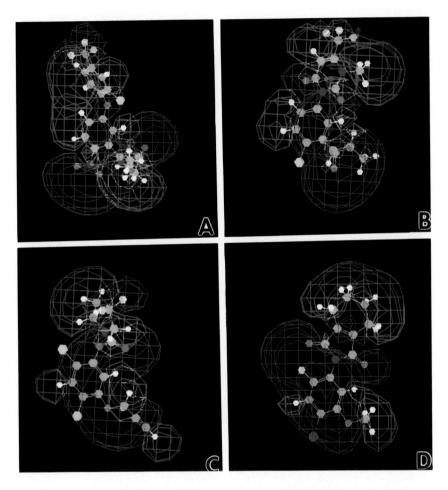

Plate 5. Molecular electrostatic potential (MEP) distribution (±10 kcal/mol) of the compounds of Figure 1. A. PPG-1013 B. UCC-C4243 C. RH-0978 D. FMC-14 E. AH-2.430 F. Oxadiazon G. F-6278 H. TNPP-ethyl I. M&B 39279 J. LS 82-556. Blue regions = negative MEP; Red regions = positive MEP. Atom color coding: green, carbon; white, hydrogen; blue, nitrogen; red, oxygen; pink, chlorine; yellow, fluorine; blue-gray, sulfur; and bromine.

Continued on next page

Plate 5. Continued.

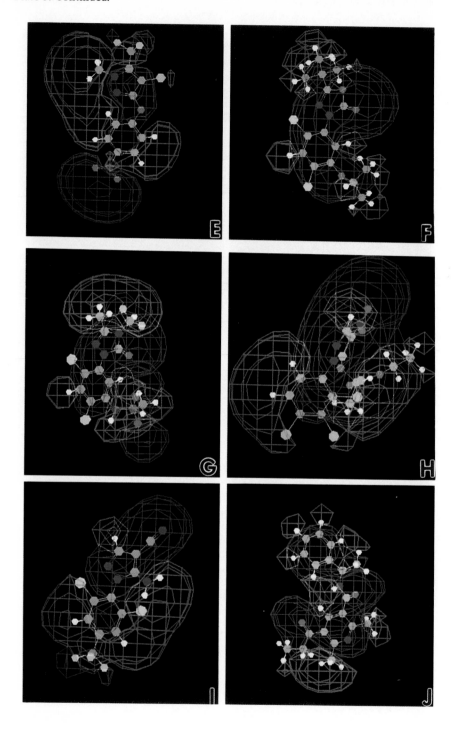

This color plate is for Chapter 16.

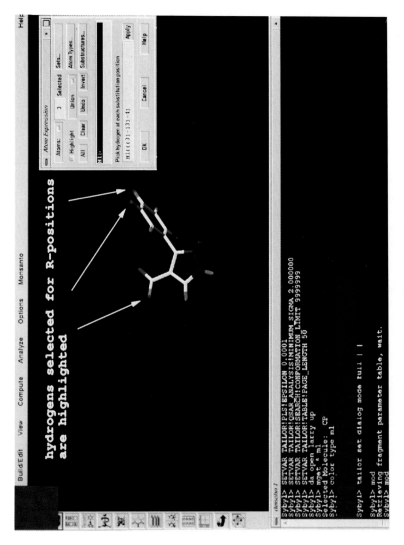

Plate 6, Selection of Substitution Pattern in MOD

This color plate is for Chapter 16.

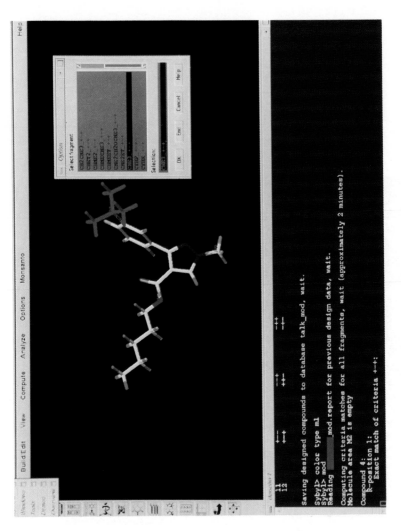

Plate 7, Selection of Structural Fragments in MOD

These color plates are for Chapter 23.

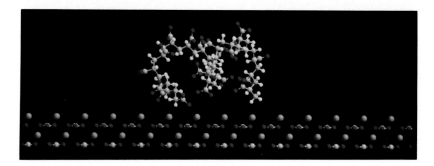

Color Plate 8. pAA on the positive 0 0 1 $CaCO_3$ surface, before molecular dynamics.

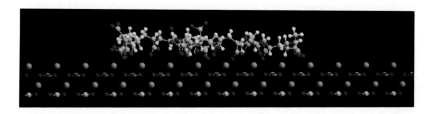

Color Plate 9. pAA on the positive 0 0 1 $CaCO_3$ surface, after 10 ps of molecular dynamics.

These color plates are for Chapter 23.

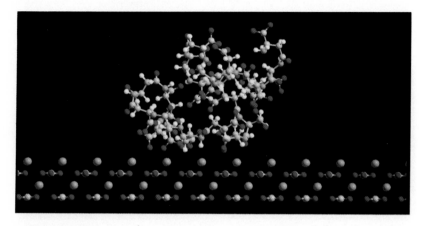

Color Plate 10. pAsp on the positive 0 0 1 $CaCO_3$ surface, before molecular dynamics.

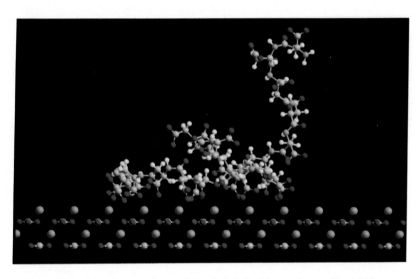

Color Plate 11. pAsp on the positive 0 0 1 $CaCO_3$ surface, after 1000 ps of molecular dynamics.

These color plates are for Chapter 23.

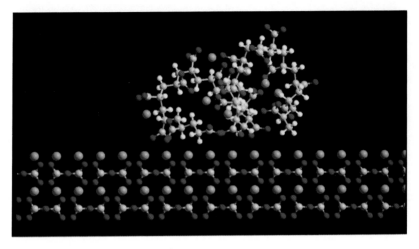

Color Plate 12. pAA + 7 Ca^{+2} ions on the positive 1 0 -2 CaCO$_3$ surface, before molecular dynamics.

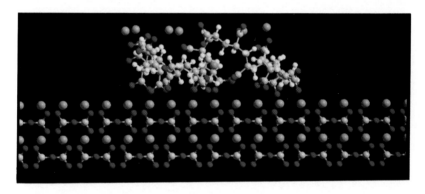

Color Plate 13. pAA + 7 Ca^{+2} ions on the positive 1 0 -2 CaCO$_3$ surface, after 1000 ps of molecular dynamics.

These color plates are for Chapter 23.

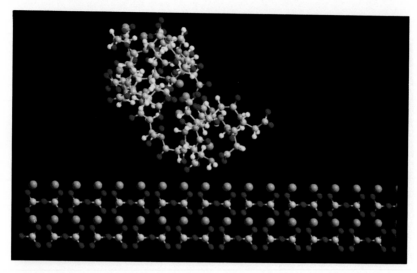

Color Plate 14. pAsp + 7 Ca^{+2} ions on the positive 1 0 -2 CaCO$_3$ surface, before molecular dynamics.

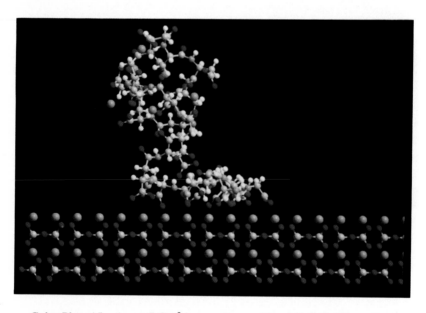

Color Plate 15. pAsp + 7 Ca^{+2} ions on the positive 1 0 -2 CaCO$_3$ surface, after 1000 ps of molecular dynamics.

Table 2

Run #	X_1	X_2	X_3	X_4	X_5	X_6	X_7	X_8	X_9
3	-	-	+	-	+	+	+	-	-
7	-	-	-	+	-	-	+	-	+
2	-	+	-	+	+	+	-	-	-
5	-	+	-	-	+	-	+	+	+
11	-	+	+	+	-	-	-	+	-
12	+	-	+	+	+	-	-	-	+
9	+	+	-	-	-	+	-	-	+
6	-	-	+	-	-	+	-	+	+
4	+	-	-	+	-	+	+	+	-
10	+	+	+	-	-	-	+	-	-
1	+	+	+	+	+	+	+	+	+
8	+	-	-	-	+	-	-	+	-

Table 3

Cpd No.	R_1			R_2			R_3		
	E_1	σ_1	π_1	E_2	σ_2	π_2	E_3	σ_3	π_3
3	-	-	+	-	+	+	+	-	-
7	-	-	-	+	-	-	+	-	+
2	-	+	-	+	+	+	-	-	-
5	-	+	-	-	+	-	+	+	+
11	-	+	+	+	-	-	-	+	-
12	+	-	+	+	+	-	-	-	+
9	+	+	-	-	-	+	-	-	+
6	-	-	+	-	-	+	-	+	+
4	+	-	-	+	-	+	+	+	-
10	+	+	+	-	-	-	+	-	-
1	+	+	+	+	+	+	+	+	+
8	+	-	-	-	+	-	-	+	-

Table 4

R	E	σ_p	π	R	E	σ_p	π
COOH	1.45	0.45	-1.30	$CHMe_2$	0.76	-0.15	1.58
COOMe	1.45	0.45	-0.96	$OCHMe_2$	0.75	-0.45	-0.45
NO_2	1.39	0.78	-1.30	CH_2OPh	0.74	0.07	1.49
CCl_3	1.38	0.33	1.38	CH_2CH_2Ph	0.70	-0.12	3.79
NEt_2	1.37	-0.90	-0.16	CH_2Ph	0.70	-0.09	2.22
CMe_3	1.24	-0.20	1.98	NH_2	0.35	-0.66	-1.70
NMe_2	0.43	-0.72	-1.00	OH	0.32	-0.37	-1.80
CN	0.40	0.66	-1.50	F	0.27	0.06	-0.61
NHMe	0.39	-0.84	-1.60	O-Pentyl	0.58	-0.34	0.83
OMe	0.36	-0.27	-1.20	$C(CH_2)Me$	0.57	0.05	1.03
CF_3	0.91	0.54	0.06	Et	0.56	-0.15	1.18
CH_2CN	0.89	0.01	-0.97	Cl	0.55	0.24	-0.17
Cyclohexyl	0.87	-0.22	2.77	CHMeOH	0.50	-0.07	-1.00
$CHCl_2$	0.81	0.19	0.68	OEt	0.48	-0.24	-0.76

parameter interactions and curvature of the response variable within the structural space examined.

It is clear that several of the suggested substitutions will produce very difficult synthetic problems and suggest molecules which are unacceptably reactive. For example one might hesitate before preparing the α-keto ester suggested by compound 9 in table 6. One would also not likely prepare acid chlorides or flourides as might be suggested by the design. In addition one must take into consideration properties of the suggested fragments which may dominate over the design parameters. For example, compound 12 would contain a methyl ester on the aromatic ring. Under many biological conditions this moiety would be subject to hydrolysis and this may compete with the response variable of interest.

It is for these reasons that these designs must not be prepared from limited lists of fragment data such as table 4. In order to provide the method with the most robust potential the design method was computerized as discussed below. The computer gives easy access to the several thousand fragment parameters available from the MEDCHEM[5] data base. The opportunity to use structural parameters other than those shown in table 4 is also in the computerized design.

The SYBYL MACRO: design generation

The underlying basis of the MOD program is the on the fly generation of a suitably sized PLACKETT-BURMAN matrix. The size of the generated matrix is based on the the number of substituent positions selected for variation. These matrices have special properties and problems. Especially useful is the property that allows creation of equally valid matrices by rotating the row positions or column positions (so long as these are done as separate operations). This allows an experimenter to generate different designs for the same number of parameters. The formulae that generate these matrices with their special properties result in 4n rows and 4n-1 columns. Thus the number of runs which are required by these designs are 4,8,12,16 etc. and the number of parameters respectively are 3,7,11,15 etc. Since the number of parameters which will be examined in the MOD strategy are usually (but not restricted to) three (E, σ and π) per substitutable position the MOD designs must accomodate 3,6,9,12 etc. parameters. To achieve this, the number of columns in next highest design is truncated in MOD. (Truncation is not possible for the number of rows or structures). Additionally, problems may result from confounding effects for certain matrices such as the 16X15. Where these are a problem, the next higher matrix is truncated.

It should be pointed out at this stage that the assumption that the parameters represent main effects and are not confounded is an expedient oversimplification that must be addressed outside the the macro. A second useful, but not necessarily valid assumption is the additivity of fragment properties. It is essential to apply standard modeling approaches to the 3D structures created by the design to verify their properties. A SYBYL database (.mdb) containing the structures is produced by the macro for this purpose and for future QSAR/CoMFA analysis[15].

To actually generate the design matrix, it is necessary to select the desired parameters and assign high and low threshold values. The SYBYL spreadsheet containing the parameter data has in fact about 58 parameters for over 1500 fragments[5].. Each parameter has a range of values. Not all fragments have data for all parameters. The parameters mentioned throughout the text are used as default parameters and assigned default threshold values that allow all possiblestrict design. While the macro does not use a three level design, it is useful from the standpoint of synthetic feasibility to have that middle level. If none of the +++ level fragments turn out to be synthetically reasonable for a given molecule, for example, having a reserve set of fragments of ++0, +0+ and 0++ gives the user more control over

Tabel 5

R	E	σ	π	R	E	σ	π
COOH	+	+	-	CHMe2	+	-	+
COOMe	+	+	-	OCHMe2	+	-	-
NO2	+	+	-	CH2OPh	+	+	+
CCl3	+	+	+	CH2CH2Ph	0	-	+
NEt2	+	-	-	CH2Ph	0	0	+
CMe3	+	-	+	NH2	-	-	-
NMe2	-	-	-	OH	-	-	-
CN	-	+	-	F	-	+	-
NHMe	-	-	-	O-Pentyl	-	-	+
OMe	-	-	-	C(CH2)Me	-	+	+
CF3	+	+	0	Et	-	-	+
CH2CN	+	0	-	Cl	-	+	-
Cyclohexyl	+	-	+	CHMeOH	-	0	-
CHCl2	+	+	+	OEt	-	+	-

Table 6

Cpd No.	R1	R2	R3
3	Et	C(CH2)Me	NEt2
7	NMe2	NEt2	CHMe2
2	OEt	CH2OPh	OH
5	OEt	F	CCl3
11	C(CH2)Me	NEt2	F
12	CMe3	COOMe	Et
9	COOMe	Et	O-Pentyl
6	Et	Et	C(CH2)Me
4	NEt2	CHMe2	COOMe
10	CH2OPh	NH2	OCHMe2
1	CH2OPh	CCl3	CH2OPh
8	NEt2	F	F

deviations from the three way combinations of + and -. The user may adjust the values as desired. The thresholds are used to sort the molecules into two or three levels. If the high and low thresholds are equal there are two levels. If they are not, a third, 0 level is also created.

Using the experimental design program

To generate an experimental design within SYBYL 6.0[16] (or later versions) the synthetic chemist needs to draw the lead molecule to be developed. Since the program requires a hydrogen to mark the positions at which substituents will be varied, editing may be needed. The chemist should consider very carefully how the substituent positions are defined in order to ensure synthetically feasible diversity. If, for example, the parent molecule contains an ester, the chemist should probably replace not the alkyl group, but the O-alkyl group instead so that the ester can be varied as different types of carbonyl derivatives. Once the parent structure is on screen, the chemist is ready to run the macro, which is accessible through a customized SYBYL menu or via command mode. The program is available as "design synthesis" in our in-house custom menu. The R-positions to be varied are selected using the mouse and are automatically highlighted (PLATE 6). At this point, the parameters of interest and their + and - threshold values are selected to generate the design matrix. The number of molecules needed for the design is reported to the user. Next, the program now puts the parent structure on screen, indicates the position to be substituted and displays the fragment set that best matches the design matrix pattern of high and low associated with that R-group for the first row (PLATE 7). The matrix pattern is appended to the fragment name.

If the user desires to choose the fragment with criteria based on parameter values, an option is provided to view these values on screen. In the case where no fragments are synthetically feasible or desirable, the chemist has two options. The first option is to deviate from the matrix high/low pattern by specifying a new one using the RESTRICT CRITERIA option from the menu. This allows the user , for example to replace +++ with another pattern such as 0++ to find more desirable substituent fragments. Alternately, the user may opt to choose a fragment using the FULL_FRAGMENT LIST option. A menu of all fragments and their parameter criteria pattern is accessed through this option. Both of these scenarios are deviations from the strict design and the exact deviation is noted by the macro in report output. Each substituent on each molecule is selected iteratively in this manner until all the molecules have been created. The report output containing the design matrix, deviations, the selected substituents, and a list of 2D structures is then generated and printed for the user. The 3D structures are saved as a SYBYL database for further modeling.

The ability to edit molecule design sets was found to be essential for synthetic feasibility. Hence, the user may edit a design when it is created or in a latter SYBYL session. The user simply selects the option of editing an "old" design and selects the molecule(s) to be edited. It is then necessary to select the position(s) on that molecule to be modified. The substituent to be replaced is highlighted in red and the program proceeds as in creation mode with the original matrix criteria and fragment menus

Conclusion

The traditional application of experimental design strategies has been in areas in which the independent variables are continuous, that is can be controlled at any value within the parametric domain. However, this is not the case with the structural parameter space. The various parameters which are used to describe chemical structures are not continuous and progress from low values to high values at irregular

NOTE: Color plates appear in color section.

intervals. A cubic section of a formal design will show all parameter levels at the corners of the parameter space. Plackett-Burman designs, when used with continuous variables, cannot reveal parameter interactions or curvature in the response surface.

In MOD designs experimental points will be distributed through parametric space with clustering near the corners of the formal design. The tightness of the clustering will depend on the threshold values used to determine the design levels for the fragments. The more restrictive the threshold values the closer the level points will be to the corners of the section. With less restrictive thresholds the design will contain points toward the center of the design. The designs, then, take on some of the character of multiple level designs. This property may give the potential for the revelation of curvature of the response surface and parameter interactions.

The set of classical parameters suggested in the examples discussed were chosen because most synthetic chemists will be familiar with them and with their meaning. These parameters are somewhat co-variant, not fully independent. The MEDCHEM data base from which they were obtained contains many other parameters. Other parameters may be more appropriate for specific lead development.

Acknlowledgements: The authors are indebted to Svante Wold and Sergio Clementi for their comments, critisisms and suggestions in the development of the MOD program. We owe a special thanks to David L. Duewer for his encouragement and technical advice since 1982.

References

1. Hammett, L.P., *Physical Organic Chemistry*, McGraw-Hill, New York, 1940.
2. Hansch, C. A.; Muir, R. M.; Fujita, T.; Maloney, P. P.; Geiger, F.; Streich, M.,*J. Am. Chem. Soc.* **1963,** *85,* 2817.
3. Charton, M., *Chemical Technolgy,* **1974,** 502
4. Taft, R. W. *Steric Effects in Organic Chemistry;* Wiley: New York, N.Y., 1956, pp 556-675.
5. Medicinal Chemical Project, Pamona College, Claremont, California.
6. Leo, A.; Jow, P. Y. C.; Silipo, C. and Hansch, C. A., *J. Med Chem.* **1975** *18,* 865.
7. Nys, G. G. and Rekker, R. F.,*Chimica TherapeuticJ,* **1973,** *9 ,* 521.
8. K.W. Ratts, U.S. Pats. 3,829,306; 3,830,841 and 3,901,685.
9. Brannigan, L.H., Duewer, D.L., *QSAR: Rational Approaches to the Design of Bioactive Compounds*; Elsevier, Amsterdam, **1991,** p 553 (8th European Symposium on QSAR, Sept. 1990)
10. Hellberg,S., Sjostrom, M., Skagerberg, B., Wilkstrom, C., Wold, S., *Acta Pharm. Jugosl.,* **1987,** 37, 53-65.
11. Plummer, E.L., *Reviews in Computational Chemistry*, VCH, New York, **1990,** Chapter 4, pp 119-168.
12. Carlson, Rolf, *Design and Optimization in Organic Chemistry,* Data Handling in Science and Technology 8, Elsevier, Amsterdam **1992.**
13. Plackett, R. L. and Burman, J. P., *Biometrica,* **1946,** 33, 305-328.
14. Clementi, S., Cruciani, G., Baroni, M., *QSAR: Rational Approaches to the Design of Bioactive Compounds*; Elsevier, Amsterdam, **1991,** p 217 (8th European Symposium on QSAR, Sept. 1990).
15. Cramer, R.D. III, Patterson, D. E., Bunce, J. D., *J. Am. Chem. Soc.,* **1988,** 110, p 5959.
16. Tripos Associates, Inc. 1699 Hanley Road, Suite 303, St. Louis MO, 63144-2913.

RECEIVED December 8, 1994

Chapter 17

Use of Predictive Toxicology in the Design of New Chemicals

Vijay K. Gombar and Kurt Enslein

Health Designs, Inc., 183 East Main Street, Number 1050,
Rochester, NY 14604

While many applications of computer-assisted
techniques involve the design of molecules with
maximum biological activity, this chapter addresses a
different criterion of molecular design. We consider
monitoring toxicity as the design process brings
structural changes because the most desirable chemical
is not necessarily the one with maximum activity but
rather the one with maximum activity-to-toxicity ratio.
The computer-assisted technique for predicting toxicity
profiles   described here relies on robust and cross-
validated quantitative structure-toxicity relationship
(QSTR) models developed from experimentally
determined animal toxicity data. The technique, as
embedded in the   software package **TOPKAT**, first
confirms whether the query chemical lies inside, near the
boundary of, or outside the optimum prediction space
(OPS) of a QSTR model in order to assess the reliability of
predictions. Currently, such toxicity measures as rodent
carcinogenicity, mutagenicity in *Salmonella typhimurium*,
teratogenicity, chronic lowest observed adverse effect
level (LOAEL) and maximum tolerated dose (MTD), acute
rat oral LD_{50} and mouse inhalation LC_{50}, and aquatic
EC_{50} and LC_{50} can be estimated by using these
techniques. The methodology is explained here using
developmental toxicity (DT) data.

Widespread applications of powerful computers with tools for
generating life-like graphics have attracted the attention of chemists
(*1*). Among other uses of computers, the possibility of assessing
certain molecular properties before even the molecule is synthesized
is, perhaps, the most beneficial. For instance, by displaying on a
high-resolution monitor a three-dimensional picture of the
established geometry and the electronic structure of a binding site
in a protein, a chemist can dock possible ligands or design *de novo*

0097–6156/95/0589–0236$12.00/0
© 1995 American Chemical Society

some favorably fitting ligands before selecting the one to synthesize
(2-4). This computer-aided technique has been applied to design
molecules possessing a desired type of activity. A chemical so
designed, however, may not be worth anything if it turns out to be
toxic. Therefore, rapid and reliable, preferably computer-assisted,
assessment of toxicity associated with any molecular structure
should be an integral part of any molecular design program. With the
advent of solid-state synthetic chemistry and with growing
applications of combinatorial chemistry to create large libraries of
small (non-protein) molecules (5), the need for large scale screening
for toxicity is pressing. Also, federal agencies responsible for
administering risk assessment should be able to conveniently and
rapidly assess toxicity for timely assessment of health and
environmental risks of existing and new chemicals, particularly in
times of shrinking resources as risk assessment is becoming a social
priority.

In principle, if the mode of action/mechanism by which a toxic
response is elicited or the structure of the site at which a toxicant will
bind were known, the computer-assisted technique, as applied for
optimizing the activity, could be used for minimizing toxicity.
Unfortunately, toxicity is a very complex endpoint resulting from a
multitude of causes. Two chemicals, for example, having extremely
low and identical LD_{50} value (acute median lethal dose) may produce
altogether different acute effects. Similarly, two carcinogenic
chemicals may produce dose-related neoplastic lesions of different
morphology, in different organ systems, at different sites.
Consequently, we have used the linear free energy related (LFER)
approach to develop QSTR models (6). These statistically significant
and cross-validated models, along with the data bases from which
these models are developed, are installed in the **TOPKAT** software
package (7) for computer-assisted assessment of toxicity associated
with a molecular structure.

Since the QSTR models in **TOPKAT**, or for that matter any
quantitative structure-property relationship (QSPR) models, are
equations, all it takes to predict the value of the response variable
(property/toxicity) is to provide the values of the set of predictor
variables (structure descriptors). Every predicted value, however, may
not be reliable. Therefore, we employed algorithms for performing
various checks to ascertain whether the query structure falls within
the optimum prediction space (OPS) of the model being used for
toxicity assessment. Some of the mammalian and aquatic toxicity
indicators which can be currently assessed by the **TOPKAT** program
are:

Toxicity Indicator	Assay Protocol
Carcinogenicity	2-year study on both sexes of rat and mice
Mutagenicity	Ames test
Teratogenicity	With or without maternotoxicity
LOAEL	Chronic, oral, rat
LD_{50}	Acute, oral, rat
LC_{50}	Acute, flow-through, fathead minnow
Skin/Eye Irritancy	Draize test

In the following sections we describe the underlying methodology used. As an example, we have chosen the QSTR model for developmental toxicity (DT). After a brief discussion of the steps in building data bases, structure quantification, and developing robust statistical models, we exemplify the operations of the software for (i) assessing developmental toxicity potential (DTP), and (ii) for validating the assessed DTP.

Data Base Building

Obviously, the first requirement for developing a QSTR model for a desired measure of toxicity is the compilation of results of bioassays, e.g., the results from the *Salmonella typhimurium* histidine reversion assay for a QSTR model predicting mutagenicity. It is, however, very critical that the answer sought from a QSTR model be clearly defined because this determines which experimental results can be included in the data set and which should be excluded. For instance, if a researcher expected to predict whether the chemical he/she is designing will be mutagenic to a particular strain of *Salmonella* then the training set for the model should include results only from assays conducted with that particular strain. On the other hand, if the answer sought is whether the new chemical would be an Ames mutagen or not, then the results from assays on any *Salmonella* strain could be included in the training set. The data bases from which the QSTR models installed in the **TOPKAT** have been developed are, therefore, compiled from experiments performed under as uniform conditions as possible.

For experimental data on developmental toxicity, we identified 5,559 open literature citations from which the 1238 rat studies were selected. 830 of these studies were found to be unusable for different reasons (Table I). For 408 unique chemicals, uniform bioassay data were available. The molecular structures of these chemicals revealed that another 34 chemicals were not suitable for QSTR studies.

For the 374 chemicals suitable for building QSTR models, we extracted, from the original research papers, the doses at which any signs of maternal toxicity (MT) or fetal anomalies (Table II) were

Table I. TOPKAT Developmental Toxicity Data Base

Total rat studies	1238
Studies not usable	830
Ambiguous results	
Multiple studies	
Post-natal studies	
Segment-I studies	
Compounds with usable data	408
Structure not suitable	34
Mixtures	
Uncertain structure	
Organometallics	
Salts and acid/base pairs[a]	
Compounds suitable for QSTR models	374

[a]When separate assays were conducted on a salt and its corresponding weak acid or base, the results from salt were not considered.

reported by the authors. Two interesting observations were made: (i) some studies were performed at one dose only and both developmental and maternal toxicity were reported at that dose, and (ii) for some studies, neither DT nor MT was reported even at the highest exposure dose. Studies in both these groups were considered cases of inadequate experimentation, at least for QSTR modeling, because selection of lower doses in group (i) and higher doses in group (ii) could alter the outcome. Recognizing the uniformity required in the response variable values, we decided to further exclude all studies falling in these two categories. 273 chemicals finally remained in the data base.

Table II. Some Signs of developmental and maternal toxicity

Developmental Toxicity	Maternal Toxicity
Reduced fetal growth	Decreased feed intake
Fetal death	Weight loss
Resorptions	Increased mortality
Abnormal brain	Alopecia
Cleft palate	Behavioral
Skeletal anomalies	Respiratory distress
Limb defects	Lethargy
External malformations	Increased water intake
Hemorrhage	Local irritation
Runting	Rough fur
Visceral defects	Vaginal bleeding

Based on the relative doses which produced developmental and/or maternal toxicity, we assigned developmental toxicity potential (DTP) score to each of the 273 chemicals using the following criteria: If no signs of DT are observed even at the highest dose, the DTP score assigned was 1, i.e. no DTP is associated with the chemical. A chemical was assigned a DTP score of 2 if both DT and MT are observed at all doses except at least one dose at which neither MT nor DT is observed. A DTP score of 3 was assigned to chemicals which induce signs of DT one dose before the dose producing signs of MT. Finally, the chemicals for which DT is observed two or more doses before the dose inducing signs of MT were assigned a DTP score of 4. The population of DTP scores 1,2,3, and 4 in the data base of 273 chemicals was 129, 59, 27, and 58, respectively.

One of the prerequisites of applying the LFER approach in developing QSTR models is that the chemicals in the training set be structurally similar and have the same mechanism of action. In the absence of the knowledge of mode of action/mechanism, we decided to divide the 273 chemicals into broadly similar structural classes. Three classes of reasonably close population size were identified: 87 aliphatics, 95 carboaromatics, and 91 heteroaromatics. The distribution of DTP scores in these three classes is given in Table III.

Due to a small number of chemicals with DTP score 3, particularly in the carboaromatic and heteroaromatic classes, and due to the skew of the distribution towards chemicals with DTP score 1, it was decided to combine the chemicals with DTP score of 2,3, and 4 into one group, called positive (POS), and to label the chemicals

Table III. Distribution of DTP Scores Among Chemical Classes

DTP Score/ Class	Chemical Class		
	Aliphatic	Carboaromatic	Heteroaromatic
1	39	41	49
2	16	22	21
3	14	8	5
4	18	24	16
NEG	39	41	49
POS	48	54	42

with DTP score 1 as NEG. Consequently, the QSTR model was developed to discriminate chemicals which will produce developmental toxicity, i.e. POS chemicals, from the ones which will not, i.e. NEG chemicals. Such a QSTR will answer the question: Will the query structure induce/not induce developmental toxicity if in a bioassay it were tested at least up to a dose which produces maternal toxicity. For obtaining any finer information, such as whether any signs of MT be observed before DT is noticed, different group boundaries would have to be set. Similarly, given all the dose levels as compiled in the data set, a model for predicting the lowest dose which will produce any fetal anomalies could also be developed.

Structure Quantification

Effective structure quantification is the basis of all structure-property relationship studies. It has been shown (8) that the structure descriptors should be able to quantify transport, bulk, and electronic attributes of a molecular structure.

Electronic Attributes. Theoretically it is possible to quantify electronic attributes at any desired level of complexity ranging from quantum mechanical to a mere count of lone pairs. In recent years a number of methods have been published which do not require the knowledge of molecular geometry and are extremely fast in computing electronic properties (9,10) such as partial atomic charges, residual electronegativity, effective polarizability, etc. These algorithms have a practical advantage, especially for QSPR studies, because they can be applied to large sets of big molecules without significantly depleting available computational resources. We have shown (11), through an example of mutagenicity of some triazenes, that the QSTRs obtained by using the electronic descriptors computed from these non-geometry-based methods are of better or equal quality than that of QSTRs developed using sophisticated molecular orbital methods.

For the developmental toxicity models we have applied a rather new approach to quantify electronic attributes of molecules. Some of the advantages of this approach over the one used for the study of triazenes are:

a) complete electro-topological information encoding,
b) greater statistical reliability,
c) easy comprehension and implementation,
d) suitable for validation of predictions.

A brief description of the two-stage method is given in the following sections.

Identification of Fragments. The first step in quantification of electronic attributes of a molecule is to parse the molecular structure for identification of 1-atom and 2-atom fragments which are present in our proprietary fragment library. If there are any atoms or hybridization states of atoms in a molecule which are not represented in the library, such molecules need to be excluded from the analysis. At present this library has about 2500 fragments consisting of atoms H, B, C, N, O, F, Si, P, S, Cl, Br, and I in a variety of hybridization states and covers a wide range of organic molecules. If needed, the library can be expanded to include atoms of particular interest.

There are two sets of fragments in the library; general and specific fragments. Unlike general fragments, the specific fragments include attached H atoms and impose restrictions on the topological environment of atom(s).

Computation of Electro-topological State Values. After having identified the unique set of fragments in a molecule, the electro-topological state values, E-values (*12*), of these fragments are computed. The E-value of a fragment encodes information about its electron content (valence, sigma, pi and lone-pair), topology, and environment. Since an E-value is computed by taking into account the effects of both intrinsic and environmental features, it changes even with remote variations in structures; of course, the magnitude of variation depends on the severity of change. The computation of the E-value of any atom begins with assignment of intrinsic state value, I, to every non-hydrogen atom i in the molecule according to:

$$I_i = \frac{\delta^v + 1}{\delta}$$

where δ is the number of σ-electrons and δ^v is the sum of sigma, pi and lone-pairs of electrons. The I value is then corrected for the field influence ΔI_i on atom i. ΔI_i may be calculated as:

$$\Delta I_i = \frac{\sum (I_i - I_j)}{r^2_{ij}}$$

where r_{ij} is the number of atoms in the shortest path between atoms i and j, both inclusive. The sum of I_i and ΔI_i then gives the E-value atom i. The E-value for a fragment is calculated by summing up the E-values of the atoms involved in the fragment. It is these E-values on all 1-atom and 2-atom fragments which are used here to quantify the electronic attributes of molecules.

In our experience, the type, count and the E-values of 1-atom and 2-atom fragments collectively define, for all practical purposes, a chemical structure uniquely. For 3-methoxy-1-propanol, for instance, four 1-atom and four 2-atom fragments are generated. They are:

1-Atom		2-Atom	
Type	Count	Type	Count
$-CH_2$	3	$-CH_2-CH_2-$	2
$-O-$	1	$-CH_2-OH$	1
$-OH$	1	$-CH_2-O-$	1
$-CH_3$	1	$-O-CH_3$	1

Collectively, the counts and E-values of these fragments will uniquely quantify the structural information of 3-methoxy-1-propanol only.

Transport Attributes. The transport attributes are generally expressed in terms of the logarithm of partition coefficient between n-octanol and water, logP (13). Since LogP is an additive-constitutive property, it can be calculated from the constant contributions of structural features of a molecule (14-16). Considering these contributions as different weights associated with various structural fragments, we decided to use the counts of 1-atom and 2-atom fragments instead of logP and left the selection of fragments and the assignment of weights of the selected fragments to be determined by the statistical method used for developing the structure-property relationship.

Since molecular shape and molecular symmetry also influence molecular transport, we included topological shape descriptors, $^m k$ (kappa), of orders 1 through 7 (17,18) and seven indices of molecular symmetry (19) for quantifying the transport attributes of molecules.

Bulk Attributes. Besides molecular weight, we used size-corrected E-values on 1-atom and 2-atom fragments for quantification of molecular bulk. The size-corrected E-values are computed from a rescaled count of valence electrons; $4\delta^v/n^2$ instead of δ^v.

Model Development

Predictor Variables. All the structural descriptors, namely, shape indices, symmetry indices, and counts, E-values, and size-corrected E-values of 1-atom and 2-atom fragments for the training set chemicals were subjected to a frequency check. Any variables having non-zero values for less than five chemicals were not considered as predictor variables. This was done to enhance the statistical reliability of the predictor variables.

In order to reduce problems due to possible collinearity of variables, the pair-wise correlations of these variables were examined. From a pair of variables with correlation coefficient of 0.9 or higher only one variable was retained in the descriptor set. The variable with a higher order (larger number) in predetermined hierarchy was generally retained. The hierarchy of selection of predictors used in this work is given in Table IV. A variable with higher order in the hierarchy, generally, is easier to compute and comprehend and is more continuous (more non-zero values) than one with a lower order.

Table IV. Hierarchy of Different Variables in Correlation Checks

Category	Order in Hierarchy
Molecular Weight	6
Symmetry Indices	5
E-values	4
Size-corrected E-values	3
Shape Indices	2
Fragment Counts	1

Method. The choice of the statistical method for developing a QSTR model, among other factors, depends on the scale on which the response variable is expressed. For a toxicity measure which is expressed in terms of the dose required to cause/inhibit a predefined effect, regression and related techniques are perfectly suitable. The examples of such response variables are: LD_{50}, MTD, LOAEL, etc. On the other hand, if the toxicity measure represents membership in certain group for a predefined dose, some classification technique is applied for developing a model capable of distinguishing members of one class from others. Classification of chemicals into four groups based on the evidence of carcinogenicity of chemicals by the National Toxicology Program (NTP), for instance, provides a polychotomous response variable.

For the chosen example of developmental toxicity, expressed as membership either to the NEG or POS group, the method of linear discriminant analysis (LDA) was used. This method produces two functions; one for the NEG class of chemicals and the other for the POS class of chemicals. Both functions use the same predictors. The coefficients of the predictors are, of course, different in the two functions and are so adjusted as to maximize the accuracy of classification. The number of descriptors in a function can be controlled by different techniques. See Gombar and Enslein (20) for more details on the method of LDA.

The BMDP 7M procedure was employed for carrying out LDA on the three data sets of aliphatic, carboaromatic, and heteroaromatic chemicals. The selection of variables in this step-wise procedure was controlled by the F-values (ENTER and REMOVE) of variables and the orthogonality of the variables in the discriminant function was checked by their multiple linear correlation (TOLERANCE). To begin with, any variable with an F-TO-ENTER value of 1.7 was included in the function. If inclusion of a subsequent variable reduced the F-TO-REMOVE value of an already included variable below 1.7, the old variable was excluded from the function. And if the inclusion of a subsequent variable would result in a multiple linear correlation coefficient of 0.95, such a variable would be marked. Under such circumstances, the order of variables in which they could be included in the function was changed so as to select the most potent variables. If the number of predictors in the discriminant function exceeded one-fifth the number of compounds in the data set, the F-TO-ENTER and F-TO-REMOVE limits were systematically increased.

A variety of diagnostics were performed on the tentative discriminant function thus obtained. For every compound, the posterior probabilities of classification were examined. Exactly equal posterior probability for both the groups, NEG and POS, is a sign of a unique variable-compound association. The offending variable was identified and removed from the analysis. This generally does not affect the quality of the function.

Secondly, the Mahalanobis Distance (MD) of each compound was examined. A conspicuously large value of MD for both groups is generally a sign of some poorly distributed variables in the function. In such cases, the offending variable was removed and the step-wise procedure was repeated.

The stability of the model was examined by comparing the performance of the function on the training set with that in the leave-one-out test. A significant reduction in the accuracy of classification is a sign of an unreliable function. Such a situation can not be improved unless restrictions on the number of predictors or degree of orthogonality are relaxed.

The stability of the model was further tested by randomly splitting the modeling set into a training set (90%) and a prediction set (10%). The discriminant function is developed from the training set and the values of response variable for the compounds in the prediction set are computed. The computed values are compared with the actual values. If the accuracy in the prediction set is of the same order as that in the original set, the model may be considered robust. The test was repeated (25 to 30 times) till every compound was placed at least once in the prediction set. The structures of the compounds mispredicted repeatedly were noted to define limitations of the applicability of the function.

The robust model obtained at this stage was validated in order to rule out the possibility of accurate classification by chance. The values of the response variables of the compounds were randomly switched till the population of NEG and POS was roughly the same as in the actual training set. If the accuracy of the function for this randomly assigned response values is not significantly different from the accuracy of the model, the function may be unreliable. A number of such validation comparisons were made to ascertain the reality of the function.

Results

Some of the important statistical parameters determining the quality of linear discriminant models developed for discriminating chemicals with no developmental toxicity potential from the rest are collected in Table V. It can be seen that the three QSTR models are statistically significant. In all cases the ratio n/p is greater than three and the F-ratios of the models are significant for the respective degrees of freedom. The F_{min} values of 2.0 and more indicate that the least contributing descriptor is also significant at $p < 0.005$. A descriptor with the largest F-TO-REMOVE value, F_{max}, is the strongest predictor in a given model. Small values of Wilk's Λ imply good separation of NEG and POS chemicals. The classification accuracies of the three models in the leave-one-out jackknife test are given in Table VI. It can be seen that the overall accuracy ranges between 86 and over 91%. It may be mentioned that the chemicals classified in the dead zone of posterior probability of classification between 0.3 and 0.69, i.e. the indeterminates, are considered mispredictions in calculating these overall accuracy numbers. In order to confirm that this accuracy of the models was not due to chance, we randomly assigned chemicals to the NEG and POS classes, keeping the populations of NEG and POS classes equal to the original population, and redeveloped a discriminant function using the descriptors of the models. For all three subsets, these functions returned no more than 53% correct classification indicating the association between the model descriptors and the actual membership to NEG and POS classes.

The robust and cross-validated models were installed in the **TOPKAT** software package to permit their interactive use for assessing the DTP associated with a given molecular structure.

Table V. Statistical Parameters of TOPKAT DTP QSTR Models

Statistical Parameter		Submodel	
	Aliphatics	Carboaromatics	Heteroaromatic
N^a	87	95	91
n^b	79	92	90
p^c	22	29	29
Electronic	15	19	21
Shape	2	0	0
Symmetry	1	1	0
Fragment count	4	9	8
F-ratio	7.70	7.59	7.96
Wilk's Λ	0.25	0.22	0.21
F_{max}	57.2	34.3	58.4
F_{min}	2.9	2.3	4.7

[a]Number of compounds in the **TOPKAT** data base.
[b]Number of compounds in the training set.
[c]Number of descriptors in the QSTR model.

Table VI. Classification Accuracies of TOPKAT DTP Discriminant Models in the Leave-one-out Jackknife Test

Statistic		Submodel	
	Aliphatics	Carboaromatics	Heteroaromatic
N-NEG	35	39	49
N-POS	44	53	41
Sensitivity	88.6%	87.0%	86.1%
Specificity	88.6%	97.4%	86.0%
Accuracy	88.6%	91.4%	86.0%
Indeterminate	2.5%	2.2%	5.5%

Using DTP Models in TOPKAT

The first step in using **TOPKAT** is to input the molecular structure of the query chemical as a SMILES string (21). If the input structure corresponds to a chemical in the data base, the program communicates that fact to the user and displays the chemical name and ID (CAS or NIOSH number), experimentally assigned DTP class along with its original source, and the **TOPKAT**-assigned DTP class. However, while designing new molecules the input structure, generally, will not be one from the training set. In order to present a real-life application of different functionalities of the program, we applied it to a number of chemicals (Figure 1) which are not in the data base but whose DTP class assignment has been determined. It is easy to be misled by the outcome of applying QSTR to any chemical structure. As mentioned earlier, QSTR, being an equation, will always result in a value for the response variable if the descriptor variables were provided. The value however, may not be meaningful or reliable. The software is equipped with algorithms for pointing out when the QSTR-assigned DTP class may not be reliable. Some of the situations are discussed below:

Unusual Atoms/Bonds. The program parses the structure to identify any atoms or bonding situations which, for some reasons, are not supported by the program. On entering the structure of diazo-oxo-norleucine (CAS 157-03-9), the program warned that the unusual bonding of N atom in the $-CHN_2$ fragment could not be processed and execution was terminated. Such a situation is not common when dealing with single organic molecules since **TOPKAT** can handle 12 atoms in 20 different hybridization states.

Groups Present in Misclassified Chemicals. When all atom- and bond-types in the query structure are present in the training set chemicals, **TOPKAT** scans for the substructures which, according to its experience from the resubstitution test, are associated with misclassified compounds. The user is warned in such a case that the **TOPKAT**-assigned class may be unreliable but the program does not stop processing. For 3-aminopropionitrile (CAS 151-18-8), the program issued this warning for the -CN group. It should be mentioned that in this particular case though the **TOPKAT**-assigned class matched the experimentally-assigned class, the accurate match may have occurred just by chance.

Fragments Unrepresented. As **TOPKAT** parses the input structure it checks whether or not the 1-atom and 2-atom fragments in the query are represented in the training set chemicals. If they are not, the fragment(s) is(are) highlighted for the user's perusal. In such a case the user, with his knowledge of the subject and experience, may over-ride the warning and accept the **TOPKAT**-assigned DTP class. By default, however, the query chemical will be labeled "not covered" and the result may not be reliable. Such a situation is not very common because (i) the QSTR modules in **TOPKAT** are, generally, developed from large sets of heterogeneous data sets, and (ii) the fragment library has been designed to cover combinations of over 100 atom-types; for instance, $>C=$, $>C<$, $-CH_3$, $-CH_2-$, etc., are different atom-types. Endosulfan (CAS 115-29-7) was found to be "uncovered"

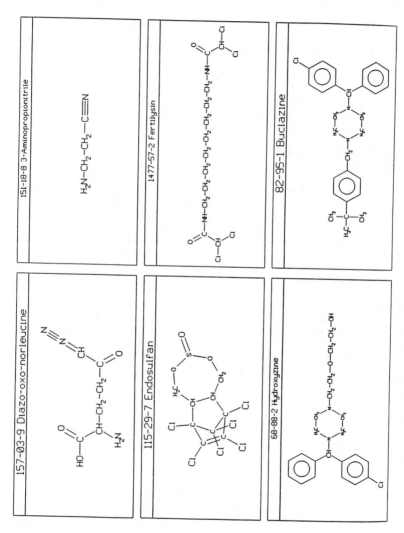

Figure 1. Structures of compounds outside the DTP database.

due to the cyclic >S=O fragment which is non-existent in the training set of 79 aliphatic chemicals.

Descriptors Outside the Range. It is important not to extrapolate the QSAR models too far outside the ranges of descriptor variables as it will result in statistically unreliable predictions. Therefore, **TOPKAT** stores the univariate statistical data such as minimum and maximum values, average, standard deviation, etc. on descriptor variables and checks that the descriptor values for the query structure are within the range of the corresponding descriptor values from the training set. For Fertilysin (CAS 1477-57-2), for example, the program identified that one of the descriptors was more than three standard deviations outside the range. A QSPR-predicted result in such a case cannot, and should not, be accepted. The experimentally-assigned DTP class for Fertilysin is POS, whereas the **TOPKAT**-assigned DTP class is NEG; but, as we have shown, the **TOPKAT**-assigned class is not reliable in this case.

Outside the Optimum Prediction Space (OPS). Sometimes the query descriptor values may be within the training set descriptor value ranges on a univariate basis but still, due to orthogonality of the descriptors, they may lie outside the OPS of the QSAR model. Under these circumstances the user is informed about (i) the descriptor(s) which are outside the OPS, (ii) their distance from the OPS, and (iii) their relative contribution to the computed value of the response variable. In case only a few variables are outside the OPS and the distance from the OPS is within one standard deviation, the **TOPKAT**-assigned value may be accepted. However, if variables are far outside the OPS and/or the variables outside the OPS contribute the most to computed value, the **TOPKAT**-computed value may be questionable.

For Hydroxyzine (CAS 68-88-2) one variable was found to be outside the OPS with a distance within one standard deviation. The **TOPKAT**-assigned DTP class was accepted. Indeed, it was confirmed to be a correct classification. Similarly, for Buclazine (CAS 82-95-1) there were many variables outside the OPS but marginally. The **TOPKAT**-assigned DTP class for buclazine was also considered validated. In this case also the **TOPKAT**-assigned and actual DTP classes matched. For Fertilysin, as mentioned above, not only there were many univariate out of range variables but these were far outside the OPS. This indicated that the DTP classifier of **TOPKAT** should not be applied to Fertilysin.

Conclusions

In the process of designing new chemicals, it is important to assess their toxicity, preferably before they are even synthesized. Robust and statistically significant QSTR models installed in **TOPKAT**, a modular computational toxicology tool, can generate, from molecular structure, a variety of mammalian, aquatic, and environmental toxicity measures, thus providing a capability to follow the toxicity profile as molecular structure is modified during the design process. The algorithms to autovalidate a computed toxicity measure are an extremely powerful and unique function of the **TOPKAT** package.

Acknowledgements

This research was partially funded through a grant (Contract No. 3857) under the Great Lakes Health Effects Program, Health Canada, Canada. We enormously admire the relentless efforts of Dr. Dieter Riedel of Health Canada, Canada, in establishing this grant. We are thankful also for his technical contributions and enthusiastic support during this research project.

Literature Cited

1. Ugi, I.; Bauer, J.; Blomberger, C.; Brandt, J.; Dietz A.; Fontain, E.; Gruber, B.; Scholley-Pfab, A.; Senff, A.; Stein, N. *J. Chem. Inf. Comput. Sci.* **1994**, *34*, 3-16.
2. DesJarlais, R. L.; Sheridan, R. P.; Seibel, G. L.; Dixon, J. S.; Kuntz, I. D.; Venkataraghavan, R. *J. Med. Chem.* **1988**, *31*, 722.
3. Goodford, P. J. *J. Med. Chem.* **1985**, *28*, 849-857.
4. Danziger, D. J.; Dean, P. M. *Proc. R. Soc. London* **1989**, *B236*, 115-124.
5. DeWitt, S. H.; Kiely, J. S.; Stankovic, C. J.; Schroeder, M. C.; Reynolds-Cody, D. M.; Pavia, M. R. *Proc. Natl. Acad. Sci. USA* **1993**, *90*, 6909-6913.
6. Hansch, C. *Drug Metabolism Reviews* **1984-85**, *15*, 1279-1294.
7. Enslein, K.; Gombar, V. K.; Blake, B. W. *Mutation Research* **1994**, *305*, 47-61.
8. Purcell, W. P.; Bass, G. E.; Clayton, J. M. *Strategy of Drug Design: A Guide to Biological Activity*; John Wiley: New York, NY, 1973.
9. Gasteiger, J.; Marsili, M. *Tetrahedron* **1980**, *36*, 3219-3228.
10. Mullay, J. *J. Am. Chem. Soc.* **1984**, *106*, 5842-5847.
11. Gombar, V. K.; Enslein, K. *Quant. Struct.-Act. Relat.* **1990**, *9*, 321-325.
12. Hall, H. L.; Mohney, B.; Kier, L. B. *J. Chem. Inf. Comput. Sci.* **1991**, *31*, 76-82.
13. Fujita, T.; Iwasa, J.; Hansch, C. *J. AM. Chem. Soc.* **1964**, *86*, 5175.
14. Rekker, R.F. In *The Hydrophobic Fragmental Constant; Its Derivation and Application, A Means of Characterizing Membrane Systems*; Nauta, W. Th.; Rekker, R. F., Eds.; Pharmacochemistry Library 1; Elsevier: Amsterdam, 1977.
15. Hansch C.; Leo, A. *Substituent Constants for Correlation Analysis in Chemistry and Biology*; John Wiley: New York, NY, 1979.
16. Ghose, A. K.; Crippen, G. M. *J. Comput. Chem.* **7**, 565-577.
17. Gombar, V. K.; Jain, D. V. S. *Indian J. Chem.* **26A**, 554-555.
18. Kier, L. B. *Quant. Struct.-Act. Relat.* **5**, 1-7.
19. Mumtaz, M. M.; Knauf, L. A.; Reisman, D. J.; Peirano, W. B.; Derosa, C. T.; Gombar, V. K.; Enslein, K.; Carter, J. C.; Blake, B. W.; Huque, K. I.; Ramanujam, V. M. S. *Regulatory Toxicol. Pharmacol.* (in press).
20. Gombar, V. K.; Enslein, K. In *Applied Multivariate Analysis in SAR and Environmental Studies*; Devillers, J.; Karcher, W., Eds.; Kluwer Academic: Dordrecht, 1991; pp 377-414.
21. Weininger, D. *J. Chem. Inf. Comput. Sci.* **1988**, *28*, 31.

RECEIVED October 13, 1994

Chapter 18

Comparison of In Vivo and In Vitro Toxicity Tests from Co-inertia Analysis

James Devillers[1] and Daniel Chessel[2]

[1]Centre de Traitement de l'Information Scientifique,
21 rue de la Bannière, 69003 Lyons, France
[2]Université Lyons-I, Unité de Recherche Associée, Centre National
de la Recherche Scientifique 1451, 43 Boulevard du 11 Novembre 1918,
69622 Villeurbanne Cedex, France

Co-inertia analysis is a multivariate method allowing to find the co-structure between two data tables from powerful statistical and graphical tools. It was used to compare toxicity results obtained with the rabbit eye test *in vivo* to those obtained with the *in vitro* eye organ test.

The aim of toxicity testing is to allow the prediction of the effects that chemicals are likely to have in man by the extrapolation of the effects observed in experimental animals or other biological systems. Safety testing is performed either to assure the safety in use of a medicine, food or consumer product, or to estimate the extent of the occupational hazard presented by an industrial chemical (*1*).

Over the last decade, there has been an increasing pressure to reduce animal experimentation and develop *in vitro* methods in pharmacotoxicology (*2, 3*). However, before any *in vitro* toxicity test can be used with any degree of reliability, a validation exercise is needed (*4*). Usually, *in vitro* data are compared with *in vivo* data by means of regression analysis (e.g.; *5, 6*). Even if this simple statistical analysis generally gives some interesting results, it is not sufficient to estimate the relevance of these different test systems, to select adequate endpoints, and to derive valuable structure-activity relationships. This study is designed to stress the usefulness of co-inertia analysis (*7*) to overcome these different problems.

0097–6156/95/0589–0250$12.00/0
© 1995 American Chemical Society

Sample and Research Methodology

During the past several years, considerable research effort has been directed towards the development of alternatives *in vitro* to the rabbit eye test *in vivo* (*8-12*). Indeed, rabbit eye test *in vivo* (*13*) has frequently been criticized due to methodological, economical, and ethical problems (*12, 14, 15*). Under these conditions, a wide variety of *in vitro* techniques for predicting eye irritancy has been described in the literature (*8, 10, 11, 16*). Among them, the enucleated eye test (*17, 18*) based on the measurement of the change in corneal thickness that follows exposure of the eye to an irritant chemical is regularly cited in the literature as an attractive alternative *in vitro* method. Indeed, although enucleated eyes are required, the animals are humanely killed before use. Two eyes are available from each rabbit, when *in vivo* only one eye is used. Eyes may be taken from rabbits that have been used for skin irritation and that are due to be killed on the completion of that test. Complex and expensive tissue culture techniques are not required and after the initial outlay on the equipment, the running costs are minimal. The *in vitro* eye organ test can be used to examine insoluble powders and acid or alkali solutions that cannot, or cannot readily, be assayed in cell cultures (*19-21*).

In a recent article, Jacobs and Martens (*22*) estimated in a first time the corneal swelling *in vivo* and *in vitro* for 34 substances. The corneal swelling data obtained *in vivo* after 4, 24, 48, and 72 hours of exposure were then compared with the corneal swelling data obtained *in vitro* after 0.5, 1, 2, and 4 hours of exposure by means of simple regression analyses. Since their data appeared insufficiently exploited, we have tried to extract more information from their matrices by means of the co-inertia analysis (*7*).

Co-inertia analysis can be viewed as a generic multivariate method to find the co-structure between two data tables (Figure 1). These two data matrices are first considered independently. They can be analyzed by means of different multivariate approaches such as principal components analysis (PCA), correspondence factor analysis (CFA), or multiple correspondence factor analysis (MCFA). These separate analyses underline the basic structure of the two data tables. In a second step, a matching analysis is performed in order to detect a co-structure between the two data matrices. This second analysis is based on the research of co-inertia axes maximizing the covariance between the coordinates of the projections of the rows of each data table.

The mathematical model of this analysis can be formulated as follows:

Let (X, D_p, D_n) and (Y, D_q, D_n) be two statistical triplets; Table X is the first data set (after an initial transformation); D_p contains the

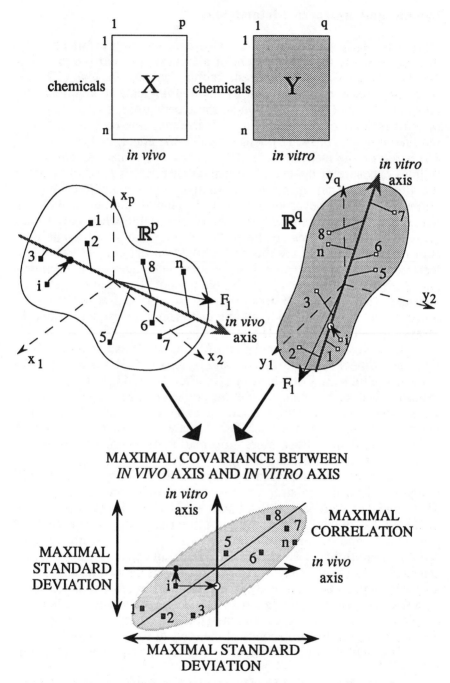

Figure 1. General principle of co-inertia analysis.

weights associated with the columns of Table X; D_n contains the weights associated with the rows of Table X. Table Y is the second data set (after an initial transformation); D_q contains the weights associated with the columns of Table Y; D_n contains the weights associated with the rows of Table Y.

The first statistical triplet (X, D_p, D_n) defines an inertia analysis of n points in a multidimensional space noted R^p and of p points in a multidimensional space noted R^n. After diagonalization, r principal axes are kept and the matrices R_r, C_r, N_r are generated. R_r contains the scores of the n rows on the r axes. C_r contains the scores of the p columns on the r axes. N_r contains the eigenvalues $(v_1...v_r)$.

The second statistical triplet (Y, D_q, D_n) defines an inertia analysis of n points in a multidimensional space noted R^q and of q points in a multidimensional space noted R^n. After diagonalization, s principal axes are preserved and the matrices R_s, C_s, M_s are generated. R_s contains the scores of the n rows on the s axes. C_s contains the scores of the q columns on the s axes. M_s contains the eigenvalues $(\mu_1...\mu_s)$.

Let u and v be a pair of vectors. The former is normalized by matrix D_p in the multidimensional space R^p and the latter is normalized by matrix D_q in the multidimensional space R^q. The projection of the multidimensional space associated with Table X onto vector u generates n coordinates in a column matrix:

$$\xi = X D_p u \tag{1}$$

The projection of the multidimensional space associated with Table Y onto vector v generates n coordinates in a column matrix:

$$\psi = Y D_q v \tag{2}$$

Co-inertia associated with the pair of vectors u and v is equal to:

$$H(u,v) = \xi^t D_n \psi \tag{3}$$

If the initial Tables X and Y are centered, then the co-inertia is the covariance between the two new sets of scores:

$$Cov(\xi,\psi) = (Iner_1(u))^{1/2} (Iner_2(v))^{1/2} Corr(\xi,\psi) \tag{4}$$

with $Iner_1(u)$ as the projected inertia onto vector u (i.e.; the variance of the new scores on u), $Iner_2(v)$ as the projected inertia onto vector v (i.e.; the variance of the new scores on v), and $Corr(\xi,\psi)$ as the correlation between the two coordinate systems. Note that the square of the latter entity Corr (ξ,ψ) is maximized via canonical correlation analysis.

To obtain co-inertia axes, one diagonalizes the following matrix:

$$W = D_p^{1/2} X^t D_n Y D_q Y^t D_n X D_p^{1/2} \tag{5}$$

Let U_z be the matrix containing the first z normalized eigenvectors of W and Λ_z be the matrix containing the first z corresponding eigenvalues (noted λ_k, $1 \leq k \leq z$). The first z co-inertia axes (associated norm D_p) in R^p are obtained as:

$$A_z = D_p^{-1/2} U_z \tag{6}$$

The first z co-inertia axes (associated norm D_q) in R^q are obtained as:

$$B_z = Y^t D_n X D_p^{1/2} U_z \Lambda_z^{-1/2} \tag{7}$$

A_z and B_z are called optimal co-inertia weights of the variables, respectively, in Tables X and Y. The co-inertia scores of Table X rows are obtained as:

$$X^*_z = X D_p A_z \tag{8}$$

The co-inertia scores of Table Y rows are obtained as:

$$Y^*_z = Y D_q B_z \tag{9}$$

Furthermore, one may compare the projected variability resulting from the separate analyses and that from co-inertia analysis by calculating the scores of the initial inertia axes projected onto the co-inertia axes. Let C^*_r and C^*_s be the resulting scores with:

$$C^*_r = N_r^{-1/2} C_r^t D_p A_z \tag{10}$$

and

$$C^*_s = M_s^{-1/2} C_s^t D_q B_z \tag{11}$$

We call the diagonal elements of matrix $(X^*_z)^t D_n X^*_z$ and matrix $(Y^*_z)^t D_n Y^*_z$ pseudo-eigenvalues. Let v_k^* be the kth pseudo-eigenvalue of Table X and μ_k^* be the kth pseudo-eigenvalue of Table

Y. Such values are useful in situating the value of co-inertia axes in comparison with inertia axes. Finally, the quantity:

$$\rho_k*^2 = \lambda_k / (\nu_k* \mu_k*) \tag{12}$$

is an expression of the correlation between the two new sets of coordinates.

Co-inertia analysis was performed with the ADE software package (*23*) on two 33 x 4 data matrices which are graphically displayed in Figure 2. Indeed, ammonia was eliminated from our statistical analysis due to an unrealistic toxicity value measured *in vivo* (*22*) after 24 hours of exposure.

Results and Discussion

Figure 2A shows that the chemicals under study can be classified according to their level of toxicity measured *in vivo* but also according to the evolution of their activity in relation with the different times of exposure. Thus, for example, sodium acetate is not toxic *in vivo* while diethylamine is very toxic. Furthermore, Figure 2A reveals that the toxicity of the former does not change with the time of exposure while the toxicity of the latter considerably changes after 4, 24, 48, and 72 hours of exposure. Analysis of Figure 2B dealing with *in vitro* experiments leads to the same type of conclusions. However, the comparison of the graphical displays of the toxicity results obtained *in vivo* and *in vitro* allows to underline some discrepancies in the toxicological behavior of the 33 studied chemicals.

From a simple visual inspection of Figure 2, it appears that there exists an obvious toxicity model within the *in vivo* and *in vitro* data matrices. These obvious models are directly linked to the information encoded by the rows and columns of the data matrices. Thus, for example, if we consider that the toxicity data obtained *in vivo* depend on the structure of the chemicals under study and the time of exposure, we can easily demonstrate (Figure 3) that these two parameters are not linked by an additive relationship (i.e.; row mean + column mean - overall mean) but are related by means of a multiplicative effect using the best estimation obtained from the first factor of a noncentered PCA (*24*).

A similar multiplicative model was obtained with the *in vitro* data (results not shown). Under these conditions, it is obvious that the toxicological responses obtained with the rabbit eye test *in vivo* are correlated to those obtained with the *in vitro* eye organ test. This is

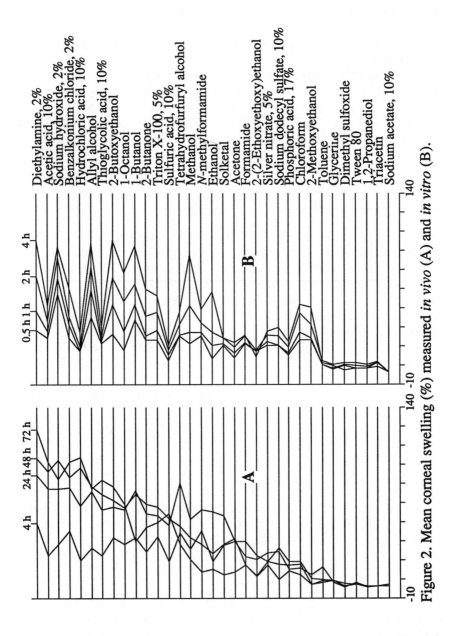

Figure 2. Mean corneal swelling (%) measured *in vivo* (A) and *in vitro* (B).

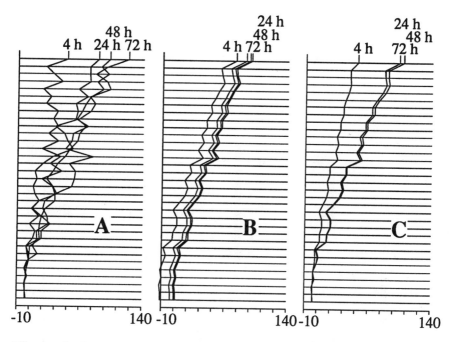

Figure 3. Mean corneal swelling (%) measured *in vivo* (A) and calculated from the additive (B) and multiplicative (C) models.

clearly underlined on Figure 4 which principally allows the comparison of the levels of toxicity in the two test systems. Thus, for example, sodium acetate, triacetin or 1,2-propanediol, which are not toxic *in vivo* are also weakly reactive *in vitro*. In the same way, diethylamine which appears very toxic in the eye test *in vivo* also presents a high degree of reactivity with the enucleated eye test. However, it is interesting to note that some chemicals such as hydrochloric acid or 2-methoxyethanol have different effects on the corneal swelling *in vivo* and *in vitro* and therefore decrease the degree of correlation between the two test systems. The study of the particular toxicity of these types of chemicals in the two test systems is particularly suitable for comparing the *in vivo* and *in vitro* assays.

Thus, it appears fruitful to try to find a common obvious model of toxicity between the two data matrices under study and then focus our attention only on the residuals (i.e.; differences between the measured toxicity data and those calculated from the model). This multiplicative model was simply derived after joining the *in vivo* and *in vitro* data tables. The graphical displays of the multiplicative model obtained *in vivo* (Figure 5B) and *in vitro* (Figure 6B) and their comparison with the graphics of the raw toxicity data (Figures 5A and

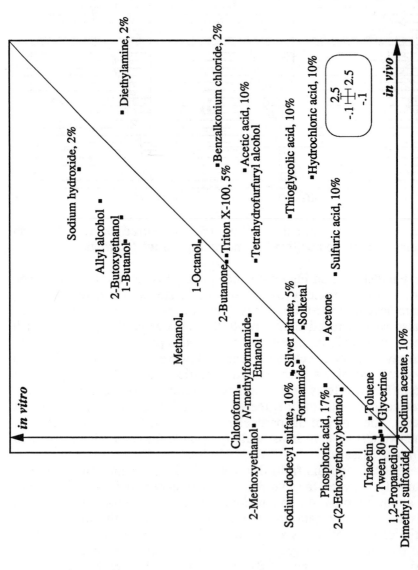

Figure 4. Plot of the toxicity levels in the two test systems. Use of the parameters of the multiplicative models.

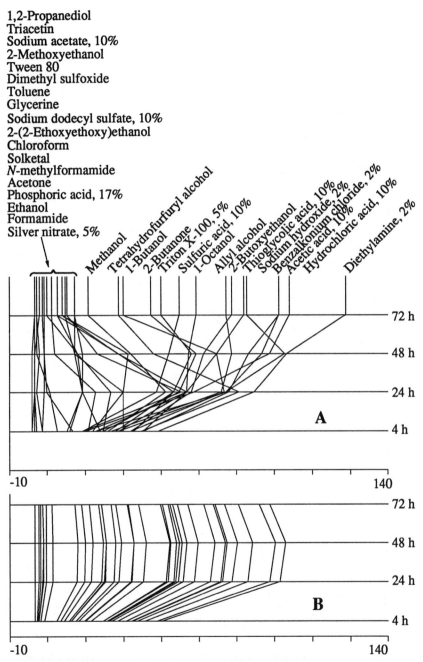

Figure 5. Mean corneal swelling (%) measured *in vivo* (A) and estimated from the common obvious multiplicative model (B).

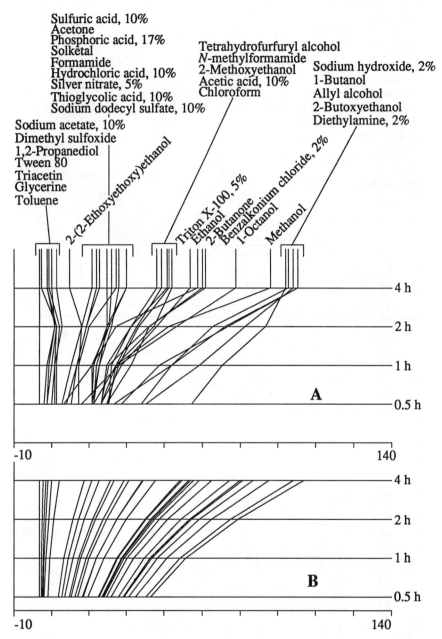

Figure 6. Mean corneal swelling (%) measured *in vitro* (A) and estimated from the common obvious multiplicative model (B).

6A) clearly show that the results recorded *in vitro* are more reproducible than those obtained *in vivo*. Furthermore, it is obvious that generally a peak of toxicity is reached after 48 hours of exposure for the rabbit eye test *in vivo* while the maximum of toxicity is only obtained after 4 hours of exposure for the enucleated eye test.

In order to extract the remaining toxicological information included in the *in vivo* and *in vitro* data matrices, we performed a co-inertia analysis on the residuals computed from the common obvious model of toxicity. The separate analyses consisted of two noncentered PCA. On the resulting maps (Figures 7 and 8), the location of the chemicals depends on the variability of their toxicological response in relation with the common model of toxicity. This methodological approach allows us to easily identify compounds presenting a particular toxicological behavior in the *in vivo* and *in vitro* test systems. Thus, for example, hydrochloric acid, acetic acid and thioglycolic acid are located at the bottom of Figure 7 since they present a high toxicity after 72 hours of exposure. At the opposite, *N*-methylformamide, methanol or 1-butanol are located at the top of the map (Figure 7) due to their high reactivity in the first hours of exposure of the tests. Analysis of Figure 8 leads to the same type of conclusions. However, the comparison of Figures 7 and 8 reveals that the time-courses of swelling development observed *in vivo* and *in vitro* for the chemicals under study are different. This is particularly well demonstrated when we perform the matching step of the co-inertia analysis and draw the bivariate graph (Figure 9) illustrating the concordance between the scores of co-inertia *in vivo* and *in vitro*. Indeed, in the upper right corner, this figure shows that chemicals with toxicity increasing during the time-course of the *in vivo* test slowly react in the *in vitro* test. This is particularly true for hydrochloric acid, acetic acid, thioglycolic acid, and sulfuric acid. At the opposite, in the lower left corner of the figure, chemicals which slowly react *in vivo* with generally a decrease in their activity with the time of exposure present an opposed toxicity behavior *in vitro* since they show an increase of their reactivity during the time-course of the test. This can be found for 2-methoxyethanol, chloroform, and so on. By contrast, chemicals located in the middle of the figure exhibit a comparable time-course for corneal swelling *in vivo* and *in vitro*.

Our study partially confirms the conclusions stated by Jacobs and Martens (*22*). Indeed, it shows that the toxicity data obtained with the rabbit eye test *in vivo* are broadly correlated to those obtained with the *in vitro* eye organ test. However, the use of co-inertia analysis clearly reveals that this degree of correlation has to be interpreted with care. Indeed, the kinetics of toxicity (i.e.; time-effect curves) of the chemicals in the rabbit eye test *in vivo* and in the *in vitro* eye organ test are not

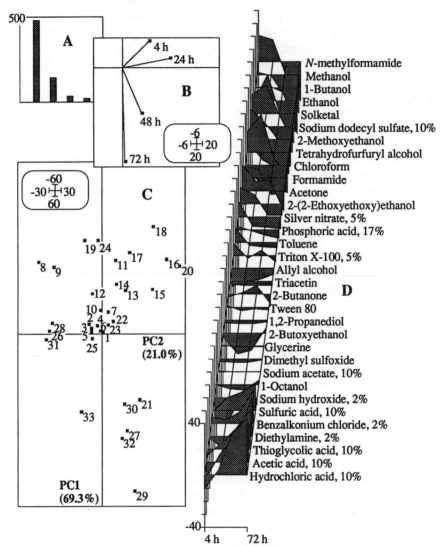

Figure 7. Noncentered PCA of the residuals (*in vivo*). Eigenvalues (A), projection of the variables (B) and chemicals (C) on PC1-PC2, and plot of the residuals for the chemicals along PC1 (D).

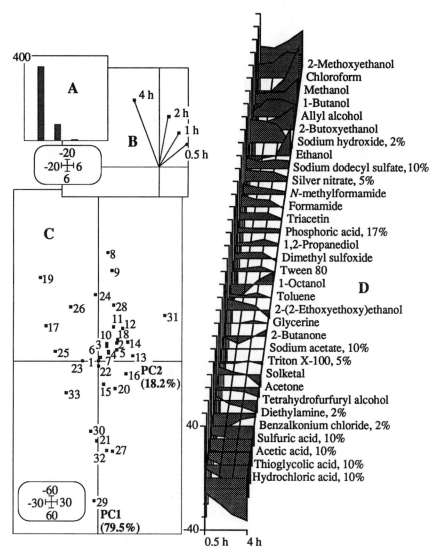

Figure 8. Noncentered PCA of the residuals (*in vitro*). Eigenvalues (A), projection of the variables (B) and chemicals (C) on PC1-PC2, and plot of the residuals for the chemicals along PC1 (D).

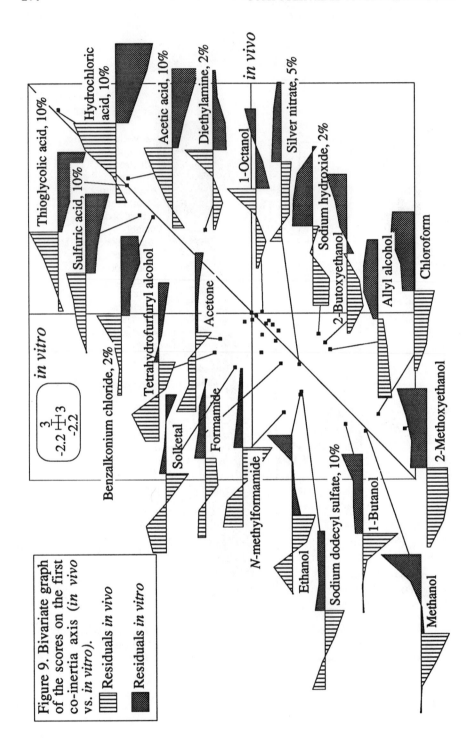

Figure 9. Bivariate graph of the scores on the first co-inertia axis (*in vivo* vs. *in vitro*).

the same. Furthermore, it is obvious that it is difficult to compare the different times of exposure in the two test systems. From a toxicological point of view, these results show that for a better comparison of two test systems and for a better description of the toxicological behavior of the chemicals tested *in vivo* and *in vitro*, it is necessary to consider all the different times of exposure. Indeed, comparisons performed after pooling the toxicity results obtained at different times of exposure obligatorily induce loss of information and misinterpretations. Consequently, the most suitable statistical approaches for the study of this kind of data are multivariate analyses rather than simple correlation analyses.

Our study also reveals that an exposure period of 4 hours in the enucleated eye test is not sufficient to correctly describe the toxicological behavior of the majority of the chemicals under study. It is interesting to note that Burton and coworkers (*18*) also use an exposure period of 4 hours for the enucleated rabbit eye test. Consequently the time-effect curves published in their article (see page 477, Figure 5) present the same characteristics as those which can be drawn from the results of Jacobs and Martens (*22*). Price and Andrews (*20*) hourly measured the corneal thickness on isolated rabbit eye until 5 hours of exposure but their results are too sparse to estimate the usefulness of their methodology. Last, our methodological approach based on the use of co-inertia analysis clearly underlines that some families of compounds (e.g.; acids) present very different toxicological behavior *in vivo* and *in vitro*.

More generally, our study reveals the heuristic potency of co-inertia analysis in pharmacotoxicology. Indeed, this statistical analysis which works on a covariance matrix is based on the mathematically coherent Euclidean model and can be universally reproduced due to its numerical stability. The method performs simultaneously the analysis of two data tables. Therefore, it is easy to underline the particular and common features of these two data matrices. This is particularly fruitful for the comparison of toxicity tests or for deriving structure-activity and structure-property relationships.

To summarize, co-inertia analysis has to be viewed as a general method allowing one to relate any kind of data sets using any kind of standard multivariate analysis.

Literature Cited

1.   Swanston, D. W. *Food Chem. Toxicol.* **1985**, *23*, 169-173.
2.   Hampson, J. E. *ATLA* **1990**, *18*, 75-81.

3. DelRaso, N. J. *Toxicol. Lett.* **1993**, *68*, 91-99.
4. Ekwall, B.; Bondesson, I.; Hellberg, S.; Högberg, J.; Romert, L.; Stenberg, K.; Walum, E. *ATLA* **1990**, *18*, 226-233.
5. Fry, J. R.; Garle, M. J.; Hammond, A. H.; Hatfield, A. *Toxic. in Vitro* **1990**, *4*, 175-178.
6. Shrivastava, R.; Delomenie, C.; Chevalier, A.; John, G.; Ekwall, B.; Walum, E.; Massingham, R. *Cell Biol. Toxicol.* **1992**, *8*, 157-170.
7. Franquet, E.; Chessel, D. *Comptes Rendus de l'Académie des Sciences Série III Sciences de la Vie* **1994**, *317*, 202-206.
8. Shopsis, C.; Borenfreund, E.; Walberg, J.; Stark, D. M. *Food Chem. Toxicol.* **1985**, *23*, 259-266.
9. Gupta, K. C. *J. Toxicol. Cut. Ocular.* **1989**, *8*, 7-16.
10. Bruner, L. H.; Kain, D. J.; Roberts, D. A.; Parker, R. D. *Fund. Appl. Toxicol.* **1991**, *17*, 136-149.
11. Bagley, D. M.; Bruner, L. H.; De Silva, O.; Cottin, M.; O'Brien, K. A. F.; Uttley, M.; Walker, A. P. *Toxic. in Vitro* **1992**, *6*, 275-284.
12. Ikarashi, Y.; Tsuchiya, T.; Nakamura, A. *J. Toxicol. Cut. Ocular Toxicol.* **1993**, *12*, 15-24.
13. Draize, J. H.; Woodard, G.; Calvery, H. O. *J. Pharmacol. Exp. Ther.* **1944**, *82*, 377-390.
14. Sharpe, R. *Food Chem. Toxicol.* **1985**, *23*, 139-143.
15. Seifried, H. E. *J. Toxicol. Cut. Ocular Toxicol.* **1986**, *5*, 89-114.
16. Watanabe, M.; Watanabe, K.; Suzuki, K.; Nikaido, O.; Ishii, I.; Konishi, H.; Tanaka, N.; Sugahara, T. *Toxic. in Vitro* **1989**, *3*, 329-334.
17. Burton, A. B. G. *Food Cosmet. Toxicol.* **1972**, *10*, 209-217.
18. Burton, A. B. G.; York, M.; Lawrence, R. S. *Food Cosmet. Toxicol.* **1981**, *19*, 471-480.
19. Parish, W. E. *Food Chem. Toxicol.* **1985**, *23*, 215-227.
20. Price, J. B.; Andrews, I. J. *Food Chem. Toxicol.* **1985**, *23*, 313-315.
21. Jacobs, G. A.; Martens, M. A. *Toxic. in Vitro* **1988**, *2*, 253-256.
22. Jacobs, G. A.; Martens, M. A. *ATLA* **1990**, *17*, 255-262.
23. Chessel, D.; Dolédec, S. *ADE Version 3.6: HyperCard® Stacks and Programme Library for the Analysis of Environmental Data.* E mail: chessel@biomserv.univ-lyon1.fr.
24. Whittle, P. *Skand. Aktuar.* **1952**, *35*, 223-239.

RECEIVED December 6, 1994

Chapter 19

Combined Use of Linear and Nonlinear Multivariate Analyses in Structure–Activity Relationship Studies

Application to Chemoreception

Daniel Domine[1,2], James Devillers[1], Maurice Chastrette[2], and Jean-Christophe Doré[3]

[1]Centre de Traitement de l'Information Scientifique, 21 rue de la Bannière, 69003 Lyons, France
[2]Université Lyons-I, Unité de Recherche Associée, Centre National de la Recherche Scientifique 463, 43 Boulevard du 11 Novembre 1918, 69622 Villeurbanne Cedex, France
[3]Musée National d'Histoire Naturelle, Unité de Recherche Associée, Centre National de la Recherche Scientifique 401, 63 rue de Buffon, 75005 Paris, France

This study deals with the combined use of correspondence factor analysis and nonlinear mapping for deriving structure-activity relationships of pheromones in Noctuidae and assessing the value of the chemical structure of female sex pheromones as a character for the classification of Noctuid moths. Graphical tools are used to facilitate the interpretation of the results.

The perception of odors and tastes depends on many factors dealing with physiology, anatomy, and psychology (1-4). The structure and physicochemical properties of the chemicals are also determinant (2, 5-9). The study of these activities requires examination of multidimensional spaces which are not perceivable by humans (10, 11). To solve this problem, numerous display or classification methods have been devised and used (12-16). These can be divided into linear methods (e.g.; correspondence factor analysis (CFA), hierarchical cluster analysis (HCA)) and nonlinear methods (e.g.; nonlinear mapping (NLM), Kohonen's self-organizing maps (KSOM)). Used separately, these methods have shown their efficiency for modeling complex biological activities and deriving structure-activity relationships (SAR) (17-20). They both present specific advantages and drawbacks. However, most published papers involve the use of only one data analysis while it is obvious that combinations of linear and nonlinear

0097–6156/95/0589–0267$12.00/0
© 1995 American Chemical Society

multivariate methods can lead to interesting supplementary and/or complementary information (*11, 21*). The aim of this paper is to stress the usefulness of combining CFA with NLM to extract SAR information from a data matrix dealing with chemoreception in Lepidoptera.

Experimental

The data used in this study (Table I) came from a paper of Renou *et al.* (*22*). These authors tried to assess the value of the chemical structure of female sex pheromones as a new character for the classification of these Lepidoptera and survey SAR of sex pheromones in seven taxa of Noctuidae. For this purpose, they compiled data from the literature and they considered the publications dealing with (i) the identification of the constituent of the natural sex pheromone of a given species of Noctuidae, (ii) the attractive effect of synthetic compounds used either alone or mixed with other compounds during field trapping experiments. Considering each constituent of the pheromone blends separately, they obtained data concerning 44 chemicals and 200 species of Noctuidae. For statistical convenience, they considered 19 subfamilies as basic units and kept the seven which contained enough data to lead to a reliable statistical analysis (Table I).

The statistical and graphical analyses used in this paper were performed with STATQSAR, MacMul, and GraphMu (*23, 24*).

Results and Discussion

In Table I, the 44 chemicals are listed using a formalism commonly used in the pheromone literature. It designates each molecule according to distinctive structural features:
- the stereochemistry of the double bond (Z or E) with its location (i.e.; 5, 7, 9, 11, 12),
- the length of the hydrocarbon chain (i.e.; 10, 12, 14, 16),
- the terminal functional group (i.e.; OH: alcohol, Ald: Aldehyde, Ac: Acetate).
Thus for example, Z7-12: OH is Z7-dodecenol.

The data in Table I indicate how many species of the taxa (columns) are attracted by a pheromone blend or a synthetic attractant comprising the molecules (rows).

The best way to treat this frequency data table (Table I) is CFA (*22, 25, 26*). A complete description of this method can be found in previous papers (*25, 26*). The main advantages of CFA are that it allows to reduce the dimensionality of data matrices and it is particularly adapted to contingency and frequency data tables. CFA is interpreted from the plots of factorial planes drawn from factorial axes

TABLE I. Frequency data table for Noctuidae taxa and pheromones (male attractants)

No.	Molecules of Pheromones and Attractants	Noctuidae						
		Noctuinae	Amphipyrinae	Plusiinae	Cuculliinae	Hadeninae	Heliothinae	Acontiinae
1	Z7-12: OH	5	–	2	–	–	–	–
2	Z7-14: OH	1	–	1	1	–	–	–
3	Z9-14: OH	2	3	–	–	11	–	–
4	Z11-16: OH	1	4	–	1	–	2	–
5	10: Ac	3	–	–	–	–	–	–
6	Z5-10: Ac	14	1	–	–	1	–	1
7	Z7-10: Ac	1	–	–	–	–	–	–
8	12: Ac	5	–	2	–	–	–	–
9	Z5-12: Ac	6	1	3	–	–	–	1
10	E5-12: Ac	2	–	–	–	–	–	–
11	Z7-12: Ac	16	3	27	4	4	–	1
12	E7-12: Ac	6	5	2	1	–	–	–
13	Z9-12: Ac	4	2	1	2	–	–	–
14	E9, 11-12: Ac	2	1	–	–	–	–	–
15	11-12: Ac	1	–	–	–	–	–	–
16	E9-12: Ac	1	–	–	–	–	–	–
17	14: Ac	2	1	1	1	1	–	–
18	Z5-14: Ac	8	–	–	1	–	–	–
19	Z7-14: Ac	10	1	2	1	5	–	1
20	E7-14: Ac	1	–	–	–	–	–	–
21	Z9-14: Ac	29	19	3	15	17	1	1
22	E9-14: Ac	–	1	–	–	2	–	–

Continued on next page

TABLE I. Continued

No.	Molecules of Pheromones and Attractants	Noctuidae						
		Noctuinae	Amphipyrinae	Plusiinae	Cucullinae	Hadeninae	Heliothinae	Acontiinae
23	Z11-14: Ac	2	3	-	1	10	-	-
24	E11-14: Ac	-	1	-	-	-	-	-
25	Z9, E11-14: Ac	-	2	-	-	-	-	-
26	Z9, E12-14: Ac	-	9	-	2	1	1	-
27	16: Ac	2	1	2	2	2	1	1
28	Z7-16: Ac	3	1	-	-	-	1	-
29	Z9-16: Ac	-	-	-	-	1	1	-
30	Z11-16: Ac	16	17	-	10	29	3	1
31	E11-16: Ac	1	-	-	1	-	-	1
32	Z7-12: Ald	2	-	-	-	-	-	-
33	14: Ald	-	-	-	-	-	1	-
34	Z7-14: Ald	2	-	-	-	-	1	-
35	Z9-14: Ald	2	-	-	1	5	2	-
36	Z11-14: Ald	-	-	-	-	3	-	-
37	16: Ald	-	1	-	-	-	2	-
38	Z9-16: Ald	-	1	-	-	1	4	-
39	Z11-16: Ald	2	6	-	4	6	8	-
40	Z9, E12-14: Ac	-	-	-	1	1	-	-
41	Z5-16: Ac	1	-	-	-	-	-	-
42	Z9, Z12-14: Ac	-	1	-	-	-	-	-
43	Z7-16: Ald	-	-	-	-	-	1	-
44	Z5-12: OH	2	-	-	-	-	-	-

considered two at a time. When most of the inertia is explained on the first two factorial axes resulting from the CFA, then the interpretation is very easy, since a simple inspection of only one plane is sufficient. However, in some instances, the percentage of inertia on the first factorial axes is not sufficient to allow a simple interpretation of the results from the sole F1F2 plane. In these cases, it is necessary to consider several maps (e.g.; F1F3, F2F3) and for all points, inspect the relative and absolute contributions which depict the goodness of fit of the points on the different planes. This is a long and tedious task but more important, when performed by non-specialists in statistics, it can lead to erroneous conclusions.

When a CFA is performed on Table I only 60% of the total inertia is extracted from the first two axes (Table II). The corresponding plots are shown in Figures 1 and 2. Inspection of these figures gives interesting SAR information but can lead to erroneous interpretation. Indeed, Table II reveals that five or six axes are necessary to obtain a full account of the information contained in the data matrix (Table I). Therefore, the study of Table I with CFA requires inspection of several factorial planes. As stated above, this kind of interpretation is long and tedious and requires training in statistics.

TABLE II. Eigenvalues and percentages of inertia of the factorial axes derived from CFA

Axis No.	Eigenvalues	Inertia %
1	0.453	37.89
2	0.267	22.34
3	0.203	16.98
4	0.157	13.17
5	0.066	5.56
6	0.049	4.07

To overcome this problem we propose the use of NLM (*27*). A complete description of this method can be found in a previous paper (*11*). Briefly, NLM tries to preserve distances between points in a display space (generally 2-D) as similar as possible to the actual distances in the original space by minimizing a mapping error (E) (*27*). In our study, NLM was used to summarize on a sole plane the information contained on the factorial axes. By taking the factorial coordinates obtained from CFA to run the NLM, one benefits from the particular data treatment obtained with CFA and from the ability of

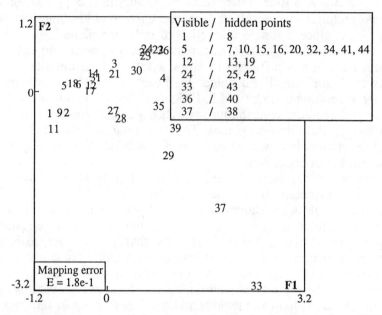

Figure 1. Projection of the chemicals on the factorial plane F1F2. For captions, see Table I.

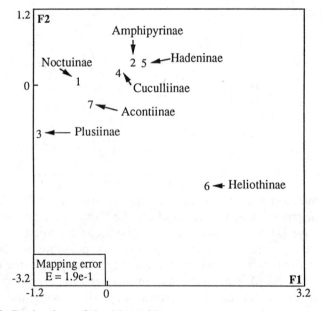

Figure 2. Projection of the Noctuidae taxa on the factorial plane F1F2.

NLM to summarize at best the information contained in a data table on a sole map.

Figure 3 shows the nonlinear map derived from the coordinates of the chemicals on the six factorial axes derived from CFA. The low mapping error (E = 3.9e-2) indicates that the main information is summarized on the map.

We have to underline that there is no criterion by which to judge that the representation on a nonlinear map is better than this obtained by a classical linear multivariate analysis. For comparison purposes, we propose to calculate the mapping error on the factorial plane F1F2 derived from CFA and compare this parameter with the mapping error obtained with NLM. Comparison of these mapping errors (Figures 1 and 3) confirms that the representation of the chemicals given by the factorial plane F1F2 is distorted due to the low percentage of inertia explained on it. It also clearly underlines the usefulness of performing the NLM analysis in order to obtain a sole easily interpretable map.

In the same way, compared with the projection of the taxa on the factorial plane F1F2 (Figure 2), a wider scattering is obtained on the nonlinear map (Figure 4) indicating that the factorial plane F1F2 only gives a distorted representation of the relations between the taxa (Table I). From a practical point of view, the wide scattering of the seven taxa on Figure 4 reveals that pheromones can be used to define new characters for classifying the Noctuidae since they are well separated on the map.

Since interpretation of the nonlinear maps obtained for the chemicals (Figure 3) and for the taxa (Figure 4) requires tedious back and forth comparisons between the maps and the information useful for their interpretation (e.g.; original data, characteristics of the chemicals), graphical tools have been used to facilitate this work (*28, 29*).

Thus, the plot of the original data (Table I) on the chemical map (Figure 3) by means of squares proportional in size to the magnitude of the original values (Figure 5) allows us to summarize all the information contained in Table I. In Figure 5, the larger the square, the larger the frequency. Figure 5.1 allows us to stress that chemicals n° 22, 24-26, 29, 33, 36-38, 40, 42 and 43 have not been cited in the literature as a component of a sex pheromone blend or an attractant for a species of the taxon Noctuinae. As the dienes of the set (except E9, 11-12: Ac (n° 14) which has a weak frequency) are also located in this area, they are not associated with the species of the Noctuinae.

Figure 5.2 shows that the most cited chemicals for Amphipyrinae are Z9-14: Ac (n° 21) and Z11-16: Ac (n° 30). Amphipyrinae is the sole taxon containing species which are associated to the dienes Z9, E11-14: Ac (n° 25) and Z9, Z12-14: Ac (n° 42) and to the E acetate E11-14: Ac (n° 24).

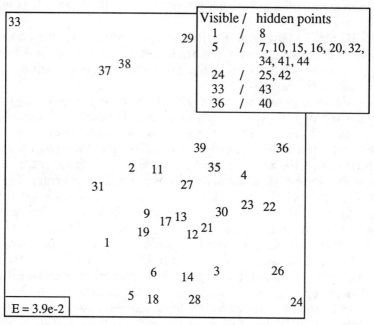

Figure 3. Nonlinear map of the chemicals. For captions, see Table I.

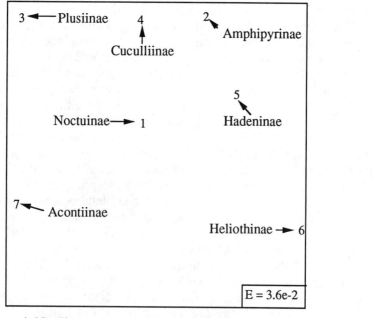

Figure 4. Nonlinear map of the Noctuidae taxa.

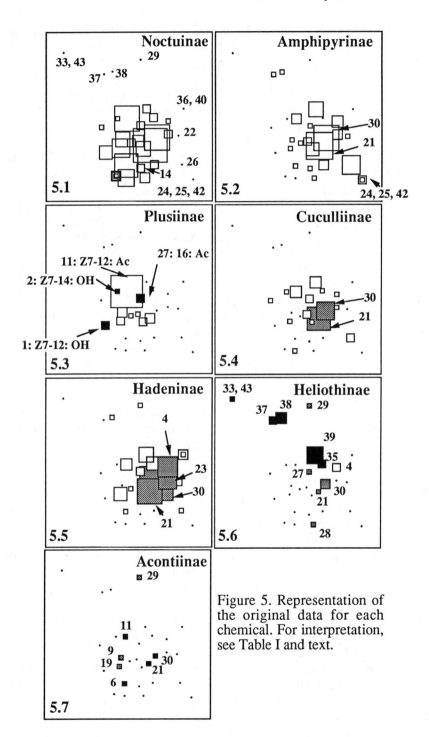

Figure 5. Representation of the original data for each chemical. For interpretation, see Table I and text.

For Plusiinae (Figure 5.3), all the chemicals cited as components of sex pheromone blends or synthetic attractants are located in the center of the figure. Except for the chemicals indicated in black (i.e.; Z7-12: OH (n° 1), Z7-14: OH (n° 2), 16: Ac (n° 27)), all chemicals are C12 or C14 acetates. No C10 chemicals and aldehydes were found to attract Plusiinae. The largest frequency is observed for Z7-12: Ac (n° 11). Indeed, the great majority of sex attractive blends for male Plusiinae contains Z7-12: Ac (22).

Figure 5.4 shows that for Cuculliinae, the chemicals which are the most cited in the literature are generally the same as those of the other taxa (i.e.; Z9-14: Ac (n° 21), Z11-16: Ac (n° 30)).

For Hadeninae (Figure 5.5), the largest squares are observed for chemicals n° 4 (Z11-16: OH), 21 (Z9-14: Ac), 23 (Z11-14: Ac) and especially for compound n° 30 (Z11-16: Ac). For Z11-16: OH (n° 4), the frequencies are lower or null for the other taxa (Figure 5).

For Heliothinae (Figure 5.6), this taxon shows a relatively exclusive response to C16 aldehydes. Z11-16: Ald (n° 39) was the main component of the five species studied up to now (22). All the acetates (colored in grey) except 14: Ac corresponding to the aldehydes are present but their frequency is generally lower. There is no chemical with a C10 or C12 hydrocarbon chain. Most of the chemicals involved contain 16 carbon atoms.

For Acontiinae (Figure 5.7), although the frequencies are weak (i.e.; small squares), it is interesting to note that the chemicals involved are only acetates. This map also allows to stress β-oxidation chain-shortening steps which are known to play an important role in the biosynthesis of pheromones and produce apparent displacement of the considered double bond through a shift of two carbon atoms (22). Thus, in Figure 5.7, one can observe the two following series: (i) Z5-10: Ac (n° 6), Z7-12: Ac (n° 11), Z9-14: Ac (n° 21), Z11-16: Ac (n° 30) indicated in black and (ii) Z5-12: Ac (n° 9), Z7-14: Ac (n° 19), Z9-16: Ac (n° 29) colored in grey.

In a similar way, the plot of the original data (Table I) on the taxon map (Figure 4) allows us to summarize on a simple collection of graphs (Figure 6), the links between the locations of the taxa on the map and the chemicals, and therefore find chemotaxonomical characters for Noctuidae. In Figure 6, the larger the square, the larger the frequency. From a general point of view, it is noteworthy that the acetates and especially the monounsaturated of general formula $Z(x-5)$-x: Ac (i.e.; Z7-12: Ac (Figure 6.11), Z9-14: Ac (Figure 6.21), Z11-16: Ac (Figure 6.30)) are the most common chemicals in Noctuidae since they are present in all taxa (except Heliothinae and Plusiinae in some instances) with relatively large frequencies (large squares). As above, β-oxidation chain-shortening steps can be stressed from these large

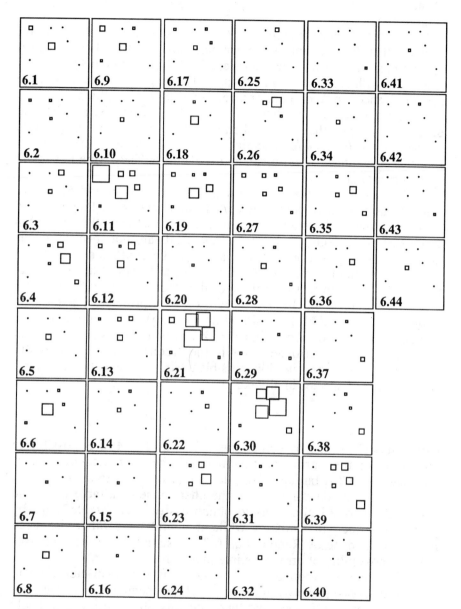

Figure 6. Representation of the original data for each taxon. The number after the dot corresponds to the chemical number (Table I).

frequencies even if Z5-10: Ac (Figure 6.6) is less present in the species of the Noctuid taxa studied and is principally associated with Noctuinae (Figure 6.6).

For the alcohols (Figures 6.1-6.4 and 6.44), it is worth noting for example that the frequencies are generally low and that they are never present in all taxa. Thus, for example, Z7-12: OH (Figure 6.1) was only associated with Noctuinae and Plusiinae. The alcohols under study were never evidenced in Acontiinae and very scarcely in Cuculliinae.

If we focus on the dienes (Figures 6.14, 6.25, 6.26, 6.40, and 6.42), it appears that these chemicals are often associated with Amphipyrinae (except for chemical n° 40) but some of them can be found in Noctuinae, Cuculliinae, and Hadeninae. They were not found in the other taxa.

If we consider the aldehydes (Figures 6.32-6.39 and 6.43), we can see that they have never been cited as attractants for Plusiinae and Acontiinae species. It is also interesting to note that Heliothinae is often associated with aldehydes but these chemicals are not specific to this taxon. The interpretation of this map could be continued in order to stress, for example, specificities for some taxa. This can present an interest in crop protection for designing traps for specific species. Thus, for example, Z5-14: Ac (Figure 6.18) is principally associated with Noctuinae, Z11-14: Ald (Figure 6.36) is associated with Hadeninae, E11-16: Ac (Figure 6.31) was exclusively found in Noctuinae and Cuculliinae even if the frequencies are not really significant. At the opposite, it is possible to stress chemicals active for a large number of species in all taxa.

Conclusions

Our results show that by combining CFA with NLM, it is possible to benefit from the particular treatment of the chemoreception data obtained with CFA but also from the ability of NLM to summarize at best the information contained in the pheromone data table on a sole map. Indeed, NLM allows us to obtain easily interpretable maps by people untrained in statistics while it would have otherwise been necessary, as done by Renou *et al.* (22), to study all the factorial axes with the possibility of erroneous interpretations.

The interpretation of the nonlinear maps is restricted here to finding relationships between the taxa and the chemicals. However, our graphical approach is open and therefore, any other information could be represented on the maps. Thus, for example, it could be possible to find links between the existing taxonomy of Noctuidae and their classification from the responses to pheromones by simply projecting anatomical information on the map of the taxa. In the same way, any information such as habitat could be represented on this map.

It is important to note that our approach does not only apply to the responses of Lepidoptera to pheromones but also to any activity related to chemoreception and more generally to any biological activity.

This paper also demonstrates that there is no single universal method to treat a data matrix and that in most cases a combination of methods allows one to obtain complementary or supplementary information on the relationships within a data matrix.

Literature Cited

1.   Walker, J.C.; Jennings, R.A. In *The Human Sense of Smell*; Laing, D.G.; Doty, R.L.; Breipohl, W., Eds; Springer-Verlag: Berlin, 1992; pp. 261-280.
2.   Walters, D.E. In *Sweeteners. Discovery, Molecular Design, and Chemoreception*; Walters, D.E.; Orthoefer, F.T.; Dubois, G.E., Eds; ACS Symposium Series 450; American Chemical Society: Washington, DC, 1991; pp. 1-11.
3.   Duchamp, A.; Jourdan, F.; Gervais, R. In *Odeurs et Désodorisation dans L'Environnement*; Martin, G.; Laffort, P., Eds; Tec & Doc - Lavoisier: Paris, 1991; pp. 1-24.
4.   Lawless, H.T.; Glatter, S.; Hohn, C. *Chem. Senses* **1991**, *16*, 349-360.
5.   Beets, M.G.J. In *Fragrance Chemistry. The Science of the Sense of Smell*; Theimer, E.T., Ed; Academic Press: New York, NY, 1982; pp. 77-122.
6.   de Saint Laumer, J.Y.; Chastrette, M.; Devillers, J. In *Applied Multivariate Analysis in SAR and Environmental Studies*; Devillers, J.; Karcher, W., Eds; Kluwer Academic Publishers: Dordrecht, 1991; pp. 479-521.
7.   Chastrette, M.; Rognon, C.; Sauvegrain, P.; Amouroux, R. *Chem. Senses* **1992**, *17*, 555-572.
8.   Schiffman, S.S. In *Computers in Flavor and Fragrance Research*; Warren, C.B.; Walradt, J.P., Eds; ACS Symposium Series 261; American Chemical Society: Washington, DC, 1984; pp. 33-50.
9.   Hopfinger, A.J.; Walters, D.E. In *Computers in Flavor and Fragrance Research*; Warren, C.B.; Walradt, J.P., Eds; ACS Symposium Series 261; American Chemical Society: Washington, DC, 1984; pp. 19-32.
10.   Kowalski, B.R.; Bender, C.F. *J. Am. Chem. Soc.* **1972**, *94*, 5632-5639.
11.   Domine, D.; Devillers, J.; Chastrette, M.; Karcher, W. *J. Chemometrics* **1993**, *7*, 227-242.

12. Devillers, J.; Karcher, W. *Applied Multivariate Analysis in SAR and Environmental Studies*; Kluwer Academic Publishers: Dordrecht, 1991.

13. Barnett, V. *Interpreting Multivariate Data*; John Wiley & Sons: Chichester, 1981.

14. Everitt, B.S. *Graphical Techniques for Multivariate Data*; Heinemann Educational Books: London, 1978.

15. Gnanadesikan, R. *Methods for Statistical Data Analysis of Multivariate Observations*; John Wiley & Sons: New York, NY, 1977.

16. Mager, P.P. *Multivariate Chemometrics in QSAR: A Dialogue*; Research Studies Press: Letchworth, 1988.

17. Stone, M.; Jonathan, P. *J. Chemometrics* **1994**, *8*, 1-20.

18. Rose, V.S.; Hyde, R.M.; Macfie, H.J.H. *J. Chemometrics* **1990**, *4*, 355-360.

19. Devillers, J. In *Quantitative Structure/Activity Relationships (QSAR) in Toxicology*; Coccini, T.; Giannoni, L.; Karcher, W.; Manzo, L.; Roi, R., Eds; Commission of the European Communities: Luxembourg, 1992; pp. 27-41.

20. Devillers, J.; Domine, D.; Chastrette, M.; Karcher, W. In *Proceedings of the Fifth International Congress: Ecotossicologia dei Pesticidi, Analitica di Laboratorio, Biotecnologie, Riva del Garda (Italy), September 22-25, 1992*; Landi, E.; Piccolo, A.; Dumontet, S., Eds; Ordine Nazionale dei Biologi, Rome, 1992; pp. 111-140.

21. Chastrette, M.; Devillers, J.; Domine, D.; de Saint Laumer, J.Y. In *Proceedings of the 13th International CODATA Conference, Beijing, 1992.*

22. Renou, M.; Lalanne-Cassou, B.; Michelot, D.; Gordon, G.; Doré, J.C. *J. Chem. Ecol.* **1988**, *14*, 1187-1215.

23. STATQSAR; CTIS: Lyon, France, 1993.

24. Thioulouse, J. *Computers and Geosciences* **1990**, *16*, 1235-1240.

25. Devillers, J.; Karcher, W. In *Practical Applications of Quantitative Structure-Activity Relationships (QSAR) in Environmental Chemistry and Toxicology*; Karcher, W.; Devillers, J., Eds; Kluwer Academic Publishers: Dordrecht, 1990; pp. 181-195.

26. Doré, J.C.; Gilbert, J.; Ojasoo, T.; Raynaud, J.P. *J. Med. Chem.* **1986**, *29*, 54-60.

27. Sammon, J.W. *IEEE Trans. Comput.* **1969**, *C-18*, 401-409.

28. Domine, D.; Devillers, J.; Chastrette, M.; Karcher, W. *Pestic. Sci.* **1992**, *35*, 73-82.

29. Devillers, J.; Thioulouse, J.; Karcher, W. *Ecotox. Environ. Safety* **1993**, *26*, 333-345.

RECEIVED December 6, 1994

Chapter 20

Comparative Quantitative Structure–Activity Relationship

Insect Versus Vertebrate Cholinesterase

Corwin Hansch

Department of Chemistry, Pomona College, Claremont, CA 91711

To construct a science of quantitative structure-activity relationships (QSAR) we must begin to do comparative studies of QSAR in many different systems. In this way we can begin to understand the similarities in different compounds interacting with the same biological system and different biological systems interacting with the same compounds. Examples are given for the inhibition of insect and vertebrate cholinesterase by a variety of chemicals. For the inhibition of vertebrate enzyme the hydrophobic properties of the inhibitors is important, but not for fly enzyme.

Since the initiation of the QSAR paradigm in 1962 (1) thousands of quantitative structure-activity relationships have been published for all kinds of organic compounds acting on all sorts of organisms or parts thereof (DNA, enzymes, organelles, etc.). In addition to these, physical organic chemists have been hard at work since 1935 formulating quantitative relationships for organic reactions in solution. Our own data bank, which is only a fraction of the literature, contains about 6,000 examples about equally divided between biological and organic chemistry. The time has come to begin the serious organization and generalization of this mass of published and still rapidly growing work. Already it is possible to make interesting comparisons and generalizations (2-5). We believe that it is particularly important to tie the less certain biological QSAR to the more firmly based relationships established by physical organic chemists (2,3). Before considering the central topic of this report, it helps to illustrate our general approach with a few diverse examples.

Many years ago (3) we derived equation 1 for the colchicine-like mitosis in onion root tips caused by simple organic solvents such as: alcohols, acetone, $CHCl_3$, xylene, ether, etc.

$$\log 1/C = 0.95 \log P + 0.63 \qquad n = 22, r^2 = 0.916, s = 0.381 \qquad (1)$$

In this equation, C is the molar concentration of chemical producing the standard aberrant mitosis. This seemed a highly esoteric study only of academic interest, still we entered it into our bank. Recently, we came across equation 2 by Onfelt

0097–6156/95/0589–0281$12.00/0
© 1995 American Chemical Society

correlating colchicine mitosis in hamster cells for a different but similar set of compounds (*3*).

$$\log 1/C = 0.85 \log P + 0.27 \qquad n = 10, r^2 = 0.901, s = 0.227 \qquad (2)$$

The plant and mammalian cells behave in very similar ways and one begins to wonder if perturbations of mitosis by solvents could be responsible for some kinds of birth defects.

Compounds of type I have been extensively studied as agents for cancer chemotherapy.

Variations of I, in which Y = Cl, Br, I, OSO_2R, have been shown to react with DNA and other nucleophilic agents. Correlation equations for reactions of I with nucleophilic agents (water, 4-nitrobenzylpyridine, calf serum) are based on σ^- constants of X. The following values of ρ (slope) were found: -1.73, -1.92, -2.02, 2.02, -2.31 (mean = -2.00) (*3*). Electron releasing groups promote the reaction with nucleophiles, presumably by way of the intermediate onium compound.

Studies of the anticancer value of these aniline mustards for mice with various forms of cancer yield more complex QSAR, but with σ^- terms having the following values of ρ: -0.96, -1.31, -1.52, -1.70 (mean = -1.37). In the case of the model systems of H_2O, pyridine and serum the substituents play a more important role (more negative) than in the in vivo reaction. This suggests that in vivo a more reactive nucleophile may be involved. This might not be DNA although DNA is usually assumed to be the target since it has been shown to be alkylated by the mustards. Since the mustards are so highly reactive it seems unlikely that they could avoid reacting with NH_2 or SH groups of important macromolecules.

Equation 3 correlates LD_{50} data for aniline mustards acting on rats (*3*).

$$\log 1/C = -1.38 \, \sigma^- + 0.41 \, I + 4.13 \qquad n = 17, r^2 = 0.865, s = 0.267 \qquad (3)$$

And equation 4 correlates LD_{50} for $X\text{-}C_6H_4SCH_2CH_2Br$ acting on red spider eggs (*3*).

$$\log 1/C = -1.42 \, \sigma^- + 1.49 \, I + 4.44 \qquad n = 14, r^2 = 0.914, s = 0.113 \qquad (4)$$

In equation 3 the indicator variable I = 1 when Y = Br. I = 0 when Y = Cl or I. In equation 4 I = 1 for examples where X = COOR. The esters probably hydrolyze to yield a more hydrophilic carboxylate ion. The lipophilic ester group may be acting as a prodrug unit. The values of ρ for equations 3 and 4 are essentially the same and agree with average value found for the four examples of antitumor activity in mice. It has been assumed that for $X\text{-}C_6H_4SCH_2CH_2Br$ the same type of onium intermediate is essential. The question arises, why are the values of ρ from the model systems higher than for the in vivo systems? It has been shown that the mustards react with lone pair electrons of the DNA bases, but these would not differ much from the benzylpyridine in nucleophilicity. We believe that other nucleophiles, such as NH_2 or SH on critical macromolecules, may also be involved.

Much more complex QSAR have been derived showing the different SAR for benzylpyrimidines inhibiting dihydrofolate reductase from vertebrate and bacterial enzyme (6). It turns out that hydrophobic substituents play a crucial role for inhibition of the vertebrate enzyme, while steric properties are more important for bacterial enzymes. These comparative QSAR point the way to the design of more selective drugs. We now consider the case of cholinesterase inhibitors where we find a similar situation.

Insect Cholinesterase

One of the advantages of the above mentioned studies of dihydrofolate reductase inhibitors is that x-ray crystallographic structures of both types of enzymes support the QSAR. In the case of cholinesterase only the structure of the enzyme from the *Torpedo Californica* has been established (7).

The following QSAR have been derived for the action of the indicated inhibitors of fly cholinesterase.

$$\log 1/C = 2.66 \ \sigma^- - 0.47 \ E_s\text{-}3 + 4.59 \qquad n = 12, r^2 = 0.953, s = 0.341 \tag{5}$$

$$\log 1/C = 2.94 \ \sigma^- - 0.54 \ E_s^c\text{-}3 + 4.78 \qquad n = 15, r^2 = 0.895, s = 0.525 \tag{6}$$

$$\log 1/C = 2.42 \ \sigma^- + 4.31 \qquad n = 7, r^2 = 0.970, s = 0.309 \tag{7}$$

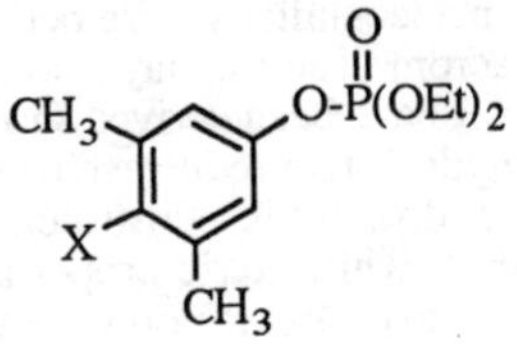

structure (11)

$$\log k_2 = 2.62\ \sigma^- + 0.77\ B5 + 0.84 \qquad n = 10,\ r^2 = 0.962,\ s = 0.435 \tag{8}$$

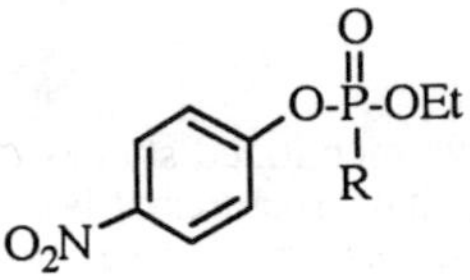

structure (12)

$$\log k_2 = 2.71\ E_s^c + 8.12 \qquad n = 16,\ r^2 = 0.893,\ s = 0.524 \tag{9}$$

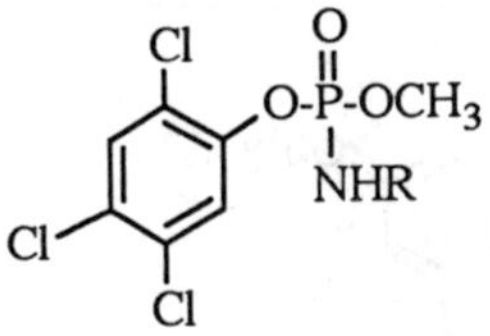

structure (13)

$$\log k_2 = 1.25\ E_s^c + 6.06 \qquad n = 7,\ r^2 = 0.947,\ s = 0.292 \tag{10}$$

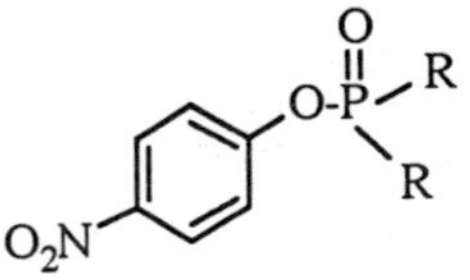

structure (14)

$$\log k_2 = 1.28\ E_s^c - 1.65\ \sigma^* + 7.00 \qquad n = 12,\ r^2 = 0.903,\ s = 0.273 \tag{11}$$

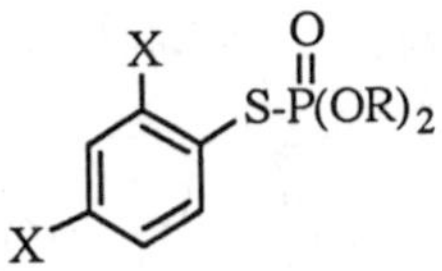

structure (15)

$$\log 1/C = 1.63\ MR_4 + 1.44\ I + 5.90 \qquad n = 12,\ r^2 = 0.876,\ s = 0.397 \tag{12}$$
$$I = 1 \text{ for } R = Et \text{ and } 0 \text{ for } R = CH_2CHMe_2$$

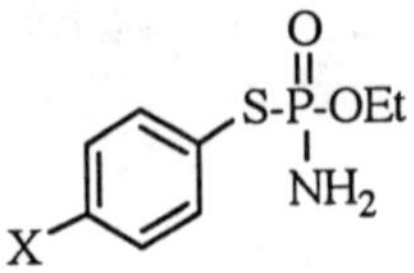

structure (16)

$$\log k_2 = 1.35\ \sigma^- + 1.45 \qquad n = 8,\ r^2 = 0.956,\ s = 0.050 \tag{13}$$

$$\text{X} \overset{\displaystyle }{\underset{\displaystyle }{\bigcirc}} - \overset{\displaystyle S}{\underset{\displaystyle OEt}{\overset{\displaystyle \|}{O\text{-}P}}}\text{-}C_6H_5 \quad (17)$$

$$\log 1/C = 2.13 \, \sigma^- + 0.69 \, MR_3 + 4.35 \qquad n = 11, \, r^2 = 0.928, \, s = 0.314 \qquad (14)$$

$$\overset{\displaystyle O}{\underset{\displaystyle X}{\bigcirc}} \overset{\displaystyle \|}{O\overset{}{C}NHCH_3} \quad (18)$$

$$\log 1/C = 0.56 \, MR_{3,4,5} + 1.56 \, MR_2 - 0.61 \, E_s\text{-}3 - 0.94 \, \sigma^2 + 1.43 \, CHG$$
$$- 0.23 \, MR^2 - 5.24 \, (F\text{-}2,6)^2 + 3.47 \, F_{2,6} + 0.66 \, RGMR - 0.62 \, HB$$
$$- 0.052 \, (MR_3)^2 - 0.56 \, E_s\text{-}2 \cdot E_s\text{-}6 + 3.46$$
$$n = 269, \, r^2 = 0.796, \, s = 0.485 \qquad (15)$$

Observe that in equations 5 through 15 no hydrophobic terms appear, only steric and electronic factors seem important. Equations 5, 6 and 15 are of interest because of the negative E_s term for meta substituents (E_s-3). The coefficients are essentially the same for these two different types of inhibitors (carbamates and phosphates). Recall that values of E_s or Hancock's modified E_s^c values are negative. This tells us that bulky groups in this position produce a <u>positive</u> steric effect. This important feature of inhibitors is also brought out by the MR for meta substituents in equations 19 and 20. The sterimol parameter B5 uncovers (equation 8) a positive steric effect for substituents flanked by methyl groups. It is difficult to interpret this term since one would expect that groups such as NO_2 would be less effective if twisted out of the ring plane. Nevertheless, σ^- is the parameter of choice and its ρ value is in the usual range. The values of ρ for the comparable inhibitors of equations 5, 6, 7, 8, 14, 16 and 19 are: 2.66, 2.94, 2.42, 2.62, 2.13, 2.87 and 2.13 with mean = 2.54.

Probably the most informative set of probes for the hydrophobic character of the fly cholinesterase is the large set of carbamates of equation 15. Despite a determined effort no evidence for a hydrophobic interaction could be uncovered. Again a negative E_s-3 term with a coefficient similar to those of equations 5 and 6 was found. This hallmark of meta substituents suggests that the carbamate binding is not unlike that of the phosphates.

Devising a mathematical relationship correlating how a set of organic chemicals affects a receptor of unknown structure and unknown flexibility is every bit as hard as it sounds. We have no absolute standards, no idea of what the best parameters are. We are still in the exploratory stage. At this point in time our best guide is not statistics, as necessary as they are, but correlating with other QSAR to confirm that what we are finding is not artifactual. The uniformity of ρ and positive steric effect of meta substituents assure us that we are on the right track. The rather poor correlation of equation 15 shows us that we have much that we still do not understand.

Now consider the action of the inhibitors on flies brought out by equations 16 through 20 for the LD_{50} against house flies.

$$\log 1/C = 2.87\ \sigma + 3.14 \qquad n = 7, r^2 = 0.955, s = 0.304 \tag{16}$$

$$\log 1/C = -0.67\ \pi + 5.94 \qquad n = 10, r^2 = 0.861, s = 0.253 \tag{17}$$

$$\log 1/C = -0.34\ \log P + 1.84\ \sigma^- + 6.87 \qquad n = 25, r^2 = 0.904, s = 0.363 \tag{18}$$

$$\log 1/C = 2.13\ \sigma^- + 0.69\ MR_3 + 4.35 \qquad n = 11, r^2 = 0.927, s = 0.314 \tag{19}$$

$$\log 1/C = 0.58\ \log P - 0.16\ (\log P)^2 + 0.40\ MR_3 + 2.73\ F\text{-}2 - 3.45\ (F\text{-}2)^2$$
$$- 0.54\ \sigma^- + 7.90$$
$$n = 47,\ r^2 = 0.870,\ s = 0.150,\ \log P_o = 1.8 \tag{20}$$

In eq 20, F-2 is field/inductive parameter for ortho substituents, while σ^- applies to other substituents.

Even with the whole fly the hydrophobic terms are weak except for the more hydrophilic carbamates of equation 20. Equations 16 and 19 lack these terms and in equations 17 and 18 the hydrophobic terms have negative coefficients. In equation 19 the MR_3 term is comparable to the E_s terms in equations 5, 6 and 15 and the MR term in equation 14. The optimum F of equation 20 is about 0.40. Smaller and larger values decrease activity. Optimum values of electronic terms in equation 15 are also low. Thus, the electronic effect of substituents on carbamates is vastly different from that on phosphates. Note that the hydrophobic terms in the whole fly equations account for inhibitor interactions with material other than enzymes (proteins, lipids).

Vertebrate Cholinesterase

There are a large number of studies of the inhibitor of cholinesterase from various vertebrates, our present data base contains 40 examples of enzymes from eel, human, rabbit, sheep and bovine plasma and erythrocytes which all contain hydrophobic terms. The following are representative examples.

Inhibition of horse erythrocyte enzyme by

$$\log k_2 = 1.75 \log P - 0.23 (\log P)^2 - 6.24$$
$$n= 12, r^2 = 0.845, s = 0.166, \log P_o = 3.0 \tag{21}$$

Inhibition of human plasma enzyme by

$$\log 1/K_i = 0.38 \, \pi + 3.58 \qquad n = 10, r^2 = 0.980, s = 0.059 \tag{22}$$

Inhibition of human plasma enzyme by

$$\log 1/C = 0.74 \, \pi + 4.12 \qquad n = 9, r^2 = 0.970, s = 0.103 \tag{23}$$

Inhibition of electric eel enzyme by

$$\log 1/K_i = 0.96 \, \pi + 0.44 \, E_s + 1.12 \, HB + 3.83 \qquad n =13, r^2 = 0.968, s= 0.190 \tag{24}$$

Inhibition of bovine erythrocyte enzyme by

$$\text{X-C}_6\text{H}_4\text{O}\overset{\overset{\displaystyle O}{\displaystyle \|}}{\text{C}}\text{NHCH}_3 \qquad (25)$$

$\log 1/K = 1.40\,\pi\text{-}2,3 + 0.30\,\pi_4 + 1.66\,\sigma_1^o - 1.78\,\sigma_2^o + 0.17\,E_s + 0.77\,F\text{-}2$
$\qquad + 1.36\,HB + 0.07$
$\qquad n = 53,\ r^2 = 0.897,\ s = 0.238$ $\qquad\qquad\qquad (25)$

The above QSAR was first published using MR (molar refractivity) instead of π (25). We have found that π yields a better correlation. The parameter σ^o is derived from phenylactic acids and applies where resonance is not involved. For all instances where positive σ^o values occur, these substituents are parameterized by σ_1^o, and for negative values σ_2^o is used. F and E_s apply only to ortho substituents and HB is a hydrogen bonding constant for certain substituents. It is interesting to compare equations 15 and 25. No evidence could be found for using hydrophobic terms in equation 15, but in equation 25 they are the most important parameters. However, there is collinearity between π and MR so that steric effects may also be involved in equation 25. Since the positive σ_1^o term tends to cancel σ_2^o, we again find electronic effects to be small with carbamates.

Following are a few examples of QSAR for vertebrate enzymes which do not contain hydrophobic terms.

Inhibition of rat brain enzyme by

$\log 1/C = 0.73\,\sigma - 0.40\,E_s\text{-}3 + 3.42 \qquad n = 10,\ r^2 = 0.884,\ s = 0.199$ $\qquad (26)$

Inhibition of horse serum enzyme by

$$\text{X-C}_6\text{H}_4\overset{\overset{\displaystyle O}{\displaystyle \|}}{\text{C}}\text{OCH}_2\text{CH}_2\overset{+}{\text{N}}(\text{CH}_3)_3 \qquad (27)$$

$\log K = 1.14\,\sigma^+ - 0.02 \qquad n = 6,\ r^2 = 0.922,\ s = 0.187$ $\qquad\qquad (27)$

Inhibition of bovine enzyme by

$$\text{X-C}_6\text{H}_4\overset{\overset{\displaystyle CF_3}{\displaystyle |}}{\text{C}}=\text{NO}\overset{\overset{\displaystyle O}{\displaystyle \|}}{\text{C}}\text{NHCH}_3 \qquad (28)$$

$\log 1/C = 1.60\,F\text{-}2 + 1.86\,MR\text{-}2 + 6.47 \qquad n = 12,\ r^2 = 0.857,\ s = 0.301$ $\qquad (28)$

Inhibition of bovine erythrocyte enzyme by

$$\begin{array}{cc} CH_3 & O \\ | & || \\ X\text{-}C_6H_4N\text{-}CH_2CH_2SP(OEt)_2 & (29) \end{array}$$

$$\log 1/K_i = 1.73\ \sigma + 4.03 \qquad n = 7,\ r^2 = 0.836,\ s = 0.191 \tag{29}$$

Why hydrophobic terms are lacking in equations 26 through 29 is not clear. The X-ray crystallographic structure of the enzyme from *Torpedo Californica* clearly shows a large hydrophobic pocket which rationalizes the many QSAR from vertebrate enzyme containing hydrophobic terms. The E_s term in equation 26 is similar to that seen in equations 5, 6 and 15. It is similar to the MR term in equations 19 and 20. However, it is different from the E_s term in equation 25. A point which must be kept in mind is that the variation in the size of the substituents in most of the studies is not large. Increasing the size of substituents will sooner or later change the shape of a particular equation.

Equation 30 illustrates the different interaction of substrates of the type $RCH_2CH_2OCOCH_3$ (*30*) with enzyme from *Electrophorous electricus*.

$$\log k_{cat}/K_m = 1.31\ \sigma_I + 1.64\ MR - 4.34 \qquad n = 14,\ r^2 = 0.941,\ s = 0.302 \tag{30}$$

Examples where R is a charged group (eg. $(CH_3)_3N^+$) or neutral groups (eg. $(CH_3)_3C$) are equally well fit. Clearly R is not binding in hydrophobic space. There may well be more than one mode of binding. In *Electrophorous electricus* enzyme there is a very large hydrophobic active site. Molecular modeling has obtained evidence which rationalizes equation 24 (*31*).

In summary, the above analysis leads us to believe that there is a very significant difference in the hydrophobic character of fly (and possibly other insects) and vertebrate cholinesterase. It seems likely that advantage could be taken of this difference in the design of more selective insecticides.

The above brief review shows that the studies of pesticides initiated by Fukuto and Metcalf (*11-14*) in the fifties using Hammett constants were quite prescient.

Parameter Glossary

The values for the various physicochemical parameters have been taken from reference 32 where there is a more detailed discussion of their origin and use.

σ: Normal Hammett constant defined from the ionization of benzoic acids:

$\sigma_X = \log K_X - \log K_H$ where X refers to a substituted benzoic acid and H to benzoic acid.

σ⁻:   Defined analogously to σ using the ionization constants of phenols. Used when there is delocalization of a negative charge between substituent and reaction center.

σ⁺:   Defined from the S_N1 solvolysis of $X\text{-}C_6H_4C(CH_3)_2Cl$. Used when there is delocalization of a positive charge between substituent and reaction center.

F and σ_I: Different versions of field/inductive constants

P: Octanol/water partition coefficients

π: Hydrophobic parameter for substituents, $\pi_X = \log P_X - \log P_H$

E_s: Taft's steric parameter based on the acidic hydrolysis of $XCOOCH_3$, $E_s = \log k_X - \log k_H$

E_s^c: Hancock's form of E_s corrected for hyperconjugation

MR: Molar refractivity, $MR = \left(n^{2-1}/n^2 + 2\right)\dfrac{MW}{d}$ where n = index of refraction, MW = molecular weight and d = density. Largely a measure of bulk because of the overriding importance of $\dfrac{MW}{d}$

HB: Hydrogen bonding indicator variable.

The numbers associated with the above parameters indicate the position of the substituent on the aromatic ring.

Literature Cited

1.   Hansch, C.; Maloney, P.P.; Fujita, T.; Muir, R.M. *Nature* **1962**, *194*, 178-180.
2.   Hansch, C. *Acc. Chem. Res.* **1993**, *26*, 147-153.
3.   Hansch, C.; Telzer, B.R.; Zhang, L. *Critical Rev. Toxicol.* in press.
4.   Debnath, A.K.; Shusterman, A.J.; de Compadre, R.L.L.; Hansch, C. *Mutat. Res.* **1994**, *305*, 63-72.
5.   Hansch, C.; Kim, D.; Leo, A.J.; Novellino, E.; Silipo, C.; Vittoria, A. *Critical Rev. Toxicol.* **1989**, *19*, 185-226.
6.   Selassie, C.D.; Li, R-L.; Poe, M.; Hansch, C. *J. Med. Chem.* **1991**, *34*, 46-53.
7.   Sussman, J.L.; Harel, M.; Frolow, F.; Oefner, C.; Goldman, A.; Toker, L.; Silman, I. *Science* **1991**, *253*, 872-879.
8.   Hansch, C. *J. Org. Chem.* **1970**, *35*, 620-621.
9.   Relimpio, A.M. *Gen. Pharmacol.* **1978**, *9*, 49-59.
10. Darlington, W.A.; Partos, R.D.; Ratts, K.W. *Toxicol. Appl. Pharmacol.* **1971**, *18*, 542-547.
11. Metcalf, R.L.; Fukuto, T.R.; Frederickson, M. *J. Agric. Food Chem.* **1964**, *12*, 231-236.
12. Fukuto, T.R.; Metcalf, R.L.; Winton, J. *J. Econ. Entomol.* **1959**, *52*, 1121-1127.
13. Fukuto, T.R.; Metcalf, R.L.; Winton, J.; March, R.B. *J. Econ. Entomol.* **1963**, *56*, 808-810.
14. Fukuto, T.R.; Metcalf, R.L.; Winton, J. *J. Econ. Entomol.* **1961**, *54*, 955-962.
15. Murdock, L.L.; Hopkins, T.L. *J. Agric. Food Chem.* **1968**, *16*, 954-958.
16. Sanborn, J.R.; Fukuto, T.R. *J. Agric. Food Chem.* **1972**, *20*, 926-930.

17. Steurbaut, W.; Dejonckhoere, W.; Kips, R.H. In *Crop Protection Agents*; McFarlane, N.R. Ed.; Academic Press, 1977, 79-85.
18. Goldblum, A.; Yoshimoto, M.; Hansch, C. *J. Agric. Food Chem.* **1981**, *29*, 277-288.
19. Fukuto, T.R.; Metcalf, R.L. *J. Am. Chem. Soc.* **1959**, *81*, 372-377.
20. Kamoshita, K.; Ohno, I.; Fujita, T.; Nishioka, T.; Nakajima, M. *Pestic. Biochem. Physiol.* **1979**, *11*, 83-103.
21. Brestkin, A.P.; Brik, I.L.; Godovikov, N.N.; Kabachink, M.I.; Teplov, N.E. *Izv. Akad. Nauk. Khim.* **1967**, 1854-1857, E.E.
22. Christian, S.T.; Gorodetzky, C.W.; Lewis, D.V. *Biochem. Pharmacol.* **1971**, *20*, 1167-1182.
23. Purcell, W.P. *J. Med. Chem.* **1966**, *9*, 294-297.
24. Miyagawa, H.; Fujita, T. *Farmaco* **1982**, *37*, 797-804.
25. Nishioka, T.; Fujita, T.; Kamoshita, K.; Nakajama, M. *Pestic. Biochem. Physiol.* **1977**, *7*, 107-121.
26. Cavallito, C.J.; Yun, H.S.; Kaplan, T.; Smith, J.C.; Foldes, F.F. *J. Med. Chem.* **1970**, *13*, 221-224.
27. Omerod, W.E. *Biochem. J.* **1953**, *54*, 701-704.
28. Rosenfeld, D.D.; Kilsheimer, J.R. *J. Agric. Food Chem.* **1974**, *22*, 926-930.
29. Brestkin, A.P.; Brik, I.L.; Ginetsinskaya, L.I.; Godovikov, N.N.; Kabachink, M.I.; Teplov, N.E. *Izv. Akad. Nauk. Khim.* **1968**, 1966-1968, E.E.
30. Hasan, F.B.; Cohen, S.G.; Cohen, J.B. *J. Biol. Chem.* **1980**, *255*, 3898-3904.
31. Selassie, C.D.; Klein, T.E. in *3D QSAR in Drug Design: Theory, Methods and Applications.* Kubinyi, H., Ed., ESCOM Science Publishers, Leiden, Netherlands, 1993, p. 257-275.
32. Hansch, C.; Leo, A. *Substituent Constants for Correlation Analysis in Chemistry and Biology*, Wiley-Interscience, 1979.

RECEIVED November 21, 1994

Chapter 21

Effect of Tautomeric Equilibria on Hydrophobicity as Measured by Partition Coefficients

Albert J. Leo

Department of Chemistry, Pomona College, Claremont, CA 91711

The tautomeric form of a solute which is preferred by any given non-polar solvent depends largely on it's hydrogen bond donor/acceptor strength and it's polarity--the same properties which determine its partition coefficient, P, between water and that solvent (1). Octanol/water and chloroform/water log P values can give useful predictions of tautomeric ratios, and help predict transport rates and bioaccumulation of pesticides and drugs that fall into this class. Alkane/water and ester/water log P values complete the "critical quartet" (2) and can describe more completely solute-solvent behavior.

It is to be expected that the tautomeric forms of a given drug or pesticide molecule will react differently at the receptor site and thus they may evoke different biological responses. Some effort has been made to determine the effect of the solvent upon the tautomeric equilibrium constant of some important bioactive molecules, such as the purine bases (3). See Figure 1A. It is not surprising that a non-polar solvent can drive the equilibrium away from the form 'normally' expected in an aqueous medium. It can thus greatly influence the tendency of the solute to self-associate. Both tautomers of the 6-dimethylamino analog can dimerize, but a higher degree of association is not favored. As is noted for almost all enol/keto tautomers, the enol is favored by non-polar solvents and the keto by polar ones. The parent isoguanine can associate beyond the dimer stage. It forms helical gels in aqueous solution, and it surely is in the keto form as it does this (Figure 1B). The capability of existing in two tautomeric forms can affect the rate of the 'random walk' process which is needed to reach the active site. Depicted in a very simplified fashion in Figure 2, a tautomeric solute might traverse the serum or cytosol largely as the keto form, but be more effective in crossing a non-polar membrane as the enol. UV and IR absorption spectra were employed to obtain the results shown in Figure 1, and, of course, these have been the preferred techniques (along with NMR) to determine tautomeric ratios. However, interpretation of these spectra are not always as straightforward as one might hope for, as will become apparent later in this paper, and so results from an entirely different methodology can be a useful addition to the picture.

0097–6156/95/0589–0292$12.00/0
© 1995 American Chemical Society

(A)

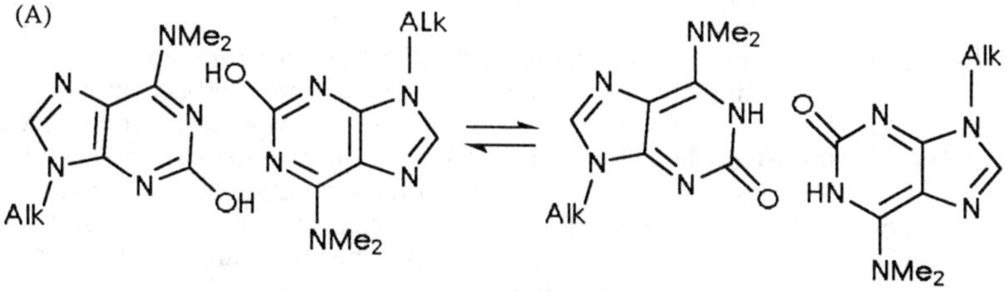

Enol favored by
Non-Polar Solvents

90% in Ethyl Acetate

Keto favored by
Polar Solvents

90% in Water

Dimerization Negligible at 10^{-4} M.

Other Solvents Studied: ethyl acetate   chloroform   DMSO
                         acetonitrile   ethanol   methanol
                         formamide

(B)

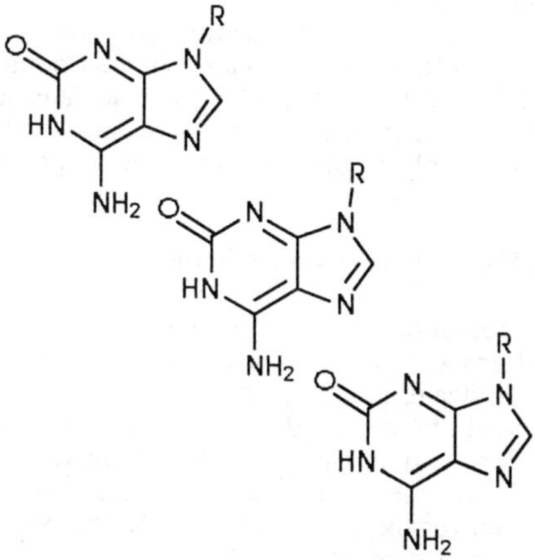

Figure 1. Tautomerism and Association in Biologically Important Solutes

Tautomers Stabilized by Internal Hydrogen Bonds: β-Dicarbonyl Solutes

There are well-established methods of assigning fragmental values to 'isolated' carbonyl, hydroxyl, amino and imino groups so that their sum contributes to an estimate of the octanol/water log P (log P_{oct}) of the solute (4,5). When these groups are present in a tautomeric solute, the observed log P_{oct} will most generally lie somewhere in between the values calculated for the two structural forms, with the enol form calculating higher due to its lesser hydrogen-bond-acceptor (HBA) strength, b. See Figure 3. In the case of acetylacetone, where spectral evidence indicates that about 83% of the solute is in the keto form in water (6), the calculated (5) log Ps bracket the observed value but seem to indicate that the wet octanol may favor the enol form. Even at relatively high concentrations the octanol/water partition coefficient is constant over a ten-fold concentration range, indicating that *intra*molecular hydrogen bonding must prevail over self-association via *inter*molecular bonds.

A reliable fragmental procedure is not available for the calculation of chloroform/water log P values. Spectral evidence indicates that the tautomeric ratio for acetylacetone in chloroform is just about the reverse of what it is in water; i.e., 86% is in the enol form.

In the case of ethylacetoacetate, spectral evidence (6) points to a very small enol content in water. It must not be much higher in wet octanol, because the CLOGP value (5) for the keto form is even a bit *higher* than the measured log P_{oct} and the CLOGP value for the enol form is much higher still. Mills and Beak (6) use NMR data to calculate an 8% enol content in chloroform. This seems a bit strange, since the chloroform/water partitioning data does show a concentration dependence, indicating that some self-association must be taking place. This could hardly occur without the enol form being present at significant levels.

The measured log P_{oct} for diethylmalonate is also very close to the calculated value for the di-keto form and far from the enol form. This indicates that in wet octanol diethylmalonate is mostly in the keto form that water also favors. There is no concentration dependence in the partition coefficients in either octanol/water or chloroform/water indicating an absence of association.

A later section will deal with the calculation of effective hydrogen bond donor strength (HBD) from octanol and chloroform log P values. At this point one can note that for all three compounds in Figure 3 the effective HBD calculates as zero or slightly negative. However, this is not a true indicator of possible enol level, because intramolecular H-bonding would effectively negate any tendency to donate to the solvent.

Tautomerism Where No Internal H-Bond is Possible (Figure 4)

4-Hydroxypyridine is known to be predominantly in the pyridone form in the vapor state and in most polar solvents. If any solvent promotes association, it must be of the head-to-tail variety. The octanol/water log P of the keto form is calculated satisfactorily, while the enol form would be expected to be over two log units more lipophilic. This is strong evidence for the dominance of pyridone in wet octanol as well as water. No data is available to evaluate the enol/keto ratio in chloroform.

Roughly the same results are seen for the 4-quinolone where the octanol/water partitioning data indicate that wet octanol supports as little as 10% as the enol form. However, based on estimation of the effective HBD strength (ea, as explained below), it would seem that the enol form dominates in chloroform. This is not unexpected, since it has been proposed that benzene ring annelation should shift the equilibrium toward the enol form (7).

Wheland (8) reports that the 9-anthrone is the more stable of the two tautomers, even though each can be separately isolated. It is difficult to measure the ratio in

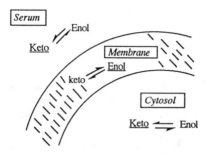

Figure 2. Tautomerism Affects "Random Walk" of Bioactive Solute to Enzyme Site

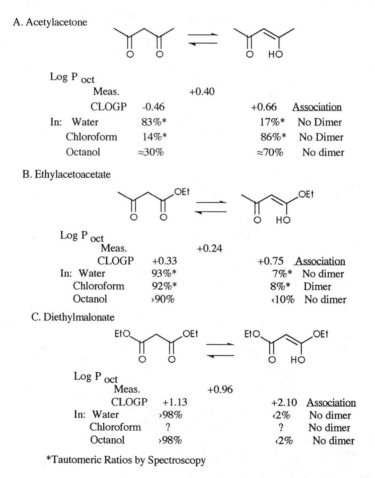

A. Acetylacetone

Log P oct

Meas.		+0.40		
CLOGP	-0.46		+0.66	Association
In: Water	83%*		17%*	No Dimer
Chloroform	14%*		86%*	No Dimer
Octanol	≈30%		≈70%	No dimer

B. Ethylacetoacetate

Log P oct

Meas.		+0.24		
CLOGP	+0.33		+0.75	Association
In: Water	93%*		7%*	No dimer
Chloroform	92%*		8%*	Dimer
Octanol	›90%		‹10%	No dimer

C. Diethylmalonate

Log P oct

Meas.		+0.96		
CLOGP	+1.13		+2.10	Association
In: Water	›98%		‹2%	No dimer
Chloroform	?		?	No dimer
Octanol	›98%		‹2%	No dimer

*Tautomeric Ratios by Spectroscopy

Figure 3. Tautomerism in β-Di-Carbonyl Solutes

water, but in the polar solvent, DMSO, both U.V. and I.R. measurements seem to show a definite preference for the *enol* form (*6,9*). This is surprising since the same investigators find that in non-polar solvents, such as carbon tetrachloride and cyclohexane, spectral data indicate a negligible fraction of enol present by these methods. This is nearly the reverse for the b-dicarbonyls where non-polar solvents definitely favor the enol form and polar solvents favor the keto. With the highly polar DMF as solvent, U.V. finds the enol slightly favored (*6*) while I.R. and NMR finds the keto highly favored (*9*). Obviously there is some inconsistency in the determination of the tautomeric ratio by spectroscopic measurements, and so it is interesting to compare them with estimation from partitioning measurements.

Estimation of the anthrol/anthrone ratio by comparison of calculated and measured octanol/water partition coefficients (Figure 4) would lead one to expect nearly 50% of each in wet octanol. This is consistent with the values found in DMSO. In chloroform, anthrone is seen by U.V. as *entirely* in the keto form, and by I.R. it is calculated as nearly that high--88%. However, partitioning between chloroform and water leads one to predict that the enol form predominates in chloroform. This is consistent with finding for acetylacetone but is at odds with the spectral data just cited.

As seen in Figure 5, the difference between log P_{oct} and log P_{clf} (log P chloroform/water) has been found to depend largely upon the solute HBD strength, a, and the solute volume (Taft, R.; Leo, A.; Anvia, F.; Vasanwala, R.; Raczynska, E. in press). Both the anthrol and anthrone tautomeric forms of the solute fit the criteria upon which Equation (1) in Figure 5 is based. The effective HBD strength of any solute, ea, is then defined by Equation (2) of Figure 5, which can easily be calculated from the difference in the two log Ps and the simple molecular volume, V_X By this method, phenol has a value of 0.60, while a_2^H from the 1:1 complex in CCl4 is 0.58. 1-naphthol is seen to have a HBD strength of 0.67, which is in reasonable agreement with the directly determined a_2^H value of 0.61. By contrast, the ea values calculated for 9-anthrol is much greater, indicating: (1) the fraction of keto form in the chloroform phase cannot be very significant; and (2) benzo-fusion enhances HBD strength. Obviously there is a discrepancy between the conclusions reached from the spectroscopic data and those deduced from chloroform/water partitioning. Further investigation would seem worthwhile to determine which method is in error.

As seen in Figure 6, the determination of HBD strength of 2-pyridinol/2-pyridone supports the prevalence of the pyridone form in chloroform. An enhanced NH is expected to have a HBD strength in the vicinity of that found, 0.40, while an enhanced phenolic OH should be much higher. In this case, NMR data also seems to support the pyridone form in (deutero)chloroform.

The picture is quite different for 4-quinolol/4-quinolone, where the HBD strength by differential partitioning is seen to be much higher than any enhancement of NH could give. This is solid evidence that the 4-quinolol is dominant in the chloroform phase.

When a chlorine is substituted at the 6- position of 2-pyridinol (*7b*), the HBD strength is increased beyond that which is expected for the enhanced NH as seen in Figure 7. This constitutes evidence for some enol form being present in chloroform. The measured octanol/water value falls midway between the calculated value for the two forms, indicating that even the relatively polar wet octanol supports considerable amount of 2-pyridinol when the electrophilic chlorine is present. This type of enol enhancement has been noted previously by Katritzky et al. (*7*).

It might be expected that only the enol form of 3-pyridinol need be considered, and it has been shown to predominate in dioxan solutions (*7*). The rather odd keto structure on the right of Figure 8 has been postulated as one component (*11*), but this has not received much support. Since the measured log P_{oct} is much lower than predicted for the simple enol (D = -0.45), it appears likely that the zwitterionic form might contribute as much as 25% to solute properties in this solvent pair. The

A. 4-Pyridinol/Pyridone

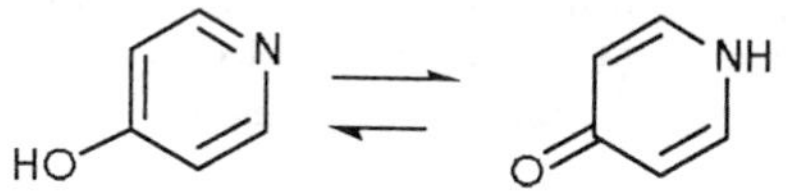

Log P$_{oct}$
    Meas.   -1.30
    CLOGP   +0.93   -1.28
In:
    Water   ‹2%   ›98%
    Octanol   ‹2%   ›98%

B. 4-Quinolol/Quinolone

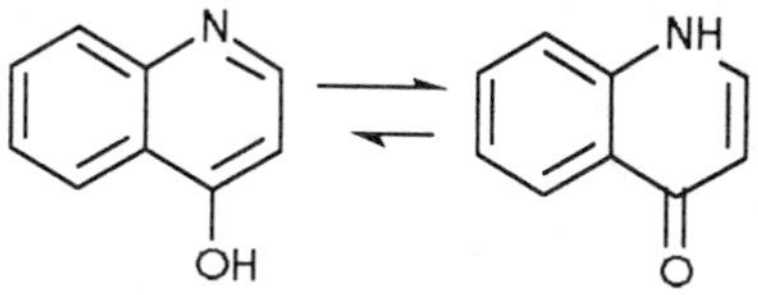

Log P$_{oct}$
    Meas.   +0.58
    CLOGP   +2.32   +0.31
In:
    Water   ‹3%   ›97%
    Octanol   10-15%   85-90%
    Chloroform   >>50%?   <<50%?

C. 9-Anthrol/Anthrone

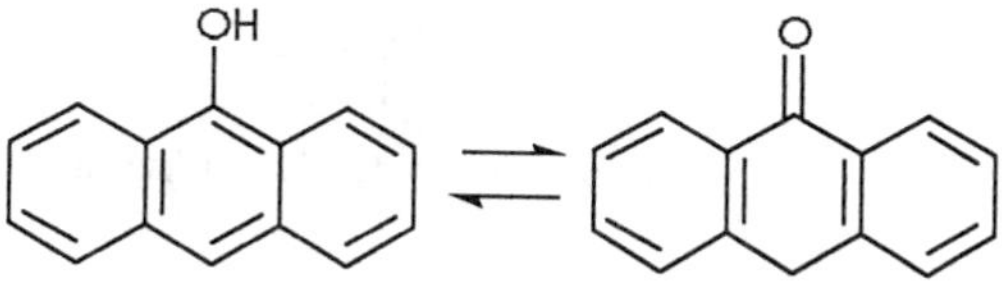

Log P$_{oct}$
    Meas.   3.66
    CLOGP   3.34   3.82
In:
    DMSO   60%*,73%**   40%*,27%**
    DMF   57%*,8%**   43%*,92%**
    Octanol   ≈50%   ≈50%
    Chloroform   12%**   88%**
                ≈100%   ≈0%

Tautomeric Ratio by: *U.V.; **I.R. or NMR

Figure 4. Tautomers with no Intramolecular Hydrogen Bond

$$\text{Log P(Oct/Clf)} = 3.23 \; \alpha_2^H - 1.00 \; V_x/100 \; - 0.03 \qquad (1)$$

$$n = 81; \quad r = 0.992; \quad s = 0.12$$

α_2^H is HBD descriptor from 1:1 complex in CCl4 using series of reference bases; $V_x/100$ is McGowan molecular volume[2] $\epsilon\alpha$ is effective HBD value applicable to multiple donor solute

36 of the solutes have single HBD sites.
45 have no HBD sites, but have single & multiple HBA sites

$$\epsilon\alpha = [\log \text{P(Oct/Clf)} + 1.00(V_x/100) + 0.03] \; / \; 3.23 \qquad (2)$$

1-Naphthol:
$$\epsilon\alpha = [2.84 - 1.82 + 1.144 + 0.03] \; / \; 3.23 = \mathbf{0.67} \qquad (3)$$

9-Anthrol:
$$\epsilon\alpha = [3.61 - 2.50 + 1.513 + 0.03] \; / \; 3.23 = \mathbf{0.82} \qquad (4)$$

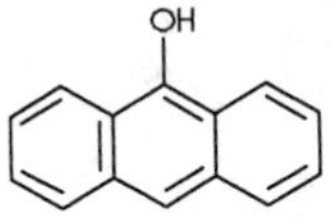

Figure 5. Estimating Effective Solute H-Donor Strength of Tautomers

2-Pyridinol/Pyridone:
$$\epsilon\alpha = [-0.58 - (-1.12) + 0.734 + 0.03] \; / \; 3.23 = \mathbf{0.40} \qquad (5)$$

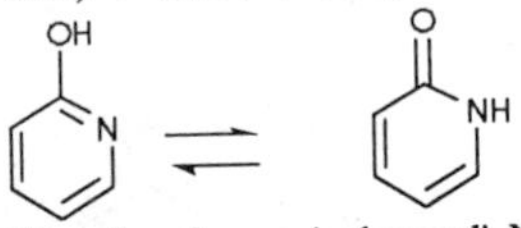

Close to the HBD value for an 'enhanced' NH since:

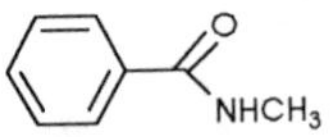

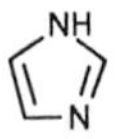

$\epsilon\alpha = $ 0.31   0.41

NMR supports pyridone form in CDCl3

4-Quinolol/Quinolone:
$$\epsilon\alpha = [0.58 - (-1.10) + 1.103 + 0.03] \; / \; 3.23 = \mathbf{0.87} \qquad (6)$$

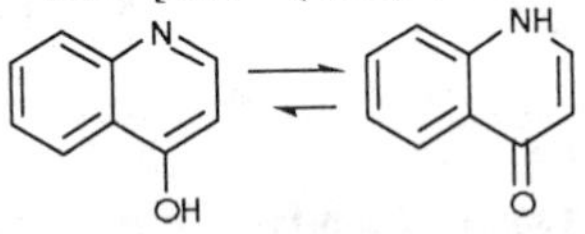

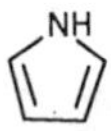

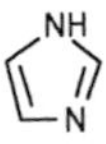

$\epsilon\alpha = $ 0.14   0.41

Supports HBD as 'enhanced' OH since $\epsilon\alpha$ is over 2X that of 'enhanced' NH; i.e. takes Quinolinol form in chloroform

Figure 6. Estimating Effective Solute H-Donor Strength (cont.)

6-Chloro-2-Pyridinol/Pyridone:

$$\varepsilon\alpha = [0.93 - 0.0 + 0.856 + 0.03] / 3.23 = \quad \mathbf{0.56} \quad (7)$$

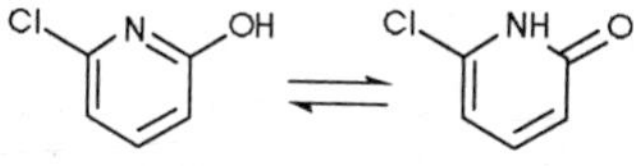

Log P_{oct}
Meas = 0.93
CLOGP   1.71   0.34

Partitioning in $CHCl_3$ shows concentration dependence; i.e.: Some association in non-polar phase.

Enol favored by electron-attracting group(s) near nitrogen atom in pyridinols.

Figure 7. Estimating Effective Solute H-Donor Strength (cont.)

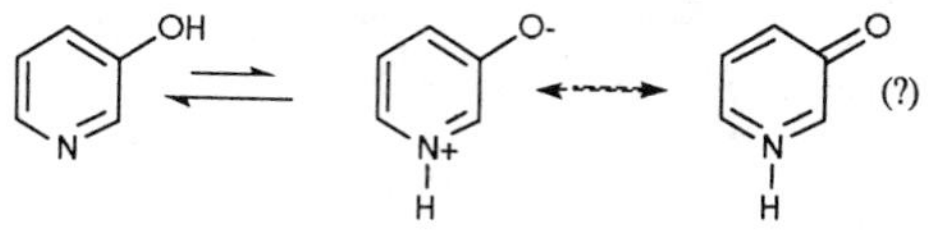

Log P_{oct} Meas. = 0.48;   CLOGP = 0.93; $\Delta = -0.45$

Full Zwitterion Correction   = -2.30

$\varepsilon\alpha = 0.82$

Figure 8. Tautomeric Forms of 3-Hydroxypyridine

Examples of R Groups:

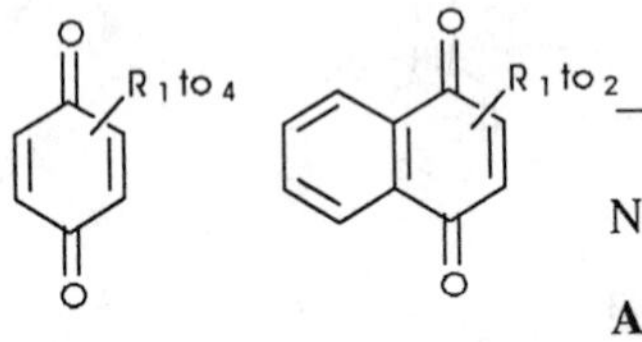

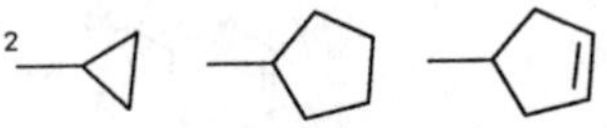

NHMe, NMe2, NHCO2Et, NHC2H4OH

Av. Dev. CLOGP for 29 Analogs = **0.12**

A. 2-Methylaminonaphthoquinone:

Log P oct.: Meas. = 1.74; CLOGP = 1.66; Dev. = +0.08

B. 2-Dimethylaminonaphthoquinone:

Log P oct: Meas. = 1.90; CLOGP = 2.07; Dev. = -0.17

Primary Amine Anomaly

2-Aminonaphthoquinone

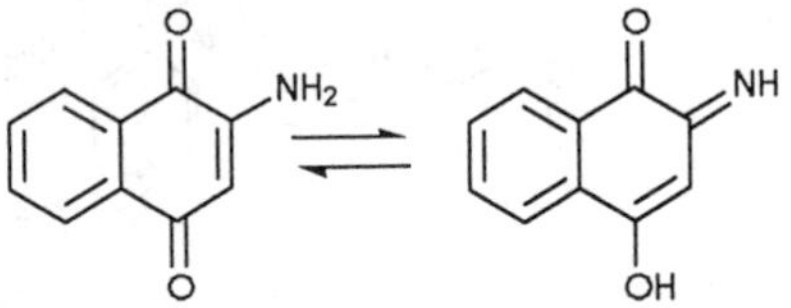

Log P oct
     Meas.   = 1.77; CLOGP   = 0.76   Dev. = +1.01

Intramol-H-bond unlikely; Spectral evidence for imine lacking.

Figure 9. Anomaly in Quinones with Primary Amine Substituents

correction factor needed for this 'tautomeric zwitterion' is much lower than for a true zwitterion, such as an amino-acid which requires a correction of -2.3.   In this case the observed high HBD strength is not much help in deciding the tautomeric ratio in chloroform, because one would predict that the N^+H would be as strong an H-donor as an enhanced phenol.

When one finds a difference between a measured log P_{oct} and a value calculated by a method based on fragments (e.g.,CLOGP), it is very risky to attribute that discrepancy to a specific phenomenon, such as tautomerism or intermolecular H-bonding, without some supporting physical evidence.   But when one knows what interactions CLOGP does and does not consider, such discrepancies can be at least thought provoking.   Figure 9 provides an example where CLOGP performs satisfactorily in predicting log P for a large number of benzoquinone and naphthoquinone analogs, including 29 specific analogs which are mono or di-substituted with secondary or tertiary amino groups.   As shown the deviation for the methylamino and dimethylamino naphthoquinone averages only 0.12, which is about the average deviation for the whole set.   By contrast, the primary amino analog is miscalculated by over one log unit.   In Figure 9, intramol-H-bonding in the amino- and methylamino- analogs would be predicted to be weak, due to the unfavorable bond angles required.   An imine tautomer could be postulated, but there seems to be no reports of spectroscopic evidence for its existence. Thus even when the partition coefficient approach does not yield clear-cut answers, it often can highlight certain anomalies which are deserving of further investigation. This is especially true when the structural features appear in biologically important compounds, such as the amine-substituted benzoquinones which are present in the promising antineoplastic maytansine analogs.

Conclusions

There is little doubt that spectroscopic measurements give the most *definitive* evidence for establishing tautomeric ratios in various solvents.   Yet quite often these methods give results which do not agree amongst themselves, and they may also appear inconsistent with physical-chemical principles.   At least in these instances, data from partition coefficients, although not as definitive, may still yield valuable information and suggest paths for further investigation.

Acknowledgments

I wish to acknowledge the important contribution to this work provided by Professors Corwin Hansch and Robert Taft in the way of guidance and suggestions, and also to the careful analytical skills of Reshma Vasanwala and Priscilla Jow in carrying out the partition coefficient measurements.   Part of the chloroform/water measurements were made under a subcontract from Molecular Solvometrics and funded by an SBIR Grant from NIH.   Financial support from the Pomona College Medicinal Chemistry Project is also gratefully acknowledged.

Literature Cited

1.   Kamlet, M.; Doherty, R.; Abraham, M.; Marcus, Y.; Taft, R. *J. Phys. Chem.* **1988**, *92*, 5244.
2.   Leahy, D.; Morris, J.; Taylor, P.; Wait, A. *J. Chem. Soc. Perkin 2* **1992**, p. 705 and p. 723.
3   (a) Sepiol, J.; Kazimierczuk, Z.; Shugar, D. *Z. Naturforsch.* **1976**, *31c*, 361.   (b) Ravindranathan, R. and Miles, H., *Biochim. Biophys. Acta*, **1965**, *94*, 606.
4.   Rekker,R. In *The Hydrophobic Fragmental Constant*; Elsevier: Amsterdam, 1977.

5.   Leo, A. *Chem. Rev.* **1993**, *93*, 1281.
6.   Mills, S.; Beak, P. *J. Org. Chem.* **1985**, *50*, 1216.
7.   (a) Elguero, J. et al. In *The Tautomerism of Heterocycles*; Katritzky, A; Boulton, A., Eds.; Academic Press: NY, 1976, pp. 87-104. (b) Gordon, A.; Katritzky, A.; Roy, S. *J. Chem. Soc. B*, **1968**, 556.
8.   Wheland, G. In *Resonance in Organic Chemistry*; John Wiley: N.Y., 1955, pp. 403-407.
9.   Sterk,H. *Monatshefte Chem.* **1969**, *100*, 916.
10. (a) McGowan, J. *J. Appl. Chem., Lond.* **1954**, *4*, 41. (b) Abraham, M.; McGowan, J. *Chromatographia* **1987**, *23*, 243.
11. Paoloni, L.; Tosato, M.; Cignitti, M. *Theor. Chim. Acta* **1969**, *14*, 221.

RECEIVED February 10, 1995

MATERIALS

Chapter 22

Structural Analysis of Carbyne Network Polymers

Scott A. Best, Patricia A. Bianconi, and Kenneth M. Merz, Jr.

Department of Chemistry, Pennsylvania State University, University Park, PA 16802

Molecular dynamics simulations of oligomers of the recently reported polycarbyne network backbone polymers indicate that calculated bond lengths between adjacent carbon backbone atoms are very long compared to the bond distance of a C-C single bond. Some degree of bond cleavage is theorized to occur between adjacent carbon atoms of the polymers' network backbones, resulting in the formation of biradicals. This theory is supported experimentally by the polymers' electronic absorption spectra, their degrees of polymerization, and their ESR spectra, which show a decreasing signal as the steric bulk between adjacent carbons is decreased.

The synthesis of the first member of a new class of carbon-based polymers, poly(phenylcarbyne) [PhC]n (1), has recently been reported (1,2). This polymer's stoichiometry is identical to that of poly(diphenylacetylene) (3,4) yet its structure is significantly different. Unlike the polyacetylenes, which are linear polymers whose backbones consist of alternating single and double bonds, poly(phenylcarbyne) has a random network backbone which consists of sp^3-hybridized carbon atoms bonded via carbon-carbon single bonds to three other backbone atoms and one phenyl substituent. This backbone structure is unique in carbon-based polymers and has been found to confer novel properties and reactivity on 1, as, for example, pyrolytic conversion to diamond or diamondlike carbon at atmospheric pressures (1). Because of this pyrolytic conversion, numerous applications for this polymer can be envisioned.

Although the network backbone microstructure of poly(phenylcarbyne) and its Group 14 congeners (1,2,5-9) has been confirmed by ^{13}C and ^{29}Si NMR, the macrostructures of this class of polymers have not been established. The seemingly completely random assembly of the polymer backbones defeats any spectroscopic or diffraction characterization technique, since no two polymer molecules in a given sample may display identical macrostructures and therefore no one macrostructure gives rise to enough characteristic signal to be unequivocally detected. Information about the macrostructure of the polymers is important, however, as polymer macrostructure appears to influence the materials' properties (see below). We have therefore used molecular modeling techniques to provide insights into possible macrostructures for the polycarbyne class of network backbone polymers, and to

0097–6156/95/0589–0304$12.00/0
© 1995 American Chemical Society

establish what effects varying the steric size of the polymers' substituents may have on preferred macrostructures. We report here a series of Molecular Dynamics (MD) simulations of poly(phenylcarbyne) and related network carbyne polymers which examine the structural characteristics of this system. We also report ESR studies on these carbyne polymers which supply further insight into the structural characteristics of this class of materials.

Experimental

The MD simulations were carried out using the all-atom AMBER force field (*10*). The simulations were done in the gas phase at a constant temperature (300K) (*11*) with a 1.0 femtosecond time step. All nonbonded interactions within the molecule were considered. The potential charges for the molecule were obtained by using the electrostatic potentials calculated from MOPAC 5.0 (*12*) using the MNDO Hamiltonian (*13*) and electrostatic fitting (*14*). A distance-dependent dielectric function was used for all simulations to mimic the effect of solvent (*15*). Initially the structures were minimized for 2000 steps using both steepest decent and conjugate gradient minimization methods. The simulations were run for a total of 160 picoseconds (60 ps equilibration, 100 ps sampling), except for the larger phenyl substituted oligomer **6** which was run for 200 ps (100 ps equilibration, 100 ps sampling). This larger oligomer, **6**, was equilibrated for 100 ps rather than 60 ps due to the increase in the size of the network. Poly(phenylcarbyne) (**1**) and poly(methylcarbyne) (**2**) were represented by small network models **3** and **4**, and the parent polycarbyne [HC]n by small network model **5**. Larger network models **6** and **7** were also constructed to represent polymer **1** and the parent polycarbyne. These models are depicted in Figure 1.

Poly(phenylcarbyne) (**1**), poly(methylcarbyne) (**2**), 99:1 poly(phenyl-co-hydridocarbyne) (**8**) and poly(phenylsilyne) (**9**) were synthesized as previously reported (*1,2,16*). ESR spectra were measured at room temperature for each polymer on solid samples using a Bruker 200 spectrometer operating at 9.70 GHz with a modulation frequency set at 100.0 kHz. Equal amounts of polymer were measured for each sample. In order to quantify the number of unpaired spins present within the samples, each sample was compared to a Bruker weak pitch standard (*17*).

Results and Discussion

We have examined models of three different polymers, poly(phenylcarbyne) [PhC]n (**1**), poly(methylcarbyne) [MeC]n (**2**), and the parent polycarbyne [HC]n (**5**). Two of these polymers, **1** and **2**, have been experimentally synthesized (*1,2*); polycarbyne [HC]n (**5**) exists only as a model and was created in order to have a reference for comparative purposes. Although the synthesized polymer networks consist of randomly assembled fused rings of varying sizes, we chose to adopt more simple, ordered models for our MD simulations in order to provide a basis for future construction of randomly oriented polymers. The models that were chosen consisted of networks of varying numbers of fused, regular, six-membered rings (Figure 1). The sizes of the model structures were varied by changing the number of fused cyclohexyl rings contained within the backbone of the network. Networks consisting of seven and nineteen fused rings were chosen for our simulations in order to provide a symmetrical system to study: the networks are constructed so that concentric rings are arranged around an innermost ring. Seven and nineteen fused rings represent fully concentric structures (Figures 1a and 1b). Model networks containing seven fused rings substituted with Ph (**3**), Me (**4**), and H (**5**) (Figure 1a) were initially studied, while larger models which contain nineteen fused rings substituted with Ph (**6**) and H (**7**) were also constructed (Figure 1b). As is illustrated in Figures 1a and 1b, in these network models, **3-7**, the substituents of the inner rings of the networks

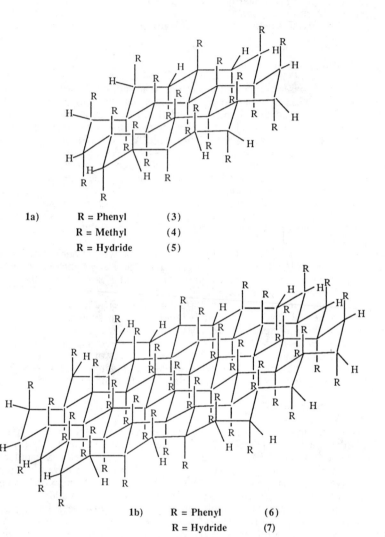

1a)   R = Phenyl   (3)
           R = Methyl   (4)
           R = Hydride   (5)

1b)   R = Phenyl   (6)
           R = Hydride   (7)

Figure 1: Schematic representing models that were created for MD simulations. The models consist of fused, regular, six-membered rings. Model networks containing seven fused rings which are substituted with Ph (3), Me (4), and H (5) are shown in Figure 1a. Figure 1b represents larger networks which were created that contain Ph (6)and H (7) as network substituents.

are arranged in alternating axial and equatorial positions, while the edges of the networks are constructed so that each carbon backbone atom is bonded to two other carbon backbone atoms, one polymer substituent, and one hydrogen atom. Although the models are oligomers and are not as large as the experimentally synthesized polymers, whose degree of polymerization is on the order of 300 monomer units (1,2), the models should provide insight into the structural arrangements which are possible in the actual polymer materials, and can serve as a basis for future modeling work.

Structural features of the oligomeric models 3-7 were analyzed by calculating average bond lengths and torsion angles. The averages take into account the symmetry found in the inner portion of the networks: all carbon backbone atoms which were located in the inner portions of the networks are bonded via carbon-carbon single bonds to three other carbon backbone atoms and one polymer substituent, thus making every carbon backbone atom equivalent. For example, the C-C bond distances of the inner cyclohexyl rings in 3-5 were calculated by taking the averages of the six individual bond lengths (atoms A-F) which define the ring (Figure 2). Other symmetry-related structural features such as torsion angles within the polymer backbone were also calculated by taking averages of individual torsion angles (for example, in Figure 2 the torsion angles consist of four consecutive carbon backbone atoms, e.g. torsion angle made by atoms A-B-C-D). Average C-C bond lengths and torsion angles (torsion angles were measured only for the interior portions of the networks) for model networks 3-7 which were obtained from these calculations are shown in Tables I and II.

For the larger network models whose backbones contain nineteen fused rings (6, 7), the calculations indicate that as the size of the polymer backbone substituent increases from hydrogen for polycarbyne (7) to the network with the larger phenyl substituents (6), the distances between adjacent carbon backbone atoms increase by approximately 0.4Å (Table I). Conversely, the torsion angles (Table II) within these larger networks are seen to decrease with increasing size of the substituent (smaller network models 3-5 also exhibit this trend and are discussed below). The larger polycarbyne model 7 has an average torsion angle of 58.6°, which is approximately equal to that found in the most stable chair conformation of cyclohexane (18). However, the larger poly(phenylcarbyne) model 6 is far more planar and its component cyclohexyl rings deviate significantly from the chair conformation.

The observed increases in bond lengths and decreases in torsion angles within the backbone of 6 are due primarily to steric factors. Furthermore, the ESR study as well as the geometrical conformations of the polymer model substituents indicate there also may be electronic factors which may be influencing the preferred conformations of 6 as well. The large phenyl substituents in 6 cause the amount of steric strain in the network to increase. Visual inspection of our model network reveals that the phenyl rings of 6 are tightly packed due to the constraints on the bond distances between carbon backbone atoms (see Figure 3). Because the phenyl substituents are in close proximity to one another, repulsive van der Waals interactions arise and cause an increase in the amount of steric strain within the polymer backbone. The lengthening of the average carbon-carbon bond distances in the network backbone is attributed to this steric strain. The smaller networks 3-5 also give evidence for the steric strain that is present in these network structures. The average bond distances for the inner portions of the networks increase with the size of their network substituents, 3 > 4 > 5 (Table I).

An examination of the torsion angles of the smaller model networks 3-5 reveals the same trend as was seen for the larger model networks 6 and 7. The polycarbyne model 5 has an average torsion angle of 58.5° while polymer models 3 and 4 have average angles of 21.9° and 34.0° respectively. The data indicate that the network backbone is becoming increasingly planar as the network substituents

Table I. Average C-C Bond Distances for Network Models 3-7

Oligomer Model	# of rings in network	C-C Dist.[a] (Å)	Std. Deviation	C-C Dist.[b] (Å)	Std. Deviation	C-R Dist. (Å)	Std. Deviation
Poly(phenylcarbyne) (3)	7	1.79	0.01	1.65	0.01	1.64	0.01
Poly(methylcarbyne) (4)	7	1.64	0.01	1.59	0.01	1.54	0.01
Polycarbyne (5)	7	1.54	0.01	1.53	0.01	1.09	0.01
Poly(phenylcarbyne) (6)	19	1.96	0.004	1.73	0.004	1.65	0.01
Polycarbyne (7)	19	1.54	0.01	1.53	0.01	1.09	0.01

[a] Distances between carbon backbone atoms in the inner portion of the networks.
[b] Distances between carbon backbone atoms on the edges of the networks

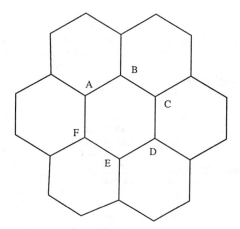

Figure 2: Schematic illustrating the determination of average bond lengths and torsion angles within the model networks.

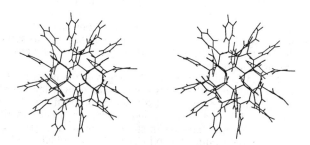

Figure 3: Stereo view of model network **3** illustrating the packing arrangement of its phenyl substituents.

Table II. Average Torsion Angles for Interior Network Backbone Atoms

Oligomer Model	# of rings in network	Torsion Angle (°)	Std. Dev.
Poly(phenylcarbyne) (**3**)	7	21.8	2.2
Poly(methylcarbyne) (**4**)	7	34.0	2.7
Polycarbyne (**5**)	7	58.5	2.7
Poly(phenylcarbyne) (**6**)	19	22.9	1.9
Polycarbyne (**7**)	19	58.6	2.2

become larger and the steric strain in the network increases. The lower torsion angle for **3** again arises from steric interaction between the phenyl substituents.

The methyl substituents of **4** are much smaller than the phenyl substituents of **3**. As is shown in Tables I and II, the average bond distances between adjacent carbon backbone atoms for **4** are less than those calculated for **3**. This decrease in bond length is attributed to the steric signature of the polymer substituents. Because the methyl groups of **4** are much smaller than the phenyl substituents of **3**, they do not require as large a separation between adjacent substituents to relieve the repulsive interactions and minimize the steric strain experienced by the polymer backbone. Hence, smaller average bond distances and larger torsion angles for the carbon backbone are observed. The same argument holds for the polycarbyne model (**5**) as well.

The degree of crowding between polymer substituents within the various network models can also be assessed by an examination of the outer portion of the networks as well. The calculated average bond length between adjacent carbon backbone atoms on the edge of the network of **6** is 1.73Å (Table II). This bond length is less than what was calculated for the inner fused rings of the network of **6** (1.96Å), yet still greater than the average bond lengths for both the inner and edge backbone distances calculated for **7**, the hydrogen-substituted model network (1.54Å). The edge carbon atoms of the backbone of **3** and **6** [R$_2$CHPh] have one less linkage to the network backbone than do the inner network atoms. There are two substituents (a phenyl ring and a hydrogen atom) for the edge carbon backbone atom rather than just the one phenyl substituent found in the inner portion of the network. The phenyl rings are not as tightly packed along the edges of the network as shown through visual inspection of our models (see Figure 3). Because of the orientation of the substituents, a decrease in the steric strain at the edges of the networks is expected. This is attributed to both the decrease in linkages to the network backbone and the presence of the smaller hydrogen atoms. Consequently the steric strain introduced by the network substituents at the edges is reduced to some degree, as evidenced by the reduction in the C-C bond lengths when compared to the inner portion of the network backbone.

This trend of bond length reduction at the edges of the network backbone also exists for the smaller models **3** and **4**. This supports our conclusion that the degree of crowding associated with the network substituents is proportional to the amount of steric strain present within the network and hence contributes to the preferred conformations of our models. Increase in steric strain with increasing substituent size has also been observed in substituted dodecahedranes, which can be regarded as oligomers of the polycarbynes (*19*).

The π-stacking adopted by phenyl rings also contributes to the conformations of the backbones of network models **3** and **6**. The strong attraction or repulsion associated with π-systems has been shown to determine the conformational preference and binding properties of polyaromatic macrocycles (*20-22*). The most important interactions between such π-systems are quadrupole-quadrupole electrostatic interactions (*23*). For aromatic rings with nonpolar substituents, π-π electronic repulsions will dominate if the aromatics are arranged in a stacked or face to face geometry. Conversely, if the substituents are arranged in a T-structure (phenyl rings arranged so that the edge of one ring is perpendicular to the face of a second ring), electronic attraction of the π-σ type will dominate (*24*). Phenyl rings will therefore try to orient themselves in a tilted T-structure in order to achieve the lowest possible interaction energy between them (*24,25*). Moreover, as the distances of separation between phenyl rings decrease, the stacked arrangements of the rings become more frequent but are still unfavorable due to unfavorable energetics. An examination of stacked and perpendicular arrangements of π-systems showed that the perpendicular arrangement displays the lower free energy due to the more favorable π-σ interactions (*26*). For our model systems **3** and **6**, visual inspection reveals that adjacent phenyl substituents are tightly packed in an orientation intermediate between the stacked and T-conformations, presumably resulting in unfavorable van der Waals interactions between them. Hence, the arrangement of the phenyl substituents contributes significantly to the calculated structural parameters which are indicated by our models.

Our calculations also indicate a difference in the average bond lengths between the backbone carbon atoms for the inner portions of networks 3 and 6 (Table I), the phenyl substituted network models, which may be indicative of an important structural feature of the polycarbyne polymer models. For the inner portion of the networks, the average bond distance between the backbone carbons of 3 is 1.79Å, while that of 6 is significantly longer, 1.96Å. This difference is attributed to the sizes of the backbone networks of the two models. The network backbone of poly(phenylcarbyne) (1) is much larger than those of our model networks (1,2). Because of the size of 1, the number of carbon backbone atoms located along the edge of the polymer is small as compared to those of the inner portion of the network. Hence, the inner portion of the polymer will have a greater effect on the physical properties of 1 than will the edge of the network. Due to its larger inner network, 6 is thought to be a better representation of the actual network backbone structure of poly(phenylcarbyne) (1) than is 3. The 1.96Å average distance between adjacent carbon atoms in the inner backbone of 6 is not feasible for C-C single bonds, but arises from the inability of the potential function used in the MD simulation to mimic bond breaking. The long average bond length calculated for 6 suggests that some degree of bond breaking may occur within the backbone of polycarbynes, resulting in the formation of biradicals. Bond cleavage within the polymer backbone would allow for maximum separation between adjacent substituents and would therefore lower the repulsive van der Waals interactions between them. It should therefore relieve steric strain and allow for more favorable backbone conformations by alleviating the crowding generated by neighboring phenyl substituents. Figure 4 illustrates schematically the bond cleavage which is proposed to occur in the backbones of polycarbynes . The bond cleavage produces biradicals which may rearrange to create areas of unsaturation within the polymer backbones. This process may occur throughout the polymers' backbones, resulting in some radical mobility. This bonding model for the backbones of polycarbynes is analogous to that proposed for amorphous silicon and germanium solid-state structures based on continuous random network models, in which some atoms are only three-coordinate and are left with a single electron in a "dangling bond" (27).

That "steric strain induced" C-C bond cleavage occurs within the backbones of polycarbyne network polymers is supported experimentally by the observation of unpaired electrons in polycarbynes (2). The experimentally-synthesized polymers poly(phenylcarbyne) (1), poly(methylcarbyne) (2), and 99:1 poly(phenyl-co-hydridocarbyne) (8) exhibit ESR signals characteristic of carbon-centered radicals, with g-values of 2.0026, 2.0033 and 2.0029 respectively (Figure 5). The number of unpaired spins for these polymers are 1.5×10^{19}, 3.7×10^{17}, and 2.8×10^{18} g / monomer units respectively. The ratio of the number of unpaired spins for poly(phenylcarbyne) (1) to 99:1 poly(phenyl-co-hydridocarbyne) (8) is 5.4 : 1 while that for poly(phenylcarbyne) (1) to poly(methylcarbyne) (2) is 40 : 1.

Because the polymer backbone is highly rigid, the steric strain in the polymer systems results in bond cleavage and in the formation of biradicals. That radical formation is driven by relief of steric strain caused by the size of the polycarbynes' substituents is also demonstrated experimentally by comparison of the ESR spectra of four different experimentally synthesized polymers, poly(phenylcarbyne) (1), 99:01 poly(phenyl-co-hydridocarbyne) (8), poly(methylcarbyne) (2), and poly(phenylsilyne) (9) (Figure 5). The intensities of the ESR absorptions are indicative of the amount of unpaired spins present within the sample. As shown in Figures 5a-5c, the intensity of the ESR signals of these polycarbynes decrease as the

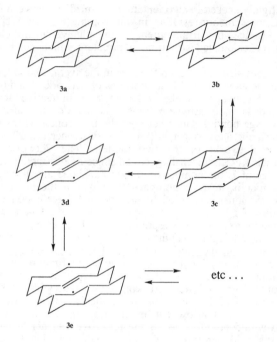

3a
3b

3d
3c

etc . . .

3e

Figure 4: Schematic representing the possible bond cleavage and radical formation within polycarbyne backbones.

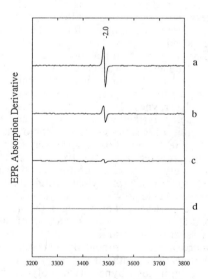

-2.0

a

b

c

d

3200 3300 3400 3500 3600 3700 3800

EPR Absorption Derivative

Figure 5: ESR spectra of a) poly(phenylcarbyne) (**1**), b) poly(phenyl-co-hydridocarbyne) (**8**), c) poly(methylcarbyne) (**2**), and d) poly(phenylsilyne) (**9**).

size of the polymers' substituents decrease. The number of radicals formed within the polymer is dependent on the size of the backbones' substituents, which our models indicate also induces steric strain within the backbone. According to our models, the larger phenyl substituents of 1 should induce a large amount of steric strain within the polymer backbone and hence greater radical formation to relive this strain. The phenyl rings may then stabilize the radicals through delocalization in the phenyl substituent. As the steric size of the substituents on the polycarbyne backbones are decreased, from polymers 8 to 2, the intensity of the ESR signal also decreases, suggesting, as predicted by our models, that the lesser degree of steric strain experienced by these polymers' backbones is the cause of a lower degree of bond cleavage in the backbone. No ESR signal is seen for the silicon-based congener of 1, poly(phenylsilyne) (9, Figure 5d). The Si-Si bonds which form this polymer's backbone are much longer than the C-C single bonds which make up the backbone of 1 (2.35Å as compared to 1.54Å) (28). A silicon-based polymer backbone substituted with phenyl groups therefore would not experience as much crowding as would the identically-substituted carbon-based polymer, and should display little or no biradical formation and ESR signal. These results support the MD simulations by illustrating the relationship between the size of the polymer substituent and the amount of steric strain within the polymer system, as evidenced by the number of biradicals formed by bond cleavage.

The electronic absorption spectra displayed by the polycarbynes (1,2) also support our models' prediction of a strained network backbone structure. These spectra are nearly identical to those of their silicon analogues, the polysilynes (5-9), consisting of intense broad absorptions with onset at approximately 450 nm, increasing gradually in intensity with decreasing wavelength to 200 nm. For polysilynes, these absorptions are attributed to Si-Si σ-conjugation over the three dimensional polymer backbone (5-9). Because of the semiconductor-like electronic spectra, the polycarbyne backbones appear to be σ-conjugated. Such σ-conjugation of C-C bonds is also seen in small molecules which incorporate strained fused rings (29-40). The strain which our calculations indicate is present in the fused rings of the polycarbyne backbones should lower the C-C bond strength and thus the energy of the bonds' σ-σ* transitions from the vacuum UV (where the σ-σ* transition of unstrained C-C bonds occurs (28)) to the observed UV and visible frequencies. The radical transfer mechanism shown in Figure 4 could also contribute to the polycarbynes' low-energy electronic absorptions: delocalization of biradicals through the polycarbynes' backbone as depicted in Figures 4c-4e should also give rise to lower-frequency electronic absorptions, as could defects (for example, OH or Cl terminations) in the polymers' structures.

Our calculations' indication that strain within polycarbyne network backbones, induced by steric crowding of the substituents (see above), increases as the network backbone increases in size suggests that the degree of polymerization which is seen in polycarbynes should decrease as the size of backbone substituents increases. This prediction is also confirmed by the experimental determination of the polycarbynes' molecular weights. The number average molecular weights, $\bar{M}_n$, for poly(phenylcarbyne) (1) (1,2), 99:1 poly(phenyl-co-hydridocarbyne) (8) (2), poly(methylcarbyne) (2) (2), and poly(phenylsilyne) (9) (16) are 3007, 4168, 4425, and 8775 daltons, respectively. These molecular weights demonstrate that, as was seen for inorganic-backbone network polymers (5-9), the degree of polymerization attained by the polycarbynes is proportional to the steric signature of their substituents, suggesting that large substituents can be incorporated only in networks which, by virtue of their smaller sizes, can minimize steric repulsions between substituents.

Conclusions

MD simulations of models for a new class of carbon-based polymers, polycarbynes, were carried out in order to provide some insight into this class of polymers' structural characteristics. Results of the simulations suggest that some degree of bond cleavage occurs within the polymer backbones that is increased by strain induced steric repulsions between the polymers' side chain substituents. The MD results are supported experimentally by the polymers' ESR spectra, electronic spectra and degrees of polymerization. Future investigations will focus on the simulation and molecular dynamics of polycarbyne model systems comprised of larger networks of varying degrees of backbone randomness that will better model the actual polymers.

Acknowledgments

We thank Dr. M. Crowder and Dr. P. M. Lanahan and Jim Yount of the Pennsylvania State University for their assistance in acquiring ESR spectra. PAB is a Beckman Young Investigator (1992-1994) and the recipient of a Camille and Henry Dreyfus Teacher-Scholar Award (1992-1997).

Literature Cited

1. Visscher, G. T.; Nesting, D. C.; Badding, J. V.; Bianconi, P. A. *Science* **1993**, *260*, 1496-1499.
2. Visscher, G. T.; Bianconi, P. A. *J. Am. Chem. Soc.* **1993**, *116*, 1805-1811.
3. Masuda, T.; Kawai, H.; Ohtori, T.; Higashimuri, T. *Polym. J.* **1979**, *11*, 813- 818.
4. Niki, A.; Masuda, T.; Higashimura, T. *J. Polym. Sci., Polym. Chem. Ed.* **1987**, *25*, 1553-1562.
5. Bianconi, P. A.; Weidman, T. W. *J. Amer. Chem. Soc.* **1988**, *110*, 2342-2344.
6. Bianconi, P. A.; Schilling, F. C.; Weidman, T. W. *Macromolecules* **1989**, *22*, 1697-1704.
7. Weidman, T. W.; Bianconi, P. A.; Kwock, E. W. *Ultrasonics* **1990**, *28*, 310-315.
8. Furukawa, K.; Fujino, M.; Matsumoto, M. *Macromolecules* **1990**, *23*, 3423-3426.
9. Szymanski, W. J.; Visscher, G. T.; Bianconi, P. A. *Macromolecules* **1993**, *26*, 869-871.
10. Weiner, S. J.; Kollman, P. A.; Nguyen, D. T.; Case, D. A. *J.. Comput. Chem.* **1986**, *7*, 230-252.
11. Berendsen, H. J. C.; Potsma, J. P. M.; van Gunsteren, W. F.; DiNola, A. D.; Haak, J. R. *J. Chem. Phys.* **1984**, *81*, 3684-3690.
12. Merz, K. M., Jr.; Besler, B. H. *QCPE Bull.* **1990**, *10*, 15.
13. Dewar, M. J. S.; Thiel, W. *J. Am. Chem. Soc.* **1977**, *99*, 4899-4907.
14. Besler, B. H.; Merz, K. M. Jr.; Kollman, P. A. *J. Comput. Chem.* **1990**, *11*, 431-439.
15. Weiner, S. J.; Kollman, P. A.; Case, D. A.; Singh, U. C.; Caterina, G.; Alagona, G.; Profeta, J., S.; Weiner, P. *J. Am. Chem. Soc.* **1984**, *106*, 765- 784.
16. Bianconi, P. A.; Smith, D. A.; Freed, C. A. *Chemistry of Materials* **1993**, *5*,   245-247.
17. Hyde, J. F. In *Sixth Annual NMR/EPR Workshop*; Palo Alto, Cal., 1962.

18. Carey, F. A.; Sundberg, R. J. *Advanced Organic Chemistry Part A: Structure and Function;* 3rd ed.; Plenum Press: New York, NY, 1990, p 802.
19. Wahl, F.; Wörth, J.; Prinzbach, H. *Angewandte Chemie* **1993**, *32*, 1722-1726.
20. Hunter, C. A.; Meah, N. M.; Sanders, J. K. M. *J. Am. Chem. Soc.* **1990**, *112*, 5773-5780.
21. Anderson, H. L.; Hunter, C. A.; Meah, N. M.; Sanders, J. K. M. *J. Am. Chem. Soc.* **1990**, *112*, 5780-5789.
22. Hunter, C. A.; Leighton, P.; Sanders, J. K. M. *J. Chem. Soc., Trans. Perkin 1* **1989**, 547-552.
23. Linse, P. *J. Amer. Chem. Soc.* **1992**, *114*, 4366 - 4373.
24. Hunter, C. A.; Sanders, J. K. M. *J. Am. Chem. Soc.* **1990**, *112*, 5525-5534.
25. Jorgensen, W. L.; Severance, D. L. *J. Amer. Chem. Soc.* **1990**, *112*, 4768-4774.
26. Linse, P. *J. Am. Chem. Soc.* **1993**, *115*, 8793-8797.
27. West, A. R. *Solid State Chemistry And Its Applications* ; John Wiley & Sons Ltd.: New York, NY, 1985, pp 82-85.
28. Greenwood, N. N.; Earnshaw, A. *Chemistry Of The Elements*; Pergamon Press: New York, NY, 1984, pp 296-385.
29. Hush, N. S.; Padden-Row, M. N.; Cotsaris, E.; Oevering, H.; Verhoeven, J. W.; Heppener, M. *Chemical Physics Letters* **1985**, *117*, 8-11.
30. Warman, J. M.; de Haas, M. P.; Padden-Row, M. N.; Cotsaris, E.; Hush, N. S.; Oevering, H.; Verhoeven, J. W. *Nature* **1986**, *320*, 615-616.
31. Oevering, H.; Padden-Row, M. N.; Heppener, M.; Oliver, A. M.; Cotsaris, E.; Verhoeven, J. W.; Hush, N. S. *J. Am. Chem. Soc.* **1987**, *109*, 3258-3269.
32. Paddow-Row, M. N.; Oliver, A. M.; Warman, J. M.; Smimt, K. J.; de Haas, M. P.; Oevering, H.; Verhoeven, J. W. *J. Phys. Chem.* **1988**, *92*, 6958-6962.
33. Verhoeven, J. W. *Pure & Applied Chem.* **1986**, *58*, 1285-1290.
34. Olah, G. A.; Surya Prukash, G. K.; Shih, J. G.; Krishnamurthy, V. V.; Mateescu, G. D.; Liang, G.; Sipos, G.; Volker, B.; Gund, T. M.; v. Rague Schleyer, P. *J. Am. Chem. Soc.* **1985**, *107*, 2764-2767.
35. Dewar, M. J. S. *J. Am. Chem. Soc.* **1984**, *106*, 669-682.
36. Gleiter, R.; Schafer, W. *Acct. Chem. Res.* **1990**, *23*, 369-375.
37. Paddow-Row, M. N.; Patney, H. K.; Brown, R. S.; Houk, K. N. *J. Am.. Chem. Soc.* **1981**, *103*, 5575-5577.
38. Balaji, V.; Jordan, K. D.; Burrow, P. D.; Paddon-Row, M. N.; Patney, H. K. *J. Am. Chem. Soc.* **1982**, *104*, 6849-6851.
39. Surya Prakash, G. K.; Fessner, W.; Olah, G. A.; Lutz, G.; Prinzbach, H. *J. Am. Chem. Soc.* **1989**, *111*, 746-748.
40. Bianconi, P. A.; Wiedman, T. W.; Kwock, E. W. In *Polymers For Lightwave and Integrated Optics: Technology and Applications*; Hornak, L., Ed.; Optical Engineering Series; Marcell Decker: New York, NY, 1991, pp 195-207.

RECEIVED November 20, 1994

Chapter 23

Computer Simulation of Polyelectrolyte Adsorption on Mineral Surfaces

Susan Fitzwater

Rohm and Haas Company, 727 Norristown Road, Spring House, PA 19477

Simulation of the adsorption of fully neutralized poly-acrylic and poly-aspartic acids on various calcite surfaces shows that both acids tend to adsorb flat against positive and growth surfaces. Poly-aspartic acid adsorbs more slowly than does poly-acrylic acid. There is no evidence for "ion bridging" to negative surfaces.

Polyelectrolytes are used in detergent builder, mineral processing, and water treatment applications. Adsorption of the polymer on selected mineral surfaces can have a significant effect on performance in all of these applications. Adsorbed polyelectrolytes can act as dispersants, providing charge and/or steric repulsion between particles. Adsorbed polyelectrolytes can also act as crystal growth inhibitors, altering the size and/or habit of crystals which form during the washing and other processes.

Molecular modeling of isolated polyelectrolytes, with and without counterions, has given us detailed information about how polyelectrolytes interact with counterions and how polyelectrolyte structure affects these interactions *(1)*. From the observed characteristics of polyelectrolyte - counterion interactions, we have inferred characteristics of polyelectrolyte - mineral surface interactions. We have successfully used these inferred characteristics to rationalize and predict polyelectrolyte structure effects on applications properties *(2, 3)*. However, our long-term goal has always been to include explicit mineral surfaces in the modeling. Recent hardware and software developments allowed us to investigate how including explicit mineral surfaces affects both modeling results and computational requirements.

In the studies described here, I examined the interaction of both poly(acrylic acid), pAA, and poly(aspartic acid), pAsp, with various $CaCO_3$ (calcite) surfaces. We have a wealth of experimental data and modeling experience on both pAA and pAsp. $CaCO_3$ surfaces are fairly well characterized; also, polyelectrolyte interaction with $CaCO_3$ surfaces is important in several commercial applications.

0097–6156/95/0589–0316$12.00/0
© 1995 American Chemical Society

Calculation Methods

The basic method I used was as follows:
1) Take an equilibrated model polyelectrolyte, with or without counterions.
2) Move it close to a model mineral surface.
3) Energy minimize to give a stable polyelectrolyte (+ counterions) + mineral surface complex.
4) Run molecular dynamics on the complex, allowing the polymer (+ counterions) to move into low-energy configurations on the mineral surface.

Some details on each step:

Model Polyelectrolytes. The model polyelectrolytes I used here are quite similar to those I have reported on before. I used 20-mers for both pAA and pAsp: this is similar to the size used in many commercial applications. Note that this means that

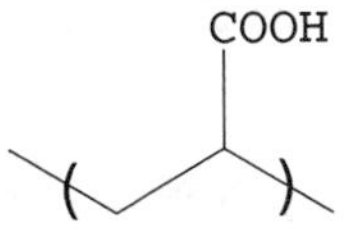

pAA 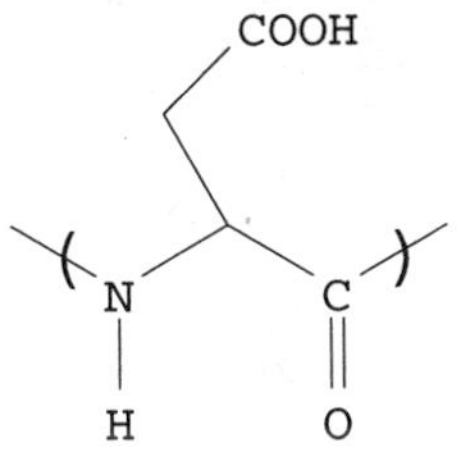   pAsp

the comparisons between pAA and pAsp in this study are done on the basis of (nearly) equal numbers of carboxylate groups, not equal molecular weight. [The model pAsp used here had 21 carboxylate groups; the extra one is at the C-terminus.]

The pAsp used in most of the modeling had 35% α- / 65% β-linkages:

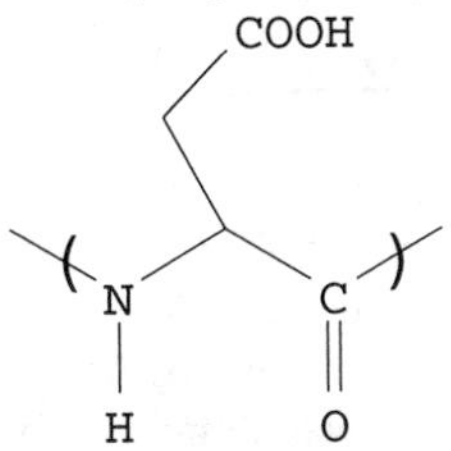

α   β

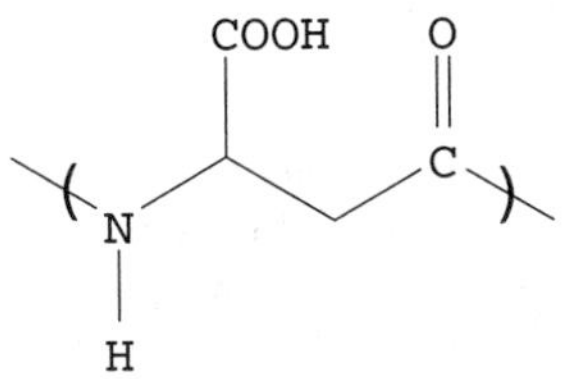

[Polyaspartate is generally synthesized by hydrolysis of polysuccinimide. Recent work *(4)* has shown that the α/β ratio is generally 25-30% α-, 70-75% β-, and is remarkably insensitive to hydrolysis conditions.] Earlier modeling work had not shown a strong effect of the α/β ratio on interaction with counterions. A couple of simulations of 100% α pAsp on the $CaCO_3$ surfaces showed minor differences from 35% α / 65% β pAsp on these same surfaces.

Since these polymers are often used in a neutral or somewhat basic environment and have pK_a's in the 4-5 range, I assumed that all of the acid groups were neutralized, that is, all of the carboxylates are in the COO^- form.

In many real applications, pAA and pAsp interact with Ca^{+2} counterion. Most of our earlier modeling studies have been done with Ca^{+2} counterions; it's usually the counterion we are most interested in. For studies in which counterion was included, I added 7 Ca^{+2} counterions to both pAA and pAsp. Experimentally, it is observed that fully neutralized pAA binds enough Na^+ or Ca^{+2} to neutralize ~2/3 of its charge *(5,6)*; one of my earlier modeling studies of pAA with Na^+ suggested the same *(7)*. Calcium sequestration studies done at Rohm and Haas suggest that both pAsp and pAA bind about 1 Ca^{+2} per 3 carboxylates.

Electrostatic interactions have strong effects on these systems. I derived charges on the atoms in the model polyelectrolytes from semi-empirical quantum calculations on oligomers. For the counterions and atoms in the model surfaces, I used "reasonable" ionic charges. The atomic charges I used are shown in the table below.

Table I. Atomic Charges for Polymers and Surfaces

Species	Atom Type	Charge	Species	Atom Type	Charge
pAA	methylene C	-0.13	pAsp	N	-0.31
	methine C	-0.16		H (N-H)	+0.26
	carbonyl C	+0.33		methine C (α form)	+0.05
	all H's	+0.055		methine H (α form)	+0.11
	carboxylate O's	-0.61		methylene C (α form)	-0.24
				methylene H (α form)	+0.09
$CaCO_3$	Ca^{+2} ion	+2.00		methine C (β form)	-0.06
	carbonate C	0.00		methine H (β form)	+0.08
	carbonate O	-0.667		methylene C (β form)	-0.18
				methylene H (β form)	+0.11
				carboxylate C	+0.32
				carboxylate O	-0.61
				carbonyl C	+0.28
				carbonyl O	-0.40

Most of the simulations were done using the Dreiding II force field *(8)* and the POLYGRAF program (Molecular Simulations Inc. of Burlington, MA, and Cambridge, UK). In these simulations, I approximated the effect of water by using a 6r dielectric constant. This is similar to the value of 4r which was found to give good results in a study of the effect of dielectric constant on molecular modeling of the structure of crambin *(9)*. In practice, D = 6r implies that charge-charge interactions are mediated by a dielectric constant in the range 25 - 60, the correct order of magnitude for water at small charge-charge separations *(10)*. I also used a 9 A cutoff for both van der Waals and electrostatic interactions. Given that electrostatic interactions are long-range, using this relatively small cutoff (or any cutoff at all) may seem questionable. However, use of the cutoff can be justified on several grounds:

1) Our major interest is in simulating qualitative behaviour in real systems, where the ionic strength is almost always appreciable. This means that the correct electrostatic potential function is

$$E_{elec} = q_1 q_2/De^{-\kappa r} r \qquad (1) \quad ,$$

where κ is the Debye screening constant, a positive number which increases as the ionic strength increases. Thus the effect of ionic strength is to wash out the charge-charge interactions, particularly at large separations. Distance cutoffs simulate this effect, albeit crudely.

2) Previous simulations done with $D = 6r$ and cutoffs agreed qualitatively with experiment.

3) A few simulations done with much longer cutoffs (20 A for the electrostatic interactions) showed the same qualitative behaviour as those done with the 9 A cutoffs.

I also ran a few simulations using the MM2 force field *(11)* in the MacroModel program, with the GB/SA *(12)* solvation model for water. These were limited because the GB/SA solvation model requires 1-2 orders of magnitude more computer time than the simple $D = 6r$ model. However, structures which were stable minima with Dreiding II / $D = 6r$ tended to be similar to stable minima with MM2 / GB/SA. Also, subjecting the Dreiding II / $D = 6r$ minima to molecular mechanics with MM2 / GB/SA induced little qualitative change in the structures. Therefore, the results presented here do not appear to depend heavily upon the specifics of the potential functions or solvation models used.

Model Mineral Surfaces. I built these using the CERIUS program (Molecular Simulations Inc. of Burlington, MA, and Cambridge, UK). There are an infinite number of possible surfaces for any given lattice type; I chose surfaces which are either present in the macroscopic $CaCO_3$ crystallites or are fast-growing, as predicted by the Bravais-Friedel-Donnay-Harker method *(13)* implemented in CERIUS. The model surfaces were large, thin slabs, significantly larger than the fully extended polyelectrolytes (to avoid edge effects), and 3-4 ion layers thick.

Here, I'll discuss results from the 0 0 1 and (very similar) 1 0 -2 surfaces, present in macroscopic crystallites, and the 2 0 3 surface, a growth surface. Formally, both the 0 0 1 and 1 0 -2 surfaces can be either positive or negative, depending on whether there is a layer of Ca^{+2} or CO_3^{-2} ions on the face. The 2 0 3 surface is formally neutral.

Obviously, the polyelectrolyte without counterions is expected to adsorb strongly on the positive surfaces and not at all on the negative surfaces. The behaviour to be expected on neutral surfaces, or when counterions are added, is not so clear. There are reports that pAA can adsorb on negative kaolin surfaces by counterion bridging *(6, 14)*, even though the typical pAA + counterion complex is formally negative.

For each polyelectrolyte - surface combination, I took the polyelectrolyte (with or without counterions) in a low-energy conformation and moved it 2 - 5 A from the model surface.

Energy Minimize. I ran the energy minimizations and the dynamics with the polyelectrolyte and counterion atoms free to move and the surface atoms fixed. There is a minor difference between "fixed" atoms in POLYGRAF and in MacroModel. POLYGRAF "fixed" atoms are truly fixed; they don't move at all. MacroModel "fixed" atoms are restrained; these atoms can move, but there is a substantial energetic penalty for doing so. There was minor, though visible, atomic movement in the surfaces during the MacroModel energy minimizations and molecular dynamics simulations. However, the surface atomic movement appeared random.

Generally, exhaustive energy minimization did not change the conformation of the polyelectrolyte significantly from the conformation of the polyelectrolyte in the original docked structure. This was true even for polyelectrolytes whose configuration on or near the surface changed dramatically during the molecular dynamics. Clearly, one cannot rely upon energy minimization alone for an adequate qualitative picture of polyelectrolyte interaction with surfaces; one must also use a molecular dynamics and/or Monte Carlo procedure which allows the polyelectrolyte to search for lower-energy configurations.

Molecular Dynamics. All POLYGRAF / D = 6r simulations were run using canonical dynamics at 300 °K for at least 1000 picoseconds using a 2 femtosecond time step. From previous work, we know that a 20-mer polyelectrolyte + divalent counterions equilibrates in less than 1000 ps. However, the polyelectrolyte + counterion complexes studied here did not equilibrate on the $CaCO_3$ surfaces within 1000 ps: the qualitative appearances of the complexes were not changing much at the end of the simulation, but the potential energy was still decreasing. The longest simulations I ran were 4000 ps; it is questionable that even this system was equilibrated at the end of the simulation.

However, gross conformational changes do occur in 1000 ps. Some qualitative differences between pAA and pAsp were very apparent in the 1000 ps simulations, as were differences in the adsorption of the polyelectrolytes with and without counterions.

I ran a few MacroModel / GB/SA Newtonian dynamics simulations; the longest of these was ~1000 ps. Results of these simulations were not qualitatively visually different from the results of the POLYGRAF / D=6r simulations.

I calculated crude binding energies from the POLYGRAF / D=6r simulations and similar simulations done without the model surfaces present. Binding energy is defined as

$$E_{bind} = \overline{E}_{surface} - \overline{E}_{nosurface} \qquad (2)$$

Here, the $\overline{E}$'s are average energies from the dynamics simulations. Energies from the first several hundred picoseconds of dynamics were not included in the averages. In assessing the binding energies, be aware that while large differences are meaningful, small differences probably are not: the energies of most of the surface simulations had not converged. Also be aware that the electrostatic contribution, which strongly dominates the binding energy, is damped by the 6r dielectric constant.

Results and Discussion

Polyelectrolyte on a Positive Surface. Color Plate 8 shows pAA on the positive 0 0 1 surface before the start of the dynamics simulation. The pAA is somewhat coiled: even though this is not the lowest-energy conformation for an isolated pAA chain, pAA + counterions is coiled and it's not clear when the counterions leave the adsorbing chain, or whether the chain extends before it adsorbs. The Color Plate 8 conformation is a stable minimum on the 0 0 1 surface.

Color Plate 9 shows pAA on the positive 0 0 1 surface after 10 ps of dynamics. The chain has flattened out on the surface. For the remaining 990 ps of the simulation, the chain twitched on the surface but did not undergo any major conformational rearrangement.

Color Plate 10 shows pAsp on the positive 0 0 1 surface before the start of the dynamics simulation. Again the chain is somewhat coiled. Color Plate 11 shows pAsp on the positive 0 0 1 surface after 1000 ps of dynamics. A significant portion of the chain has flattened out on the surface, but a good part of the chain is not in contact with the surface at all. Some longer simulations suggest that, given more time, the pAsp would have flattened out more.

Binding energies on the positive surfaces are shown below:

Table II. Binding Energies of pAA and pAsp on Positive Surfaces

Polymer	Surface	E_{bind}	E_{bonded}	E_{vdw}	E_{elec}	E_{H-B}
pAA	0 0 1	-495.67	-9.31	-39.26	-447.12	
pAsp		-352.29	13.41	-38.38	-325.99	-1.33
pAA	1 0 -2	-704.66	8.82	-18.06	-695.55	
pAsp		-562.85	23.06	-25.98	-561.85	1.92

The energy units are kcal/mole. E_{bonded}, E_{vdw}, E_{elec}, and E_{H-B} are the bonded (bond + angle + torsional), van der Waals, electrostatic, and hydrogen bonding contributions to the total binding energy.

The pAA simulation results are in good qualitative agreement with experimental data on pAA adsorption on calcite (15). The adsorption is observed to be high affinity, which implies a strong interaction between the pAA and the surface. This in turn implies that the surface on which the pAA adsorbs is positive; both the pAA and the pAsp bind most strongly to the positive surfaces. Most of the pAA segments are observed to be in close contact with the $CaCO_3$ surface.

Polyelectrolyte + Counterions on a Positive Surface. Color Plate 12 shows pAA + 7 Ca^{+2} on the positive 1 0 -2 surface before dynamics. The counterions are largely in the polymer coil. Color Plate 13 shows the pAA + counterions on the same surface after 1000 ps of dynamics. The pAA has flattened out considerably, and the counterions have moved away from the surface; they are now outside the polymer coil.

NOTE: Color plates appear in color section.

Color Plate 14 shows pAsp + 7Ca^{+2} on the positive 1 0 -2 surface before dynamics. The general appearance is similar to Color Plate 12. Color Plate 15 shows the pAsp + counterions on the same surface after 1000 ps of dynamics. The chain appears to have separated into two segments with roughly equal length: the first is adsorbed flat against the surface, and the second forms a coil several A from the surface. Six of the seven counterions are within or close to the coil.

Binding energies on the positive surfaces are shown below:

Table III. Binding Energies of pAA and pAsp with Counterions on Positive Surfaces

Polymer	Surface	E_{bind}	E_{bonded}	E_{vdw}	E_{elec}	E_{H-B}
pAA	0 0 1	-336.73	-36.35	-37.33	-263.05	
pAsp		-239.06	13.46	-29.22	-219.56	-3.74
pAA	1 0 -2	-471.12	-16.92	-24.78	-429.43	
pAsp		-298.60	28.62	-4.52	-320.62	-2.08

Note that these are less negative than those for the polymer without counterion on the positive surfaces. However, they are still appreciably negative.

Results of other simulations suggest that the type of counterion binding to polymer observed here is not stable (7). Water tends to remove Ca^{+2} ions which are not held within a chain; the Ca^{+2} within the Color Plate 15 pAsp coil is too concentrated to be stable. Both chains will lose counterions eventually; as this occurs, the adsorption will become like that of the polyelectrolyte alone on the positive surface, discussed above.

Adsorption may proceed as follows: A polyelectrolyte chain, with accompanying counterions, approaches a positive CaCO$_3$ surface. Part of the chain comes in contact with the surface; counterions shift into a portion of the chain which is still separated from the surface. As more of the chain contacts the surface, the counterions become too concentrated; one or more moves away from the chain, to lower the overall energy. This frees more of the chain to contact the surface. The counterions shift again into that (decreased) portion of the chain which is still separated from the surface; with the decreased chain volume, they are too concentrated, and another moves away from the chain. In short: adsorption drives the chain to lose counterions, loss of counterions drives further adsorption, which in turn drives further adsorption, and so on.

In this adsorption picture, the shift of the counterions away from the surface during adsorption and their subsequent release could create short-lived regions of high counterion concentration. It is conceivable that particle nucleation could occur in these regions, particularly if the polyelectrolyte molecular weight is high. Longer chains would be more likely to create larger regions, perhaps longer-lived and with higher counterion concentration. These regions may also be active in flocculation.

Polyelectrolyte + Counterions on a Negative Surface. Both pAA and pAsp + 7 Ca^{+2} complexes had stable energy minima near the negative 0 0 1 surface. However,

NOTE: Color plates appear in color section.

when the dynamics was started, neither complex approached the surface. The pAA complex rapidly deposited 3 out of the 7 Ca^{+2} ions on the surface. The chain + 4 remaining Ca^{+2} ions remained near the surface for the remainder of the simulation, elongating somewhat. The pAsp complex changed much less quickly. During 1000 ps of dynamics, it moved away from the surface slightly (significantly less than the pAA complex), possibly leaving 1 Ca^{+2} at the surface.

Both the pAA and pAsp complexes moved more quickly on the negative 1 0 -2 surface. The pAA deposited 4 out of the 7 Ca^{+2} ions on the surface. The chain + remaining ions remained near the surface for a few hundred ps, the chain elongating during this time; then the chain + 3 Ca^{+2} moved away from the surface. During 1000 ps of dynamics, the pAsp complex appeared to deposit 3 Ca^{+2} on the surface and move away slightly. A subsequent simulation showed that after ~5000 ps, the pAsp complex also moves away from this surface.

These simulations suggest that counterion bridging does not occur between either pAA or pAsp and the negative 0 0 1 and 1 0 -2 calcite surfaces.

Binding energies on the negative surfaces are shown below:

Table IV. Binding Energies of pAA and pAsp with Counterions on Negative Surfaces

Polymer	Surface	E_{bind}	E_{bonded}	E_{vdw}	E_{elec}	E_{H-B}
pAA	0 0 1	-163.87	-49.49	-9.84	-104.53	
pAsp		-30.41	12.02	-7.62	-34.70	-0.11
pAA	1 0 -2	-263.85	-46.05	-1.56	-216.25	
pAsp		-161.17	8.12	-6.82	-161.65	-0.82

Given that visual inspection of the dynamics trajectories showed no real chain binding to the surfaces, why are the binding energies negative? Apparently for 7 Ca^{+2} per chain, partitioning between the chain and the surface results in a lower energy state than all counterions staying on the chain. This is consistent with my earlier finding that the first counterions to bind to a polymer chain have more negative binding energies than do counterions which bind after several other counterions are bound to the chain (7). Apparently a negative E_{bind} doesn't necessarily indicate binding; it may result from an energetically favorable rearrangement of the polymer and counterions.

Polyelectrolyte, with or without Counterions on a Neutral Surface. Binding of both polyelectrolytes, with and without counterions, to the 1 0 4 surface (the neutral surface present in macroscopic crystallites) was extremely weak. Binding to the 2 0 3 growth surface was stronger, though not so strong as binding to the positive surfaces. Polyelectrolyte binding to the 2 0 3 surface can have a broad influence on $CaCO_3$ crystal growth: kinetics, crystal habit, ultimate particle size distribution.

Binding energies for the polymers with and without counterions on the neutral surfaces are shown on the next page:

Table V. Binding Energies of pAA and pAsp on Neutral Surfaces

Polymer	Surface	E_{bind}	E_{bonded}	E_{vdw}	E_{elec}	E_{H-B}
pAA	1 0 4	-109.98	-39.02	-33.26	-37.70	
pAsp		-78.88	9.70	-29.36	-58.03	-1.18
pAA,Ca^{+2}		-150.70	-47.18	-25.55	-77.98	
pAsp, Ca^{+2}		-34.89	13.47	-15.26	-35.33	-2.08
pAA	2 0 3	-202.51	-27.16	-35.06	-140.29	
pAsp		-260.02	14.54	-28.69	-245.77	-0.10
pAA, Ca^{+2}		-269.06	-46.82	-23.78	-198.47	
pAsp,Ca^{+2}		-165.58	12.70	-29.92	-146.53	-2.13

Visual inspection of the simulation results suggests that both pAA and pAsp are able to find a positive cleft on the surface and snuggle down into it. After 1000 ps, most of the pAA segments are in contact with the surface, though the chain does not adsorb as flat as it does on a positive surface. After 1000 ps, less than half of the pAsp segments are in contact with the surface: much more of the polymer is in loops and tails which project away from the surface.

The polyelectrolyte + 7 Ca^{+2} complexes appear to snuggle into the same positive cleft as do the polyelectrolytes alone. After 1000 ps of dynamics, the pAA complex appears to lose 3 out of the 7 Ca^{+2} counterions; these remain near the chain but in position on the surface to form an additional ion layer. After 1000 ps of dynamics, the pAsp shows no signs of losing any counterions. This apparent difference between the pAA and pAsp may not be real - it would be interesting to see what would happen in longer simulations.

Examination of the binding energies in the light of the observed behaviour suggests the following: counterions lost by pAA are more likely to bind on the surface, while counterions lost by pAsp are more likely to be released back into solution. However, this inference must be regarded as tentative, since it is not clear that the chains were truly equilibrated at the end of the dynamics run.

Conclusions

The simulations done here suggest two major conclusions:
1) Both polyelectrolytes show a strong tendency to adsorb flat against positive surfaces and the 2 0 3 growth surface.
2) The pAA and pAsp show similar conformational changes on adsorption, but the pAsp moves into its final conformations on the surface much slower than does the pAA.

The pAsp is slower to assume a flat conformation than the pAA, and slower to move away from negative surfaces.

Flat adsorption is particularly interesting to us for the following reasons. Theoretically, the adsorption geometry, flat vs. nonflat, can influence polyelectrolyte performance in detergent builder, mineral dispersant, and water treatment applications. In all of these applications, a major function of the polyelectrolyte is to act as a dispersant. Generally, polymeric dispersants can contribute both electrostatic and steric stabilization. In the pAA - calcite system, the stabilization is probably predominantly electrostatic: higher MW pAA's flocculate calcite. To provide electrostatic stabilization, the polyelectrolyte provides a charged layer around the particle; formally, the layer is located not at the particle surface but at the adsorbed polyelectrolyte - solution interface *(16)*. The exact position of this charged layer can have a strong effect upon electrostatic repulsion between particles *(17)*. Generally, moving the charged layer away from the particle surface increases the electrostatic repulsion. This in turn suggests that flat adsorption does not give the best dispersancy; dispersancy can be improved if the chain can be inhibited from adsorbing flat against the particle surface. Experimentally, we have had good success in improving the dispersancy of pAA by introducing a small amount of bulky co-monomer into the chain *(3)*.

If chain adsorption is slow enough, and if there is significant desorption and re-adsorption of chains, the fact that pAsp is slower than pAA to assume a flat adsorbed conformation may mean that it is a better $CaCO_3$ dispersant, at least on a per-COO^- basis. However, dispersancy is usually measured on a per-weight basis in practical applications. Whether the per-COO^- basis would translate into a per-weight basis is unknown.

Literature Cited

1. Fitzwater, S. J.; Freeman, M. B. *Proc. Am. Chem. Soc. Poly. Mat. Sci. Eng.* **1993**, *68*, 53.
2. Hann, W. M.; Robertson, S. T. *Industrial Water Treatment* **1991**, *23(6)*, 12.
3. Fitzwater, S. J.; Freeman, M. B. Presentation at the 82nd American Oil Chemist's Annual Meeting, 1991.
4. Mosig, J.; Gooding, C. H.; Atencio, A. H.; Ballino, B.; Wheeler, A. P. Presentation at the 207th National Meeting of the American Chemical Society, 1994.
5. Ander, P.;Kardan, M. *Macromolecules* **1984**, *17*, 2436.
6. Stenius, P.; Jarnstrom, L.; Rigdahl, M. *Colloids and Surfaces* **1990**, *51*, 219.
7. Fitzwater, S. *Polymer Preprints* **1992**, *1*, 710.
8. Mayo, S. L.; Olafson, B. D.; Goddard III, W. A. *J. Phys. Chem* **1990**, *94*, 8897.
9. Whitlow, M.; Teeter, M. M. *J. Am. Chem. Soc.* **1990**, *112*, 7903.
10. Warshel, A. *J. Phys. Chem.* **1979**, *83*, 1640.
11. Allinger, N. L. *J. Am. Chem. Soc.* **1977**, *99*, 8127.
12. Still, W. C.; Tempczyk, A.; Hawley, R. C.; Hendrickson, T. *J. Am. Chem. Soc.* **1990**, *112*, 6127.
13. Donnay, J. D. H.; Harker, D. *Am. Mineral.* **1937**, *22*, 463.
14. Jarnstrom, L.; Stenius, P. *Colloids and Surfaces* **1990**, *50*, 47.
15. Lamarche, J-M.; Persello, J.; Foissy, A. *Ind. Eng. Chem. Prod. Res. Dev.* **1983**, *22*, 123.
16. Miklavic, S.; Marcelja, S. *J. Phys. Chem.* **1988**, *92*, 6718.
17. Marra, J.; Hair, M. L. *J. Phys. Chem.* **1988**, *92*, 6044.

RECEIVED January 19, 1995

Chapter 24

Simulating the Behavior of Organic Molecules in Zeolites

C. M. Freeman[1], D. W. Lewis[2], T. V. Harris[3], A. K. Cheetham[4],
N. J. Henson[4], P. A. Cox[5], A. M. Gorman[1], S. M. Levine[1,4],
J. M. Newsam[1,4], E. Hernandez[2], and C. R. A. Catlow[2]

[1]BIOSYM Technologies, Inc., 9685 Scranton Road, San Diego, CA 92121
[2]Davy Faraday Research Laboratory, The Royal Institution,
21 Albemarle Street, London W1X 4BS, England
[3]Chevron Research and Technology Company, P.O. Box 1627,
Richmond, CA 94802–0627
[4]Materials Department, University of California,
Santa Barbara, CA 93106
[5]Department of Chemistry, University of Portsmouth, Saint Michaels
Building, White Swan Road, Portsmouth P01 2DT, England

The interactions between microporous materials and organic molecules
are critical in many industrial processes. Molecular modeling is widely
employed in understanding these interactions and in making predictions
for systems yet to be explored experimentally. Computational
techniques suitable for the analysis of zeolite-sorbate interactions are
summarized and recent studies described. Applications highlighting the
impact of molecular modeling in rationalizing zeolite templating, shape
selectivity and transport properties are provided.

The application of molecular modeling techniques to the study of hydrocarbons in
zeolites has grown substantially in recent years. The stimulus for this growth stems
from many sources: the evolution of appropriate simulation techniques, the
demonstration of successful predictions, and the importance of industrial zeolite
applications have all been important incentives. Additionally, zeolites, unlike many
other heterogeneous catalysts, are often well characterized structurally because of their
high crystallinity. Reasonable validation of simulated model systems is, therefore,
possible. Furthermore, through careful control of the conditions of synthesis and
subsequent chemical modification it is possible to exert considerable influence over
the properties of a zeolitic system. Modeling provides the ability to rationalize
empirical observations and predict the effect of known and proposed modifications.

Background

The industrial applications of zeolites provide the strongest impetus in their study.
Catalysts figure in the production of trillions of dollars worth of fuels and chemicals
worldwide each year and zeolites represent a large and growing proportion of these
catalytic materials. Although they are ubiquitous, and critically effect gross national
products, catalyst-based industrial chemical processes are usually sophisticated, but

0097–6156/95/0589–0326$12.00/0
© 1995 American Chemical Society

practical compromises. Complex and contrasting factors, ranging from thermodynamics to economics, must be balanced and harnessed to yield a successful commercial activity. Flexibility, coupled with fine control are desired of all the components of such a system, including at its core the catalyst or separating agent. The diverse demands on the system over its lifetime typically require that much effort be dedicated to optimizing the performance of commercial catalysts and sorbents. Specific requirements are varied and sometimes seemingly contradictory. A class of material which can satisfy such demands, therefore, must possess unusual properties. Zeolites, although crystalline, often possess very high effective surface areas because of their microporous structures. Zeolites are solid, though they can display high Lewis or Brønsted acidity, and furthermore, their acidity can be controlled. Zeolites maintain their structures at elevated temperatures, yet they can flex to admit and accommodate large molecules. The contrasting properties of zeolites make them match many industrial processes.

Although zeolites do possess special properties they also present particular problems. Single crystals may elude production; indeed, synthesis even of powders may be difficult and desirable properties may require extensive post-synthesis modification. Molecular modeling methods can provide insight into the varied nature of zeolite chemistry and thereby assist in the solution of these problems. The role of molecular modeling techniques in structure determination and the rationalization of structure and stability has been reviewed extensively elsewhere. In this chapter we examine the modeling of zeolite hydrocarbon interactions. This is a key area of zeolite chemistry as it is these host-guest interactions which govern the majority of commercial zeolite applications. Modeling can be applied in the selection of suitable template molecules for synthesis, the rationalization of observed binding affinities and the analysis of dynamical behavior. Applications in these areas have expanded greatly in the last ten years with reported studies sampling the extensive range of computational chemistry's simulation techniques. Here, we summarize the methodology of such simulations and overview some of their recent applications to problems in zeolite chemistry.

Zeolite Structural Chemistry. Zeolites are crystalline microporous materials; zeolite pores typically have dimensions from 3 to 13 Å and imbue the resulting structures with unique properties. Zeolite crystal chemistry has been described in detail elsewhere (*1*). Here we simply note the basic unit of their construction, the TO_4 tetrahedron, where T may be a range of atom types, and the fact that by linking TO_4 tetrahedra through bridging oxygen atoms a diverse family of cage-like structures can be formed. Figure 1 summarizes some of the key structural characteristics of zeolitic materials. Over 90 topologically distinct zeolite frameworks, each characterized by a unique three letter IUPAC code, are presently known.

Although zeolites are crystalline, it may be difficult to obtain crystals of sufficient size, L~50μm, for their efficient structural analysis by conventional single crystal diffraction. Instead it is often necessary to make use of a range of experimental methodologies, including powder diffraction, to elucidate structures. Molecular modeling techniques play a significant role in the construction of atomic descriptions compatible with all available experimental evidence (*2-8*). Furthermore once a detailed description of atomic structure has been achieved, a diverse range of information can be derived using computer simulation techniques.

Synthesis. Synthesis of microporous materials usually involves crystallization from a pre-formed gel under hydrothermal conditions. The gels used are multicomponent systems containing, in zeolite synthesis, various aluminosilicate species. Organic bases are often used during the synthesis of microporous materials, a particular motivation being their apparent templating effect (*9*), as we shall see, this is an area where modeling is having a significant impact. However, the complex and varied conditions of such syntheses have given rise to several theories as to the role of such organic cations or molecules in the crystallization process (*10*).

Zeolite Applications. Zeolites are exploited broadly in industry. Zeolite ion exchange properties allow application as water softening additives for washing powder, and in the treatment of contaminated water. Zeolitic sorbents are used in several industrially important processes, such as hydrocarbon and air separations. The most important industrial area of catalytic application is in fluid catalytic cracking (FCC) (*11, 12*) whereby crude oil is broken down to lower molecular weight fragments used in gasoline or as feedstock for the petrochemical industry. Here an acidic form of Y zeolite which has same framework structure as the mineral faujasite is employed as the primary acidic catalyst. Zeolites are also widely exploited for their shape selectivity. Reactants, products and intermediates are sterically constrained by a particular zeolite's pore system leading to many specific and shape selective catalyst applications.

Simulation Methodologies

Molecular simulation techniques are extensively reviewed elsewhere in this volume. In this section we provide illustrations of certain of these general procedures applied specifically to zeolite problems.

Potential Energy Functions for Zeolite Systems. Computational simulation techniques have at their core expressions for the potential energy of the molecular system as a function of its configuration. In general the potential energy, $E_{potential}$, of the system can be represented as a sum of terms corresponding to many-body terms of increasing complexity:

$$E_{Potential} = \sum_{ij} \frac{q_i q_j}{r_{ij}} + \sum_{ij} \Phi_{ij}(r_{ij}) + \sum_{ijk} \Phi_{ijk}(r_{ijk})$$
$$+ \sum_{ijkl} \Phi_{ijkl}(r_{ijkl}) + \dots \qquad (1)$$

The first term on the right hand side of equation 1 describes the Coulombic energy; for periodic systems this conditionally convergent sum may be evaluated using the Ewald technique, though there are subtleties for non-centrosymmetric space group symmetries (*13*). Subsequent terms are sums over all pairs, triads and higher ordered arrangements of particles. A generalized expression such as equation 1 provides a convenient form for the evaluation of the potential energy of the zeolitic system. In most practical applications, however, the sums over interacting particles will extend only as far as 'four body' terms, and even here the summation is always carried out only over a selected subset of atoms determined by the chemical connectivity of the system. The chemical interpretation of such 'four body' terms might then be as means of describing torsional potentials in hydrocarbon molecules, for example, rather than a more abstract, but generally applied, four body interaction potential. Equation 1, and the assumptions made to approximate it, provide a useful point of analysis for a given potential model. For example, in non-polar systems the long range Coulombic contribution may be small in comparison with the short range component of interaction energies and it may be practical to ignore this term. In the simulation of zeolite frameworks without hydrocarbon sorbates it may be possible to truncate equation 1 simply at the two body term (see, for example, *14, 15*).

The simulation of zeolite frameworks and zeolite frameworks containing extra-framework ions has received considerable attention and many different potential models have been described for such systems(*15-17*). The simulation of organic molecules using empirical energy functions, such as equation 1 above, has developed into a standard technique for probing the structure and energetics of organic systems generally known as 'Molecular Mechanics'(*18*). However, the incorporation of hydrocarbon molecules into models designed to describe zeolite frameworks is not without difficulty. Electrostatic interactions in particular may require special treatment. Molecular mechanics typically ignores electrostatic interactions between bonded atoms, with the Coulombic term essentially used to describe intermolecular rather than the intramolecular interactions. However, many zeolite energy functions are based on a highly ionic description, where the nearest neighbor bonding is represented as largely electrostatic in nature, and consequently atomic charges are set to formal values. Such a potential may exert an extreme influence over a polar hydrocarbon sorbate necessitating either a special treatment of host-guest interactions or a more fully integrated approach to potential development. A range of potential functions suitable for simulating zeolite systems have been reported (*14-20*).

Static Simulation. Minimization of the potential energy computed by a suitable energy function with respect to appropriately chosen variables provides a means of optimizing zeolite structural models. Such procedures, which yield energy minimum or low temperature structures, have been successfully applied to zeolite framework structures, extra-framework cation positions, and modeling sorbate locations. Energy minimizations, based on efficient numerical procedures (*21-23*), provide information on low energy sorption sites rather than probing the thermodynamic ensemble of states accessed under thermal conditions. However, minimization can probe favorable binding locations for guest molecules in systems whose size complicates more detailed calculations. The quality of all computational simulations is determined by the quality of the potential energy function employed. Energy minimization relies on an accurate representation of minima and therefore places lower demands on the accuracy and shape of the potential than vibrational, dynamical or Monte Carlo calculations, all of which also require information on the curvature about minima and even details of the energy surface distant from minima.

Vibrational Studies. At a minimum of the chosen potential energy function (equation 1), with respect to suitably chosen coordinates, the second derivative of the potential provides information about the cumulative vibrations of the system. Diagonalizing the mass weighted second derivative matrix yields the deformation coordinates of the zeolite system about the minimum studied. Such simulated data relates directly to the optical spectroscopic modes sampled by IR or Raman spectroscopy (see reference (*24*), for example) and can also form the basis of a means for determining the free energy of the system (*25*). Molecular dynamics calculations, described below, sample the vibrational degrees of freedom of the zeolite system, by virtue of their intrinsic inclusion of thermal variations, and can also yield vibrational information.

Molecular Dynamics. The first derivative of the energy function yields the forces experienced by the system components. This, combined with appropriate mass information provides acceleration, which, from a suitable starting point, allows the dynamics of the system to be calculated. Finite time steps are employed in integrating the equations of motion and a wide range of observable characteristics, from diffusion coefficients to sorption enthalpies, can be extracted from the simulations (*26*). However, there are practical limits to the lengths of time which can be modeled using molecular dynamics calculations. This upper limit currently extends to the nanosecond domain. Such a simulation will entail of the order of 10^6 energy and derivative calculations and represent a considerable computational investment. Large sorbates can, however, diffuse more slowly than can be accessed by even such lengthy calculations.

Monte Carlo. Since equation 1 and its analogs provide the potential energy of the zeolite system for a given configuration we can use computational statistical mechanics to obtain thermodynamic information for a particular system. In practice the necessary configurational integrals are performed numerically using the Metropolis (27) importance sampling scheme. Such procedures have been extensively employed in the simulation of zeolite-sorbate interactions (28-33). Monte Carlo procedures within the Grand Canonical ensemble also provide a means of estimating the loading favored by zeolites interacting with potential sorbates at a particular temperature and pressure. Here rather than simply displacing sorbate particles, as in the standard Metropolis scheme, particle insertions and deletions are also made so that the number of sorbate molecules can evolve to equilibrium values for the specified sorbate chemical potential or partial pressure.

Hybrid Studies. The complexity of many zeolite systems prevents direct application of traditional modeling methods: Potential sorbates may be flexible, too large to diffuse within simulation periods, and local minima may hamper the routine application of minimization procedures. These problems can sometimes be circumvented by applying the methodologies described above in consort. Thus molecular dynamics may be employed to search hydrocarbon conformational flexibility, Monte Carlo methods used to place sorbate configurations within the host lattice, and finally energy minimization applied to locate low energy binding sites (34). Similarly Monte Carlo searching techniques have been employed as an adjunct to constrained energy minimization in probing low energy diffusion pathways in zeolite channels (35).

Applications

The use of these various methodologies in simulating zeolite hydrocarbon interactions is growing rapidly. In this section we provide illustrations of recent topical applications. Zeolite synthesis, the first step along the path to an industrial process, has for decades been principally an empirical art but molecular modeling is beginning to contribute to our understanding of the mechanics of the formation of different zeolite structures. Many zeolite systems exert pronounced shape selectivity on the reactions which they mediate, a direct result of the steric constraints of their open porous structures. Molecular modeling is proving valuable in rationalizing and predicting not only shape selectivity but also subtle variations on this classical theme such as 'inverse shape selectivity'. Modeling is a valuable adjunct to experimental investigation in probing hydrocarbon location and mobility within zeolite systems, and we highlight recent molecular dynamics results in this area. We also provide recent examples of the modeling of densely packed systems and the modeling of the thermodynamic properties of hydrocarbon-zeolite systems through Monte Carlo procedures.

Different hydrocarbons interact with their zeolite hosts in different ways and a full range of simulation techniques are, therefore, regularly deployed in probing these interactions. If the interaction between host and guest is weak, and the guest molecule is small, diffusion will be rapid and molecular dynamics techniques can yield valuable information. Alternatively, if the hydrocarbon can interact strongly with the pores of its zeolite hosts, diffusion may be slow, direct dynamical simulation will, therefore, be inappropriate. Instead, constrained energy minimization, vibrational analysis, Monte Carlo methods, or for translational diffusion, transition state theory will be preferred. It may be that information concerning the transport of hydrocarbons through the zeolite lattice is not required of the simulation, sorption energetics and likely binding sites being of principal interest, in this case the Monte Carlo based procedures will be appropriate. In practice, then, each of the simulation techniques outlined in the previous section can be exploited in the rationalization of molecule-zeolite interactions.

Subtle changes in either host or hydrocarbon can have dramatic effects on intra-zeolite processes. For example, trimethylbenzene diffuses around 1 million fold more slowly in zeolite ZSM-5 than the *para*-dimethylbenzene molecule (36). The origin of this dramatic difference is readily apparent when the size and shape of both guest molecules are compared with the pore dimensions of ZSM-5 using interactive molecular graphics. Trimethylbenzene is unable to fit within the lattice without significant strain, and, furthermore, this strain must be increased in the process of diffusion. In contrast the *para*-dimethylbenzene (*p*-xylene) molecule fits neatly into the channel structure of the zeolite and it can be seen that diffusion does not entail significant steric overlap.

Although molecular graphics provides a qualitative rationalization of the observed differences in diffusion, a quantitative description would demand a detailed molecular dynamics study. As the diffusion rates span six orders of magnitude, we should anticipate that the simulations themselves, in order to capture the observed diffusion differences, should encompass similarly protracted time periods. At present classical dynamical simulations of the methylbenzenes are not practical as simulation runs rarely exceed 10^6 time steps. Hence the differences that can be probed by molecular dynamics must be compressed into, at most, the nanosecond time frame. In practice, therefore, we cannot rely on molecular dynamics to yield desired dynamical information for every system. However, by making assumptions about possible diffusion pathways it is possible to use constrained minimization and vibrational analysis methods in investigating the transport properties of zeolites (*35, 37*).

There are procedural challenges, then, to be overcome in the simulation of some zeolite-hydrocarbon systems. However, in practice, useful information can usually be obtained for even the most taxing systems. Recent studies have shown how molecular modeling can play an important role in understanding the process of zeolite formation.

Templates in Zeolite Synthesis

The syntheses of many novel zeolites rely on the use of organic additives. These 'template' molecules are occluded inside the zeolite as it crystallizes and can then be removed via calcination. Experimental determination of the locations adopted by these molecules inside the zeolite is difficult; X-ray diffraction studies, for example, are hindered by the polycrystalline nature of most zeolite samples. As a result, only a limited number of experimental investigations have been performed. However, as we shall see, the agreement between available experimental structures and predicted structures is excellent.

We are still some way from understanding the mechanism of framework formation during hydrothermal synthesis under the influence of structure directing templates. Indeed the mechanism of framework crystallization in the absence of organic additives remains itself uncertain. This section will highlight the role which computer modeling techniques can play in advancing our understanding of templating in the synthesis of novel microporous materials. The recognition of 'goodness of fit' between template and its framework is well established (*10*) and has parallels in the Fischer's famous 'lock and key' analogy for enzyme-substrate specificity. Quantification of the underlying determinants of this relationship has recently been reported (*38-40*).

In these studies trends in template-host interaction are analysed in terms of non-bonded interaction energies and the shape (as revealed by moments of inertia (*39*)) of template molecules. In assessing the energy of interaction between template and framework it proves necessary to account for the flexibility of the organic molecule. This has been achieved by making use of a hybrid simulation procedure (*34*) where a library of conformations is generated for the potential template using molecular dynamics procedures and representative configurations are then docked into the framework using a Monte Carlo procedure. Finally energy minimization is applied to yield a range of feasible low energy configurations for the template within the lattice.

Figure 2 illustrates the relationship between computed template-framework interaction energy and the number of non-hydrogen atoms of the template molecule.

The apparent trend implies that all atoms within the template contribute approximately equally to the interaction energy and hence occupy similar positions in relation to the framework. This is not an unreasonable view of the tight binding between template and its framework and, indeed, the trend might be used predictively in the selection of templates for synthesis of a particular material. The trend is most marked within a class of template molecule such as diamines (Lewis, D. W.; Freeman, C. M.; Catlow, C. R. A. 1994 *submitted for publication*). As is inevitably the case with molecular mechanics calculations the absolute template-zeolite interaction energies of these studies are dependent on the parameterization of the energy function (equation 1). However, it encouraging to note that similar trends are observed using different parameterizations. The general observation that favorable host-guest interactions are optimized for 'good' templates is, therefore, reasonable and provides quantitative evidence that templating is, in some way, the result of a 'good fit' between host and template. The measured vibrational characteristics of tetramethylammonium cations occluded within aluminosilicates provide experimental evidence of substantial template-framework interactions (*41,42*).

The formation of a zeolite in the presence of templating molecules is, in practice, a kinetically controlled process. Indeed, Bell and Chang have presented a model of zeolite formation as an event which proceeds from clathrate organization (*43*). Harris and Zones have demonstrated a striking relationship between host-guest interaction energy and crystallization rates for the zeolite-like cage structure nonasil and the zeolite chabazite and a range of organic template molecules (*38*). Figure 3 illustrates the relation between the calculated template nonasil-interaction and the length of time taken to yield the crystalline product for several templates. The relationship can be used predictively; poor stabilization energy for a particular template leads to long crystallization times. Clearly favorable template-nonasil interactions are associated with higher rates of crystallization. Since crystal growth is significantly faster than nucleation for the nonasil system the influence of the template is probably most significant at the nucleation stage. We can speculate that the template is able to stabilize the cluster configuration(s) leading to viable nuclei which in turn hastens crystallization. If this is indeed the critical role of the organic templates, the trend in Figure 3 implies that the nucleating center resembles the cage of the nonasil framework; as it is the crystallized framework structure of nonasil which is used to evaluate the framework-template interaction, rather than a hypothetical model of the clathrate structure. It is important to emphasize the fact that entropic terms have to date been neglected in studies of template-host interactions. Recent combined theoretical and experimental investigations of the nonasil system have illustrated circumstances where this approximation may not be valid. We note that the incorporation of entropic terms into zeolite host-guest molecule system simulations is possible through either vibrational calculations or Monte Carlo methods, though computationally highly demanding.

Recent studies have focussed on probing the interactions of template molecules with the frameworks in which they induce crystallization. In the absence of models that incorporate the effects of gel chemistry and composition this is a reasonable initial approximation. The availability of crystal structure data for zeolite frameworks containing occluded template molecules allows us to test the structural accuracy of the simulation procedures. For example, Figure 4 illustrates the experimentally observed locations of two templates, 1-aminoadamantane and the N-methylquinuclidinium cation, in the LEV framework (*44*) and the corresponding locations determined using Monte Carlo docking. Clearly there is excellent agreement between experiment and calculation.

The configuration of templates within the crystallized zeolite lattice and hence, presumably, their structure-directing role may also be altered by template-template interactions. Thus although, the tetrabutylammonium cation in isolation enjoys favorable interactions with the ZSM-5 framework, the butyl chains are large enough to

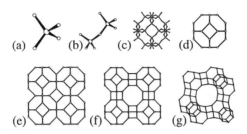

Figure 1. Some elements of zeolite structural chemistry (not drawn to scale). TO_4 tetrahedra (a) linked through shared oxygen atoms (b) to form cages, chains and sheets. Illustrated here the sodalite cage (c) and showing framework cation positions only, (d). Also shown are the framework structures of sodalite (e), zeolite A (f), and faujasite (g).

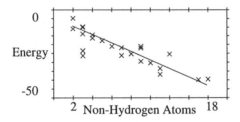

Figure 2. Computed interaction energy (kcal mol[-1]) for experimental framework-template combinations as a function of the number of non-hydrogen template atoms.

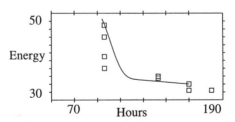

Figure 3. Comparison of computed stabilization energies (kcal mol[-1]) with nonasil crystallization times for a range of template molecules.

hinder one another once the complete zeolite framework is established and it is tetrapropylammonium cation which is the preferred template in ZSM-5 synthesis (*40*). Apparently two complementary criteria need therefore to be satisfied for an organic molecule to template successfully a zeolite. First, the favorable non-bonded interaction between template and framework must be optimized. Second, the template molecules must be able to pack efficiently within the zeolite framework. Additional constraints are certainly imposed by the mechanics and kinetics of crystal growth.

The Monte Carlo docking procedure has been extended to account for the effects of template interactions within the zeolite lattice (Cox, P. A.; Stevens, A. P.; Freeman, C. M.; Gorman, A. M. *In preparation*). The Monte Carlo stage of the calculation selects at random a location and orientation for the guest molecule within the confines of the host structure. As for the isolated guest molecule case, this configuration is retained only if its interaction energy falls below a specified threshold level. Further molecules are automatically packed into the cell subject to appropriate periodic boundary conditions. Simulated annealing (*45*), in the form of high temperature and quenched molecular dynamics, is then applied to obtain low energy packed configurations. The procedure is summarized schematically in the series of diagrams of Figure 5 illustrating the use of the procedure in generating a packed model for *p*-xylene molecules in ZSM-5. Trial applications of this new procedure to model systems, such as N-methylquinuclidinium in the LEV framework, have yielded excellent agreement with experiment.

Shape Selectivity and Inverse Shape Selectivity

The important industrial role of zeolites provides a strong incentive for the calculation of catalytic reactivities and selectivities, as a means of both rationalizing experimental results and focussing experimental effort. For many catalytic systems in the petrochemical and refining industries an interplay of complex factors govern selectivity. Simulation is often highly informative, especially when one factor, such as rate of diffusion or preferred binding, dominates.

Such a situation arises in the industrial exploitation of shape selectivity (*46*) where the steric constraints of the zeolite can exert a direct influence over the reactions catalysed within its framework. As in pharmaceutical modeling, molecular graphics and molecular modeling techniques are an effective tool for studying the shape selectivity imposed by a particular zeolite structure (*47-56*). However, the influence exhibited by the host lattice can be subtle. For example 'inverse shape selectivity' (*57*) is manifested for C_6-C_7 paraffinic hydrocarbons by molecular sieves containing void spaces of around 7 Å in diameter (*58*). Here molecules whose size and shape match the void space are preferentially sorbed over smaller or isomerically related molecules. Hence the shape selectivity is attractive in its action as opposed to the normal, repulsive or excluding selectivity.

Inverse shape selectivity has been observed in materials of the AFI structure type in a 'pore probe' test (*59*) in which 2,2-dimethylbutane (DMB) is selectively absorbed from a 1:1:1 mixture of DMB, 3-methyl-pentane (3MP) and n-hexane at 130C. In hexadecane hydrocracking experiments using Pd-zeolite catalysts (*58*), the acid forms of Al-SSZ-24, MAPO-5 and SAPO-5 also showed increased DMB production in the hydrocracked C_6 products compared to 3MP and n-hexane. These two separate lines of experimental evidence establish a strong link between the inverse shape selectivity observed in sorption and catalytic experiments (Figure 6).

Since the C_6-C_7 fraction of 'light naphtha' from hydrocracking is blended directly into the gasoline pool without reforming, high octane is a highly desirable characteristic. One method of increasing the octane rating is to increase the fraction of DMB and multiply-branched C_7 isomers produced by the hydrocracking catalyst. The realization of zeolite based catalyst systems capable of giving increased DMB yield would, therefore, be of significant interest.

One means of determining DMB:nC$_6$ and 3MP:nC$_6$ ratios sorbed in zeolite pores requires the calculation of low occupancy sorption heats. These can be obtained using Metropolis Monte Carlo methods (*60*). Experimentally the sorbed ratios of DMB:nC$_6$ and 3MP:nC$_6$ are measured; these may be estimated from the calculated sorption energy differences by the following relation:

$$\Delta E \cong -RT \ln K_{eq} = -RT \ln ([A_{zeo}]/[B_{zeo}]) \qquad (2)$$

where $[A_{zeo}]/[B_{zeo}]$ is the ratio of the concentrations of hydrocarbons A and B sorbed within the zeolite pores. The expression is approximate as the calculated energy difference is not a free energy, and terms related to pressure and entropic differences are assumed to be equal for both A and B. Calculated and observed ratios are presented in Figure 6. It can be seen that the agreement is good; providing confirmation of the validity of the simplifying assumptions employed.

Figure 6 also highlights the variation of the calculated and experimentally observed DMB:nC$_6$ ratio as a function of void space size for a range of molecular sieve structures. This figure shows the effect of inverse shape selectivity for structures with void spaces of about 7 Å which results from the preferential DMB sorption. This modeling study demonstrates that it is, in principle, now possible to examine the potential inverse shape selectivity of a given zeolite in advance of experimental investigation. Indeed, it is possible to screen hypothetical frameworks (*61,62*) for desirable properties using simulation procedures prior to their synthesis. The companion successes in the zeolite templating area suggest the possibility of attempting a directed synthesis of a desirable target framework structure using a template selected or designed by modeling.

Diffusion

In view of the considerable interest in the diffusivity of organic molecules in zeolites there has been much activity in the dynamical simulation of such systems (*63-71*). Molecular dynamics studies can be informative. For example, in the recent study of Yashonath and Santikary (*72*) two distinct, modes of xenon diffusion, one surface mediated and the other accessing pore central regions, were highlighted in sodium zeolite Y. An additional example is provided by the recent work of Kawano and coworkers in the calculation of diffusion coefficients of methane in a silicalite (*68*) and, for a range of larger hydrocarbons, by Hernandez (Hernandez, E. PhD Thesis, University of London, 1993) (see also table I). The calculated diffusion coefficients obtained appear to accord well with experimetal data.

Table I. Calculated Diffusion Coefficients of Silicalite Adsorbates at 300K and Loadings of 1 and 2 Molecules per Unit Cell

Adorbate	$D \times 10^7$ cm^{2}s^{-1} 1 molecule/cell	$D \times 10^7$ cm^{2}s^{-1} 2 molecules/cell
Ethane	692	620
Propane	56	68
n-Butane	7	14
n-Hexane	14	9

In the investigation of transport properties molecular dynamics calculations do not require *a priori* mechanistic assumptions. To probe a particular hypothesis the system can be allowed simply to evolve for a sufficient length of time rather than be laboriously probed for a sequence of alternative hypothetical models. With the

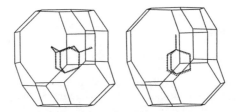

Figure 4. Comparison of calculated (dashed line) and experimental (solid line) locations for template molecules within the LEV framework. Left 1-aminoadamantane and right N-methylquinuclidinium cation (only one of several equivalent crystallographic configurations is shown).

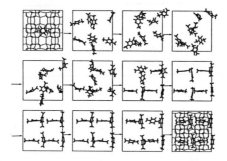

Figure 5 Stages in the combined Monte Carlo docking/simulated annealing packing of p-xylene in ZSM-5. Sequence starts top left (framework not drawn in intermediate steps for clarity).

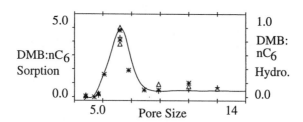

Figure 6. Experimental and calculated DMB:nC6 ratios as a function of pore size in Å. Sorption experiments labeled x, hydrocracking data Δ, and calculated values *.

assistance of modern molecular graphics procedures (*73*), diffusion mechanisms can be extracted from the temporal evolution of the system. However, molecular dynamics calculations are constrained by two important factors: firstly the potential model must provide an accurate description of interatomic interactions at any separation accessed in a simulation, and, secondly, the small size of the time steps employed in the integration of the equations of motion dictate that large amounts of computer time must be expended to simulate comparatively modest amounts of real time. Several strategies are evolving to alleviate these problems. For example, sophisticated numerical integration schemes permit the use of longer time steps for zeolite systems (*74*). We note also that transition-state theory Monte Carlo methods have been employed to determine the diffusion coefficients of large molecules in zeolite systems (*75*). Here it is possible to probe transport too slow to be accessible to molecular dynamics calculations at the expense of detailed mechanistic information. Several workers have removed the structural flexibility of the zeolite from the simulation (*69*). This has the effect of focusing computational effort on the diffusing entity rather than the host lattice. However, there is evidence that framework flexibility influences the motion of sorbed species (*71*).

Future Developments

The results cited in this chapter illustrate some of the potential of modeling techniques in the study of sorbed molecules in microporous solids. We believe, moreover, that there is an exciting future in this field in three respects. The first concerns the accuracy that will be achievable by simulation techniques; the second the detail and complexity of the simulated systems; and the third, the growing possibility of simulating reactive processes.

More accurate simulations require the development of improved interatomic potentials: Approaches based on a combination of *ab initio* calculations and empirical fitting procedures, which have been used for several years with success in the study of molecules of biochemical importance (*76-78*) are now finding widespread applications in materials modeling (*19, 79-82*). Improved interatomic potentials will contribute to the modeling of details and complexity, but of ever greater significance is the growth in computer power, especially the increasing availability of massively parallel systems which are appropriate for many of the types of simulations described here. In the greatest challenge, that is the study of zeolite reactivity, hardware improvements will extend the applicability of quantum chemical methods and new developments, such as embedded cluster techniques (*83, 84*), promise to permit increasingly reliable calculations on problems of real chemical importance.

Literature Cited

1.   Newsam, J. M. In *Solid State Chemistry: Techniques;* Cheetham, A. K., Day, P., Eds.; Oxford University Press: Oxford, UK, 1992, Vol. 2, pp 234-280.
2.   Newsam, J. M.; Deem, M. W.; Freeman, C. M. In *Accuracy in Powder Diffraction II* (NIST Special Publication No. 846); Prince, E., Stalick, J. K., Eds.; National Institute of Standards and Technology: Bethesda, MD, 1992, pp 80-91.
3.   Nenoff, T. M.; Harrison, W. T. A.; Stucky, G. D.; Nicol, J. M.; Newsam, J. M. *Zeolites* **1993**, *13*, 506.
4.   Deem, M. W.; Newsam, J. M. *Nature* **1989**, *342*, 260.
5.   Deem, M. W.; Newsam, J. M. *J. Am. Chem. Soc.* **1992**, *114*, 7189.
6.   Catlow, C. R. A.; Thomas, J. M.; Freeman, C. M.; Wright, P. A.; Bell, R. G. *Proc. Roy. Soc. Lond. A* **1993**, *442*, 85.
7.   Thomas, J. M.; Vaughan, D. E. W. *J. Phys. Chem. Solids* **1989**, *50*, 449.
8.   Thomas, J. M.; Catlow, C. R. A. *Prog. Inorg. Chem.* **1987**, *35*, 1.

338   COMPUTER-AIDED MOLECULAR DESIGN

9.   Barrer, R. M. *Hydrothermal Chemistry of Zeolites;* Academic Press: London, UK, 1982.
10. Lok, B. M.; Cannan, T. R.; Messina, C. A. *Zeolites* **1983**, *3*, 282.
11. Venuto, P. B.; Habib, E. T. *Fluid Catalytic Cracking*; Marcel Dekker: New York, 1979.
12. Sapre, A. V.; Katzer, J. R. In *Computer Aided Design of Catalysts*; Becker, E. R., Pereira, C. J., Eds.; Marcel Dekker: New York, 1993.
13. Deem, M. W.; Newsam, J. M.; Sinha, S. K. *J. Phys. Chem.* **1990**, *94*, 8356.
14. de Boer, K.; Jansen, A. J. P.; van Santen, R. A. In *Zeolites and Related Microporous Materials: State of the Art 1994;* Studies in Surface Science and Catalysis, Vol. 84; Weitkamp, J.; Karge, H. G.; Pfeifer, H.; Holderich, W. Eds.; Elsevier Science BV:1994, pp. 2083-2087.
15. Jackson, R. A.; Catlow, C. R. A. *Mol. Simul.* **1988**, *1*, 207.
16. No, K. T.; Seo, B. H.; Jhon, M. S. *Theor. Chim. Acta* **1989**, *75*, 307.
17. Bell, R. G.; Jackson, R. A.; Catlow, C. R. A. *Zeolites* **1992**, *12*, 870.
18. Burkett, U.; Allinger, N.L.; *Molecular Mechanics*; ACS Monograph 177, Am. Chem. Soc.: Washington DC, 1982.
19. Kramer, G. J.; Farragher, N. P.; van Beest, B. W. H.; van Santen, R. A. *Phys. Rev. B* **1991**, *43*, 5068.
20. Tsuneyuki, S.; Tsukada, M.; Aoki, H. *Phys. Rev. Lett.* **1988**, *61*, 4792.
21. Powell, M. J. D. Ed. *Nonlinear Optimization*; Academic Press: New York, 1982.
22. Gill, P. E.; Murray, W.; Wright, M. H. *Practical Optimization*; Academic Press, London, 1981.
23. Fletcher, R.; Powell, M. J. D. *The Computer Journal* **1963**, *6*, 163.
24. Creighton, J. A.; Deckman, H. W.; Newsam, J. M. *J. Phys. Chem.* **1994**, *98*, 448.
25. Jackson, R. A.; Parker, S. C.; Tschaufeser, P. In *Modelling of Structure and Reactivity in Zeolites*; Catlow, C. R. A., Ed.; Academic Press: London, UK, 1992.
26. Allen, M. P.; Tildesley, D. J. *Computer Simulation of Liquids*, Oxford University Press: Oxford, UK, 1987.
27. Metropolis, N.; Rosenbluth, A. W.; Rosenbluth, M. N.; Teller, A. H.; Teller, G. *J. Chem. Phys.* **1953**, *21*, 1087.
28. Stroud, H. J. F.; Richards, E.; Parsonage, N. G. *J. Chem. Soc. Faraday Trans. I* **1976**, *72*, 942.
29. Kretschmer, R. G.; Fiedler, K. Z. *Z. Phys. Chem.* **1977**, *258*, 1045.
30. Yashonath, S.; Thomas, J. M.; Nowak, A. K.; Cheetham, A. K. *Nature* **1988**, *331*, 601.
31. Smit, B.; den Ouden, C. J. J. *J. Phys. Chem.* **1988**, *92*, 7169.
32. Woods, G. B.; Rowlinson, J. S. *J. Chem. Soc. Faraday Trans. 2* **1989**, *85*, 765.
33. Soto, J. L.; Meyers, A. L. *Mol. Phys.* **1981**, *42*, 971.
34. Freeman, C.M.: Catlow, C.R.A.; Thomas, J.M.; Brode, S. *Chem. Phys. Lett.* **1991**, *186*, 137.
35. Horsley, J.A.; Fellmann, J.D.; Derouane, E.G.; Freeman, C.M.; *J. Catal.* **1994**, *147*, 231.
36. Haag, W. O.; Chen, N. Y. In *Catalyst Design, Process and Perspective*; Hegedus, L. L., Ed.; Wiley: New York, 1987, pp. 163.
37. Schroder, K.-P.; Sauer, J. *Z. Phys. Chemie, Liepzig* **1990**, *271*, 289.
38   Harris, T. V.; Zones, S. I. In *Zeolites and Related Microporous Materials: State of the Art 1994;* Studies in Surface Science and Catalysis, Vol. 84; Weitkamp, J.; Karge, H. G.; Pfeifer, H.; Holderich, W. Eds.; Elsevier Science BV, 1994, pp. 29-36.
39. Cox, P. A.; Stevens, A. P.; Banting, L.; Gorman, A. M. In *Zeolites and Related Microporous Materials: State of the Art 1994;* Studies in Surface Science and Catalysis, Vol. 84; Weitkamp, J.; Karge, H. G.; Pfeifer, H.; Holderich, W. Eds.; Elsevier Science BV, 1994, pp. 2115-2122.

40. Bell, R. G.; Lewis, D. W.; Voigt, P.; Freeman, C. M.; Thomas, J. M.; Catlow, C. R. A. In *Zeolites and Related Microporous Materials: State of the Art 1994;* Studies in Surface Science and Catalysis, Vol. 84; Weitkamp, J.; Karge, H. G.; Pfeifer, H.; Holderich, W. Eds.; Elsevier Science BV, 1994, pp. 2075-2082.
41. Newsam, J. M.; Brun, T. O.; Trouw, F.; Iton, L. E.; Curtiss, L. A. In *Novel Materials in Heterogeneous Catalysis*, Baker, R. T. K., Murrell, L. L., Eds., ACS Symp. Ser. No. 437, ACS Washington DC, 1990, pp. 25-37.
42. Brun, T. O.; Curtiss, L. A.; Iton, L. E.; Kleb, R.; Newsam, J. M.; Beyerlein, R. A.; Vaughan, D. E. W. *J. Am. Chem. Soc.* **1987**, *109*, 4118.
43. Chang, C. D.; Bell, A. T. *Catal. Lett.* **1991**, *8*, 305.
44. McCusker, L. B. *Materials Science forum* **1993**, *423*, 133.
45. Kirkpatrick, S.; Gelatt, C. D.; Vecchi, M. P. *Science* **1983**, *220*, 671.
46. Thomas, J. M. *Angew. Chem. Int. Ed. Engl.* **1988**, *27*, 1673.
47. Wright, P. A.; Thomas, J. M.; Cheetham, A. K.; Nowak, A. K. *Nature* **1985**, *318*, 611.
48. Nowak, A. K.; Cheetham, A. K.; In *New Developments in Zeolite Science and Technology*; Murakai, Y.; Iijima, A.; Ward, J. W. Eds.; Kodansha and Elsevier; Tokyo and Amsterdam, 1986, 49; pp 475-479.
49. Nowak, A. K.; Cheetham, A. K.; Pickett, S. D.; Ramdas, S. *Mol. Simul.* **1987**, *1*, 67.
50. Pickett, S. D.; Nowak, A. K.; Cheetham, A. K.; Thomas, J. M. *Mol. Simul.* **1989**, *2*, 353.
51. Pickett, S. D.; Nowak, A. K.; Thomas, J. M.; Cheetham, A. K. *Zeolites*, **1989**, *9*, 123.
52. Newsam, J. M., Silbernagel, B. G.; Melchior, M. T.; Brun, T. O.; Trouw, F. In *Inclusion Phenomena and Molecular Recognition*; Atwood, J. Ed.; Plenum Press: New York, pp. 325-337.
53. Titiloye, J. O.; Parker, S. C.; Stone, F. S.; Catlow, C. R. A. *J. Phys. Chem.* **1991**, *95*, 4038.
54. Mentzen, B. F.; Sacerdote-Peronnet, M.; Berar, J.F.; Lefebvre, F. *Zeolites* **1993**, *13*, 485.
55. de Vos Burchart, E.; Jansen, J. C.; van de Graaf, B.; van Bekkum, H. *Zeolites* **1993**, *13*, 216.
56. Blanco, F.; Urbina-Villalba, G.; Ramirez de Agudelo, M. M. In *Zeolites and Related Microporous Materials: State of the Art 1994;* Studies in Surface Science and Catalysis, Vol. 84; Weitkamp, J.; Karge, H. G.; Pfeifer, H.; Holderich, W. Eds.; Elsevier Science BV, 1994, pp. 2155-2162.
57. van Nordstrand, R. A.; Santilli, D. S.; Zones, S. I. In *Synthesis of Microporous Materials*; Occelli, M., Robson, H. E., Eds.; Vol. 1; Van Nostrand: New York, 1992, Chapter 25.
58. Santilli, D. S.; Harris, T. V.; Zones, S. I. *Microporous Materials* **1993**, *1*, 329.
59. Santilli, D. S. *J. Catal.* **1986**, *99*, 335.
60. Insight II, Discover & Catalysis and Sorption Project Software Packages BIOSYM Technologies, Inc., 9685 Scranton Road, San Diego, CA 92121.
61. Smith, J. V. *Chem. Rev.* **1988**, *88*, 149.
62. Akporiaye, D. E. *Z. Kristallogr.* **1989**, *188*, 103.
63. Bull, L. M.; Henson, N. J.; Cheetham, A. K.; Newsam, J. M.; Heyes, S. J. *J. Phys. Chem.* **1993**, *97*, 11776.
64. Nicholas, J. B.; Trouw, F. R.; Mertz, J. E.; Iton, L. E.; Hopfinger, A. J. *J. Phys. Chem.* **1993**, *97*, 4149.
65. Henson, N. J.; Cheetham, A. K.; Peterson, B. K.; Pickett, S. D.; Thomas, J. M. *J. Comp.-Aided Mat. Des.* **1993**, *1*, 41.
66. Demontis, P.; Fois, E. S.; Suffriti, G. B.; Quartieri, S. *J. Phys. Chem.* **1990**, *94*, 4329.

67. Catlow, C.R.A.; Freeman, C.M.; Vessal, B.; Tomlinson, S.M.; Leslie, M.: *J. Chem. Soc. Faraday Trans.* **1991**, *87*, 1947.
68. Kawano, M.; Vessal, B.; Catlow, C. R. A. *J. Chem. Soc., Chem. Commun.* **1992**, *12*, 879.
69. Pickett, S. D.; Nowak, A. K.; Thomas, J. M.; Peterson, B. K.; Swift, J. F. P.; Cheetham, A. K.; den Ouden, C. J. J.; Smit, B.; Post, M. F. *J. Phys. Chem.* **1990**, *94*, 1233.
70. June, R. L.; Bell, A. T.; Theodorou, D. N. *J. Phys. Chem.* **1992**, *96*, 1051.
71. Demontis, P.; Suffritti, G. B. In *Modelling of Structure and Reactivity in Zeolites*; Catlow, C. R. A. Ed.; Academic Press: London, UK, 1992.
72. Yashonath, S.; Santikary, P. *J. Phys. Chem.* **1993**, *97*, 3849.
73. Freeman, C. M.; Levine, S. M.; Newsam, J. M.; Sauer, J.; Tomlinson, S. M.; Brickmann, J.; Bell, R. G. In *Modelling of Structure and Relativity in Zeolites*; Catlow, C. R. A., Ed.; Academic Press: London, UK, 1992.
74. Deem, M. W.; Newsam, J. M.; Creighton, J. A. *J. Am. Chem. Soc.* **1992**, *114*, 7198.
75. June, R. L.; Bell, A. T.; Theodorou, D. N. *J. Phys. Chem.* **1991**, *95*, 8866.
76. Dinur, U.; Hagler, A. T. In *Reviews in Computational Chemistry*, Lipkowitz, K. B., Boyd, D. B., Eds.; VCH Publishers: New York, 1991, Vol. 2, pp 99-164.
77. Maple, J. R.; Hwang, M. J.; Stockfisch, T. P.; Dinur, U.; Waldman, M.; Ewig, C. S.; Hagler, A. T. *J. Comp. Chem.* **1994**, *15*, 162.
78. Hwang, M. J.; Stockfisch, T. P.; Hagler, A. T. *J. Am. Chem. Soc.* **1994**, *116*, 2515.
79. Hill, J.-R.; Sauer, J. *J. Phys. Chem.* **1994**, *98*, 1238.
80. Kramer, G. J.; Farragher, N. P.; van Beest, B. W. H.; van Santen, R. A. *Phys. Rev. B* **1991**, *43*, 5068.
81. Purton, J.; Jones, R.; Leslie, M. ; Catlow, C. R. A. *Phys. Chem. Minerals* **1993**, *19*, 392.
82. Gale, J. D.; Catlow, C. R. A.; Mackrodt, W. C. *Modeling Simul. Mater. Sci. Eng.* **1992**, *1*, 73.
83. Grimes, R. W.; Catlow, C. R. A.; Shluger, A. L. Eds. *Quantum Mechanical Cluster Calculations in Solid State Studies*; World Scientific Publishing Co. Pte Ltd.: Singapore, 1992.
84. Sauer, J. In *Modelling of Structure and Reactivity in Zeolites*; Catlow, C. R. A. Ed.; Academic Press, London, UK, 1992, pp. 183-216.

RECEIVED November 20, 1994

Chapter 25

Valence Bond Charge Transfer Theory for Predicting Nonlinear Optical Properties of Organic Materials

William A. Goddard III, Daqi Lu, Guanhua Chen, and Joe W. Perry

Division of Chemistry and Chemical Engineering, Materials and Molecular Simulation Center, Beckman Institute, California Institute of Technology, Pasadena, CA 91125

A simple theory (VB-CT) is developed for predicting nonlinear optical properties of organic materials. Application of this theory to donor-acceptor charge-transfer molecules leads to *analytic* formulae for the absorption frequency, hyperpolarizabilities, and bond length alternation. Derivative relationships between hyperpolarizabilities (with respect to bond length alternation) are derived. Using a continuum description of the solvent in the VB-CT framework leads to the VB-CT-S model which gives results for solvent shifts in good agreement with experiment. To predict the saturation behavior of polarizability and hyperpolarizability with respect to polymer length, we developed the VB-CTE model which is applied to nine polymeric materials.

There is a great deal of industrial interest in nonlinear optical (NLO) materials for use in

    (i) optical processing of data/images,
    (ii) optical storage of data/images,
    (iii) optically based telecommunications, and
    (iv) optically based computers.

The important properties for these applications are the hyperpolarizabilities. The effect on the energy (E) of applying an external electric field $(\mathcal{E})$ is

$$E = E_0 - \mu \cdot \mathcal{E} \tag{1}$$

where μ depends on $\mathcal{E}$ as in (2)

$$\mu_z(\mathcal{E}) = \mu_{0z} + \alpha_{zz}\mathcal{E}_z + \beta_{zzz}\mathcal{E}_z^2 + \gamma_{zzzz}\mathcal{E}_z^3 + \delta_{zzzzz}\mathcal{E}_z^4 \tag{2}$$

The polarizability is defined as

$$\alpha_{zz} = \left(\frac{\partial \mu_z}{\partial \mathcal{E}_z}\right), \tag{3}$$

the first hyperpolarizability as

$$\beta_{zzz} = \frac{1}{2!}\left(\frac{\partial^2 \mu_z}{\partial \mathcal{E}_z^2}\right), \tag{4}$$

0097–6156/95/0589–0341$12.00/0
© 1995 American Chemical Society

the second hyperpolarizability as

$$\gamma_{zzzz} = \frac{1}{3!}\left(\frac{\partial^3 \mu_z}{\partial \mathcal{E}_z^3}\right), \tag{5}$$

and the third hyperpolarizability as

$$\delta_{zzzzz} = \frac{1}{4!}\left(\frac{\partial^4 \mu_z}{\partial \mathcal{E}_z^4}\right), \tag{6}$$

where each is evaluated at $\mathcal{E} = 0$.

The properties of most current interest are β and γ which control
  (i) frequency doubling (better focus, more data),
  (ii) changes in refractive index (electro-optical switch for telecommunications), and
  (iii) frequency mixing.

Currently $LiNbO_3$ is the material of choice for such applications. However, polymers would provide great advantages in ease of processing and for tailoring the properties to match precise requirements.

Recent advances in developing new high β, γ organic materials [S. Marder, J. Perry, and coworkers (3-5)] include the development of such materials as (4)

$$\tag{7}$$

$$\Psi_{VB} \qquad\qquad\qquad \Psi_{CT}$$

and (3)

$$\tag{8}$$

The general pattern for good NLO organics is

$$\tag{9}$$

We report here a simple method, VB-CT, for predicting the NLO properties of these materials (6-7).

A typical approach for predicting polarizabilities involves summing over intermediate states formed from molecular orbitals. Thus for a laser frequency

ω the polarizability and hyperpolarizabilities have the form (8)

$$\alpha_{ij}(-w,w) = 2I_{-w,w}\left(\frac{e^2}{\hbar}\right)\sum_n{}' \frac{r_{gn}^i r_{ng}^j}{w_{ng} - w} \tag{10}$$

$$\beta_{ijk}(-w_\sigma; w_1, w_2) = 3K(-w_\sigma; w_1, w_2)\left(\frac{e^3}{\hbar^2}\right)I_{-\sigma;1,2}$$

$$\sum_{m,n}{}'\left\{\frac{r_{gn}^i \bar{r}_{nm}^j r_{mg}^k}{(w_{mg} - w_\sigma)(w_{ng} - w_1)}\right\} \tag{11}$$

$$\gamma_{ijkl}(-w_\sigma; w_1, w_2, w_3) = 4K(-w_\sigma; w_1, w_2, w_3)\left(\frac{e^4}{\hbar^3}\right)I_{-\sigma;1,2,3}$$

$$\left[\sum_{m,n,p}{}'\frac{r_{gp}^i \bar{r}_{pn}^j \bar{r}_{nm}^k r_{mg}^l}{(w_{pg} - w_\sigma)(w_{ng} - w_1 - w_2)(w_{mg} - w_1)}\right.$$

$$\left. - \sum_{mn}{}'\frac{r_{gm}^i r_{mg}^j r_{gn}^k r_{ng}^l}{(w_{mg} - w_\sigma)(w_{ng} - w_1)(w_{ng} + w_2)}\right] \tag{12}$$

Here
(i) g indicates the ground state and $\sum'$ indicates that g is excluded from the sum over excited states.
(ii) $\bar{r}_{kl}^i = r_{kl}^i - r_{gg}^i$, where r_{kl}^i is component i for the dipole matrix element between states l and k.
(iii) $w_\sigma = \sum_i \omega_i$
(iv) $K(-\omega_\sigma; \omega_1, \omega_2, \omega_3)$ is a numerical factor determined by the nature of the NLO process.
(v) $I_{-\sigma;1,2,3}$ denotes the average of all terms generated by permuting $\sigma, \omega_1, \omega_2, \omega_3$.

Given a good description of the excited states this sum-over-states approach can be used to predict accurate values of α, β, γ. However there are two problems:
(i) it rapidly becomes very complicated and expensive as system size increases
(ii) there is no obvious relationship between α, β, γ or between these properties and other properties (D, A, linker length) of the system.

VB-CT Theory

We have developed a new approach (denoted VB-CT) (6) for predicting NLO properties. Valence bond charge transfer (VB-CT) is based on a valence bond description, using only the two states, Ψ_{VB} and Ψ_{CT}, corresponding to the left and right sides of (7) or (8). VB-CT theory determines all NLO properties ($\alpha, \beta, \gamma, \delta$) and establishes relationships between them. It involves two main parameters t and V which can be extracted from the experimental λ_{max} (or from theory). It can be used to predict the solvent dependence and should be useful for designing new materials by tuning the donor (D), acceptor (A), polymer linkers, and solvent.

No Solvent. The VB-CT model assumes that the wavefunction of the molecule and all properties can be described as a linear combination

$$\Psi_{gr} = \sqrt{1 - f}\,\Psi_{VB} + \sqrt{f}\,\Psi_{CT} \tag{13}$$

of the two valence bond configurations, Ψ_{VB} and Ψ_{CT}. The optimum charge transfer (CT) fraction, f, in (13) is determined by the relative energy of Ψ_{VB} and Ψ_{CT}, the coupling between them, the change in the dipole moments, and the solvent polarity.

Without solvent the Hamiltonian is

$$H_0 = \begin{pmatrix} E_{VB} & -t \\ -t & E_{CT} \end{pmatrix}, \tag{14}$$

where

$$\langle \Psi_{CT} | \Psi_{VB} \rangle = 0, \tag{15}$$

$$-t = \langle \Psi_{CT} | \mathcal{H} | \Psi_{VB} \rangle, \tag{16}$$

and

$$V = E_{CT} - E_{VB}. \tag{17}$$

This leads to a bandgap of

$$E_g = \sqrt{V^2 + 4t^2} \tag{18}$$

and to a CT fraction of

$$f = \frac{1}{2} - \frac{V}{2\sqrt{V^2 + 4t^2}} = \frac{\partial E_{gr}}{\partial V}, \tag{19}$$

where E_{gr} is the energy of the ground state.

Bond Length Alternation. Since Ψ_{CT} and Ψ_{VB} involve alternate resonant descriptions of the intervening polyene unit, the increase of f from 0 to 1 will change each double bond ($R = 1.33$Å) of the polyene to a single bond ($R = 1.45$Å) and vice versa. Thus the bond length alternation (BLA) coordinate changes from $q = -0.12$Å $= q_{VB}^0$ to $q = +0.12$Å $= q_{CT}^0$ as the CT fraction f goes from 0 to 1, leading to a one-to-one relationship. The contributions to the Hamiltonian (14) are

$$E_{VB} = \frac{1}{2}k(q - q_{VB}^0)^2, \tag{20}$$

$$E_{CT} = V_0 + \frac{1}{2}k(q - q_{CT}^0)^2. \tag{21}$$

where V_0 is referred to as the *adiabatic* excitation energy.

The equilibrium structure q_{opt} obtained by solving

$$\frac{dE_{gr}}{dq} = 0 \tag{22}$$

leading to

$$q_{opt} = \frac{1}{2}\left(q_{VB}^0 + q_{CT}^0\right) + \frac{1}{2}\left(q_{VB}^0 - q_{CT}^0\right)\frac{V}{\sqrt{V^2 + 4t^2}} \tag{23a}$$

$$= q_{VB}^0 - f\left(q_{VB}^0 - q_{CT}^0\right)$$

$$= -0.12 + 0.24f \tag{23b}$$

Thus f and q_{opt} are linearly related to each other.

Figure 1 illustrates the dependence of the BLA on V_0 using

$$t = 1.1 eV, \qquad (24a)$$

a force constant in (20), (21) of

$$k = 33.55 \ eV/Å^2 = 773.7 \ kcal/molÅ^2 = 5.38 \ mdyn/cm, \qquad (24b)$$

from UFF (*10*), and $V_0 = 1$ eV. This leads to $q_{opt} = -0.069$Å. If VB and CT are degenerate ($V_0 = 0$) then $q_{opt} = 0$. Further stabilization of CT to $V_0 = -1$ eV reverses the BLA to $q_{opt} = +0.069$Å.

Application of an Electric Field. For systems such as (7) and (8), the polarizability and hyperpolarizability are dominated by the z component (chain axis), and we will ignore all other components. Assuming that only Ψ_{CT} contributes to the dipole moment, we write

$$\mu_{CT} = Q e R_{DA} \qquad (25)$$

where Q (expected to be between 0.5 and 1.0) is the net charge transfer for Ψ_{CT}. In an applied external electric field, $\mathcal{E}$, the Hamiltonian (14) becomes

$$H = \begin{pmatrix} E_{VB} & -t \\ -t & E_{CT} - \mu_{CT}\mathcal{E} \end{pmatrix}. \qquad (26)$$

Equations (13), (17) and (19) apply also for finite fields but with V replaced by

$$V_{\mathcal{E}} = V - \mu_{CT}\mathcal{E}. \qquad (27)$$

In particular the change in f due to the applied field is

$$\frac{df}{d\mathcal{E}} = \frac{df}{dV_{\mathcal{E}}} \frac{dV_{\mathcal{E}}}{d\mathcal{E}} = \frac{2t^2 \mu_{CT}}{(V_{\mathcal{E}}^2 + 4t^2)^{3/2}} = \frac{2t^2 \mu_{CT}}{E_g^3}. \qquad (28)$$

Polarizabilities. Given the dependence of the ground state energy on the external electric field, the dipole moment of the ground state, μ_z, is obtained from

$$\mu_z(\mathcal{E}) = -\frac{dE_{gr}}{d\mathcal{E}} = f\mu_{CT} \qquad (29)$$

and the polarizability and hyperpolarizabilities are obtained from (3)-(6), leading to

$$\alpha_{zz} = -\mu_{CT}^2 \frac{df}{dV_{\mathcal{E}}} \mid_{\mathcal{E}=0} = \frac{2t^2 \mu_{CT}^2}{E_g^3} \qquad (30)$$

$$\beta_{zzz} = \frac{\mu_{CT}^3}{2} \frac{d^2 f}{dV_{\mathcal{E}}^2} \mid_{\mathcal{E}=0} = \frac{3t^2 \mu_{CT}^3 V}{E_g^5} \qquad (31)$$

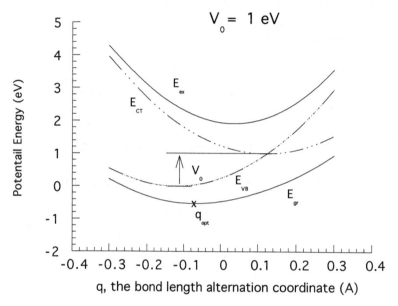

Figure 1. Relation between the energy curves for (i) pure VB and (ii) pure CT states for $V_0 = 1$ eV. The ground state and excited state resulting from interaction of VB and CT. $E_{gap} = hc/\lambda_{max}$ is the observed transition energy.

$$\gamma_{zzzz} = -\frac{\mu_{CT}^4}{6}\frac{d^3 f}{dV_{\mathcal{E}}^3}\ |_{\mathcal{E}=0} = \frac{4t^2\mu_{CT}^4[V^2 - t^2]}{E_g^7}, \tag{32}$$

$$\delta_{zzzzz} = \frac{\mu_{CT}^5}{24}\frac{d^4 f}{dV_{\mathcal{E}}^4}\ |_{\mathcal{E}=0} = \frac{5t^2\mu_{CT}^5 V[V^2 - 3t^2]}{E_g^9}. \tag{33}$$

(The following discussions will omit the z subscripts.)

Thus VB-CT leads to *analytical equations* for all hyperpolarizabilities in terms of the atomistic parameters t and V! This contrasts with the usual result (10)-(12).

Predictions of μ, α, β, and γ From VB-CT Theory. As f increases from $f = 0$ to $f = 1$, the VB-CT model leads to an alternation in which the polyene double bonds for Ψ_{VB} change to polyene single bonds in Ψ_{CT} and vice versa. Since there is a linear relation (23b) between f and q_{opt} [the change in bond length alternation (BLA)] and since f determines the polarizability and all hyperpolarizabilities, then a single BLA parameter, q_{opt}, determines the polarizability and all hyperpolarizabilies. This has been anticipated by Marder *et al*, (*9*) who pointed out that BLA is a useful parameter for examining the structure-property relationships of NLO materials. They showed that the β and γ values can be tuned by varying BLA. In addition they carried out finite-field AM1 calculations (*9*) and showed the relationships of α, β and γ to q_{opt}. Their results provide a good test of VB-CT theory.

In order to illustrate the relationships, we used (19), (23), and (30)-(33) to calculate f, q_{opt}, α, β, γ, and δ as a function of V, all with the fixed values of t and k from (24). This allowed us to obtain α, β, γ, and δ as a function of f, Figure 2. The shapes of the polarizability curves are insensitive to the value t. Thus the salient factor for polarizability and hyperpolarizability is the bond length alternation.

General observations from these relations are as follows:

(i) α has a maximum for $f = \frac{1}{2}$,

(ii) β is the derivative of α with respect to f, leading to a maximum in $|\beta|$ at $f = 0.276$ and 0.724 and zero at $f = \frac{1}{2}$.

(iii) γ is the derivative of β with respect to f, leading to the largest magnitude (a minimum) at $f = \frac{1}{2}$, with secondary maxima (1/4 the magnitude) at $f = 0.173$ and $f = 0.827$. Where $|\beta|$ is a maximum, $\gamma = 0$.

(iv) δ is the derivative of γ with respect to f, leading to maxima in $|\delta|$ at $f = 0.357$ and $f = 0.643$ and secondary maxima at $f = 0.117$ and $f = 0.883$. Where $|\gamma|$ is a maximum, $\delta = 0$.

Special cases are:

(a) When $V = 0$ (VB and CT states degenerate), we have

$$f = \frac{1}{2}, \quad q_{opt} = 0, \quad t = E_g/2. \tag{34}$$

At this point, α is a maximum, $\beta = 0$, $|\gamma|$ is a maximum, and $\delta = 0$.

(b) When $|V| = |t|$, we have

$$f = 0.276 \text{ or } 0.724, \quad q_{opt} = \pm 0.0538\text{Å}, \quad t = \frac{E_g}{\sqrt{5}}. \tag{35}$$

Thus $|\beta|$ is a maximum and $\gamma = 0$.

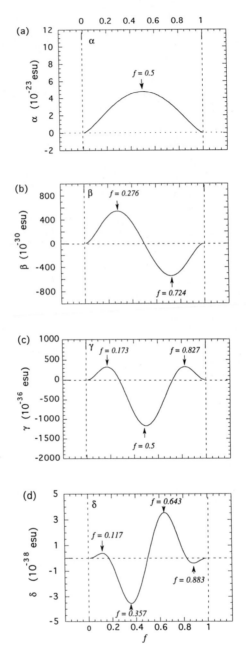

Figure 2. Predicted properties $(\alpha, \beta, \gamma, \delta)$ as a function of CT fraction, f.

(c) When $|V| = \sqrt{3}|t|$, we have

$$f = 0.173 \; or \; 0.827, \quad q_{opt} = \pm 0.0785\text{Å}, \quad t = \frac{E_g}{\sqrt{7}}. \qquad (36)$$

Thus $|\gamma|$ is a maximum and $\delta = 0$.

Comparison with AM1 Calculations. The VB-CT results are compared with AM1 calculations (*9*) (dots) in Figure 3. In making these comparisons we used $Q = 0.69$ for α, γ, and δ but $Q = 0.51$ for β. The AM1 results for β and γ agree quite well with VB-CT theory. VB-CT has α go to zero as $f \to 0$ or 1, whereas AM1 calculations lead to about half the maximum. This is probably because the current VB-CT calculations ignore the polarizability for a fixed VB or CT structure (it could have been included).

Solvation Effects. Placing a CT molecule into a polar solvent leads to reorientation of both the solvent and solute molecules. This changes the relative energy of Ψ_{VB} and Ψ_{CT}, (17), which through (14) changes the optimum fraction, f, of CT character in the ground state (13). Assuming that only CT contributes, the dipole moment of the ground state becomes

$$\mu = f\mu_{CT} = fQeR_{DA} = fQ\mu_{DA} \qquad (37)$$

In the VB-CT-S model (*7*) we assume that electronic states other than Ψ_{VB} and Ψ_{CT} have much higher energies and that t (the coupling between Ψ_{VB} and Ψ_{CT}) is independent of solvent. Thus quantitative evaluation of the solvation effects requires only the change in relative energy, (17), due to the presence of solvent, and we write

$$V_S = V + \Delta V_S \qquad (38)$$

Here V is the energy difference between Ψ_{VB} and Ψ_{CT} *without* solvent and ΔV_S is the change in the relative energy caused by the addition of solvent. Including BLA the final V is obtained from (19) using V_S in place of V.

To evaluate ΔV_S we approximate (*11*) the donor and acceptor by two spheres of radius r_D and r_A with charges distributed symmetrically as in (39)

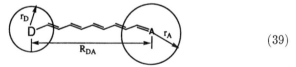

(39)

Donor   Acceptor

The net result is

$$\Delta V_S = -\frac{e^2}{4\pi\epsilon_0}\left(1 - \frac{1}{\epsilon}\right) fQ^2 S_F, \qquad (40a)$$

where

$$S_F = \frac{1}{2r_D} + \frac{1}{2r_A} - \frac{1}{R_{DA}}, \qquad (40b)$$

depends only on the geometry.

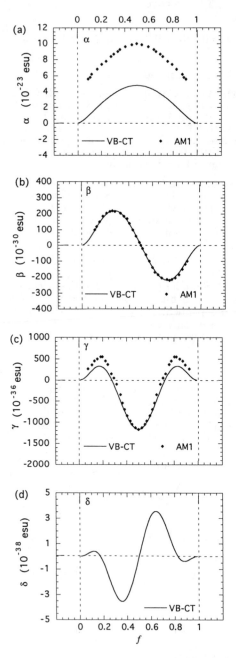

Figure 3. Comparison of predicted properties $(\alpha, \beta, \gamma, \delta)$ with AM1 calculations (reference 9).

The polarizabilities have the form (30)-(33) where μ_{CT} is given by (37) and V is replaced by V_S, (38).

Measurements in solution lead to the rotationally averaged values of the polarizabilities,

$$\alpha = \frac{1}{3}\alpha_{zz} = \frac{2t^2\mu_{CT}^2}{3E_g^3} \tag{41}$$

$$\beta = \frac{\mu_z\beta_{zzz}}{\mu} = \frac{3t^2\mu_{CT}^3 V}{E_g^5} \tag{42}$$

$$\gamma = \frac{1}{5}\gamma_{zzzz} = \frac{4t^2\mu_{CT}^4(V^2 - t^2)V}{5E_g^7} \tag{43}$$

Comparison with Experiment. The dots in Figure 4c show the experimental values (*4*) of the second hyperpolarizability γ for molecule (*7*) in a variety of solvents

To compare VB-CT-S theory with experiment, we must evaluate six parameters: t, V_0, S_F, R_{DA}, Q, and k. Using the Universal Force Field (*10*) (UFF) in conjunction with Charge Equilibration theory (*12*) to predict the charges, we obtain

$$R_{DA} = 7.30\text{Å} \tag{44}$$

for molecule (*7*). Similarly UFF (*10*) leads to the value of k in (24b). The remaining parameters t, V_0, S_F, and Q are each intrinsic parameters of CT molecules and can be determined directly from experiment (*4*).

From (35), γ is zero when $|V| = |t|$. Experimentally (*4*) $\gamma = 0$ for a solvent polarity of $\epsilon = 2.209$, leading to $E_g = 2.648$ eV for this polarity. Thus from $|V| = |t|$ and (7), we can evaluate t for (7),

$$t = E_g/\sqrt{5} = 1.184eV. \tag{45}$$

V_0 and S_F can be obtained by fitting absorption peaks in two different solvents. We chose dioxane ($\epsilon_1 = 2.209$ with absorption energy $E_{g1} = 2.648$ eV), with CH_3CN ($\epsilon_2 = 37.5$ and $E_{g2} = 2.604$ eV). This leads to two equation of the form

$$V_0 + \frac{1}{2}k\left[(q_i - q_{CT}^0)^2 - (q_i - q_{VB}^0)^2\right] - \frac{e^2}{4\pi\epsilon_0}\left(1 - \frac{1}{\epsilon_i}\right)f_iQ^2S_F = \sqrt{E_{gi}^2 - 4t^2}, \tag{46}$$

where f_i depends on t and E_g. Solving equation (46) leads to

$$Q^2S_F = 0.0373\text{Å}^{-1} \tag{47}$$

$$V_0 = 0.833eV \tag{48}$$

To separate out Q from S_F, we can fit to the magnitude of γ at some ϵ. We chose to do this for CH_3CN ($\epsilon = 37.5$). The experimental value (*4*) is $\gamma_{static} = -35$ esu, whereas the calculated value would be $\gamma = -118$ esu for $Q = 1$. This leads to

$$Q^4 = 0.297, \tag{49a}$$

or

$$Q = 0.738, \tag{49b}$$

Substituting into (47) leads then to

$$S_F = 0.0685\text{Å}^{-1} \qquad (50)$$

Given t, V_0, S_F, Q, R_{DA}, and k from (45), (48), (50), (49), (44), and (24b) we can calculate α, β, γ, and δ for all solvent polarities, ϵ. The resulting averaged values are shown in Figure 4.

Currently only γ is available from experiment, Figure 4c. Despite the simplicity of this model, VB-CT-S fits reasonably well with experiment (4). It will be valuable to measure the α and β for this molecule in various solvents in order to further test the model.

Use in Design and Prediction. The VB-CT-S model is quite simple. It involves

i. two properties (k and R_{DA}) which can be obtained from the force field (spectroscopy or theory)

ii. three electronic parameters (V_0, t, Q) characteristic of the isolated molecule, which can be derived from theory or experiment on the isolated molecule or from experiment in solution [as illustrated in above]

iii. one solvent independent parameter, S_F, which must be obtained from an experimental value of the α, β, or γ in a polar solvent.

Given these six parameters one can predict the properties [λ_{max}, α, β, γ, δ, and q_{opt}] as a function of solvent polarity.

In designing new nonlinear materials, one might consider replacement of the donor, of the acceptor, or of the linker. The value for V_0 should depend strongly on the ionization potential (IP) of donor (D) and the electron affinity (EA) of acceptor (A). These in turn might be related to the change in redox potentials for some solvent. Similarly the differential charge transfer, Q, can be estimated from IP_D and EA_A. The effect of changing the length of the linker or of replacing the polyene linker in (7) with other polymers is discussed below.

Linker Excited States

The VB-CT model assumes that all other excited states are much higher than the VB and CT states. In particular the resonant states involving the bridge or linker must be much higher. For octatetraene the absorption maxima is about 4 eV indicating that the resonance state of the linker in equation 51 is more than 4 eV above the VB state.

(51)

Since the donor-acceptor molecules considered here have the CT state about 1 eV above the VB state, neglect of the resonance state should be a good approximation. When the energy of the resonance structure is similar to those of VB and CT, the contributions from the linker resonance state must be included. This complicates the theory so that the results are no longer analytic.

To predict the dependence of α, β, γ on polymer length, we developed valence bond charge transfer exciton (VB-CTE) theory (8). Here the excited states are considered as charge transfer excitons with an electron removed from the HOMO on one monomer and added into the LUMO on another monomer p sites away. With some additional approximations this leads to an analytic result and to saturation behavior in good agreement with experiment. Thus for polythiophene (52) we obtain the results in Figure 5.

(52)

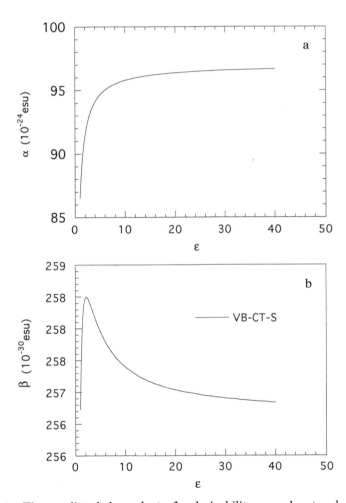

Figure 4. The predicted dependent of polarizability on solvent polarity (expressed in terms of the static dielectric constant ϵ). (a) Polarizability, α. (b) Hyperpolarizability, β. (c) Second hyperpolarizability, γ. (d) Third hyperpolarizability, δ_{zzzz}. The values plotted are the static averages values. For γ in (c) a comparison is made between theory (solid line) and experiment (dots).

Continued on next page

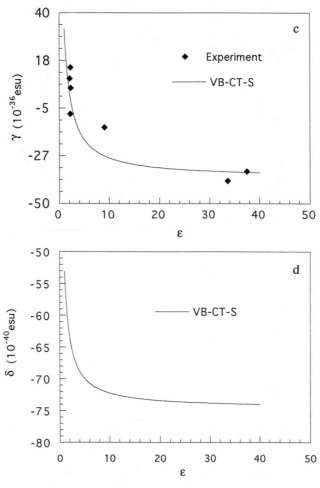

Figure 4. Continued.

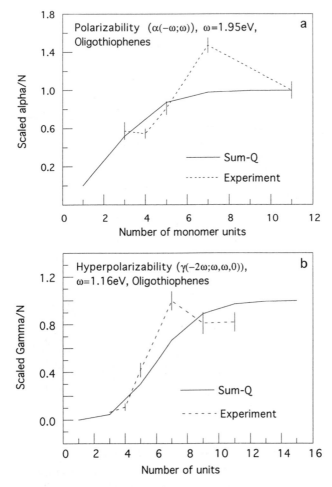

Figure 5. Comparison of theory and experiment for the saturation behavior of oligothiophenes. (a) Polarizability α ($\omega = 1.95$ eV) and (b) hyperpolarizability γ ($\omega = 1.16$ eV). The dashed line connects experimental data (with error bars). The solid line connects theoretical predictions.

VB-CT-E theory involves just two parameters t and $V = IP - EA$. They are related to bandwidth B and bandgap E_g by (53) and (54),

$$B = 4t \tag{53}$$

$$E_g = V - 2t \tag{54}$$

The saturation length for polarizability is (55),

$$L_\alpha \sim 1 + 20\frac{t}{V} \tag{55}$$

and the saturation length for second hyperpolarizability is (56),

$$L_\gamma \sim 1 + 38\frac{t}{V}. \tag{56}$$

Thus applying VB-CT-E to other polymers require only two pieces of information, say a bandwidth and a bandgap. Table I shows the predictions for the polymers in Figure 6.

Table I. Saturated values for α and γ and saturation lengths (L_α, L_γ) predicted from VB-CT-E theory. These values assume $\omega = 0$.

Quantity	Band Width B	Band Gap[a] E_g	t^b	V^b	Saturation Length		α_{zz}/N	γ_{zzzz}/N
Polymer	(eV)	(eV)	(eV)	(eV)	L_α	L_γ	(10^{-23}esu)	(10^{-34}esu)
6a	3.9	2.1	1.0	4.1	5.8	10.1	2.41	11.3
6b[c]	3.5	3.4	0.9	5.1	4.4	7.5	1.13	2.54
6c	2.8	3.0	0.7	4.4	4.2	7.1	2.49	15.9
6d	3.8	3.2	0.95	5.1	4.8	8.2	1.13	2.55
6e	2.5	2.2	0.6	3.4	4.6	7.8	3.88	44.9
6f	3.3	3.3	0.8	5.0	4.2	7.1	0.40	0.51
6g	2.7	5.4	0.7	6.8	3.0	4.8	0.083	0.020
6h	1.3	3.1	0.3	3.8	2.6	4.0	1.09	3.45
6j	-	2.77	0.83	4.09	5.0	9.0	1.77	6.87
6i	-	1.8	0.83	2.85	6.8	12.0	5.65	82.1

[a]From theory.
[b]From experiment values of B, E_g.
[c]Twisted $(22°)$.

First Principles Prediction of Solvent Effects

We are using PS-GVB/SOLV (13,14) to predict *ab initio* solvation effects for molecules such as (7) and (8). PS-GVB/SOLV considers that there is a dielectric continuum with dielectric constant ϵ surrounding the molecule as in Figure 7. Here the interface is the van der Waals surface around each atom. The procedure is as follows:

1. The PS-GVB program (14) is used to calculate the HF or GVB wavefunction and the optimum orbitals are used to determine the electron density.
2. The density from step 1 is converted to point charges (using electrostatic potential derived charges).

a   Polydiacetylene(Acetyleneic)

b   Polyparaphenylene

c   Polyparaphenylene vinylene

d   Polyprrole

e   Polythiophene vinylene

f   Polyvinylene sulfide

g   Polymethineimine

h   Polybenzothiophene

i   Polyacetylene (trans)

j   Polythiophene

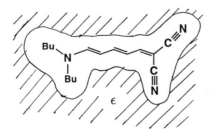

Figure 6. Polymers considered in Table I.

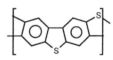

Figure 7. Illustration for PS-GVB/SOLV calculations.

3. DelPhi (from B. Honig of Columbia) calculates the solvent response to the charges from step 2, using the Poisson-Boltzmann equation. The net result is a set of charges at the interface representing the effects of the polarized solvent.
4. The interface charges from step 3 are input to PS-GVB which solves for a new HF or GVB wavefunction consistent with the solvent polarization.
5. The results in step 4 are used in step 2 and the process is iterated until convergence.

The net result is a wavefunction and structure self-consistently adjusted to the solvent. This should provide the ability to consider entirely new systems.

Acknowledgements

We thank Seth Marder of the Beckman Institute for helpful discussions. This research was funded by NSF (CHE 91-100289 and ASC 92-17368). The facilities of the MSC are also supported by grants from DOE-AICD, Allied-Signal Corp., Asahi Chemical, Asahi Glass, Chevron Petroleum Technology, Hughes Research Laboratories, BF Goodrich, Vestar, Xerox, and Beckman Institute.
        Some calculations were carried out on the NSF-Pittsburgh supercomputer and on the JPL CRAY.

Literature Cited

*To whom correspondence should be addressed.
1. *Introduction to Nonlinear Optical Effects in Molecules and Polymers*; Prasad, P. N.; Williams, D. J., Eds.; Wiley, New York, NY, 1991.
2. *Nonlinear Optical Materials*; Kuhn, H.; Robillard, J., Eds.; CRC Press, Inc. 1992.
3. Dagani, R.; *C and E News*; **1994**, pp 26.
4. Marder, S. R.; Perry, J. W.; G. Bourhill; Gorman, C. B.; Tiemann, B. G. *Science* **1992**, *261*, 186.
5. Marder, S. R.; *et al. Science* **1994**, *263*, 511.
6. Lu, D.; Chen, G.; Perry, J. W.; Goddard III, W. A. *J. Am. Chem. Soc.*, submitted.
7. Chen, G.; Lu, D.; Goddard III, W. A. *J. Chem. Phys.*, in press.
8. Lu. D.; Chen, G.; Goddard III, W. A. *J. Chem. Phys.* **1994**, *101*, 4920.
9. Marder, S. R.; Beratan, D. N.; Cheng, L. T. *Science* **1991**, *252*, 103; Marker, S. R.; Gorman, C. B.; Tiemann, B. G.; Cheng, L. T. *J. Am. Chem. Soc.* **1993**, *115*, 3006; Marder, S. R.; *et al.*, *J. Am. Chem. Soc.* **1993**, *115*, 2524; Gorman, C. B.; Marder, S. R. *Proc. Natl. Acad. Sci.* **1993**, *90*, 1129.
10. Rappé, A. K.; Casewit, C. J.; Colwell, K. S.; Goddard III, W. A.; Skiff, W. M. *J. Am. Chem. Soc.* **1992**, *114*, 10024.
11. Marcus, R. A. *J. Chem. Phys.* **1956**, *24*, 996.
12. Rappé, A. K.; Goddard III, W. A. *J. Phys. Chem.* **1991**, *95*, 2260.
13. Tannor, D. J.; Marten, B.; Murphy, R.; Friesner, R. A.; Nicholls, A.; Honig, B.; Ringnalda, R.; Goddard III, W. A. *J. Am. Chem. Soc.*, submitted.
14. PS-GVB and PS-GVB/SOLV are from Schrödinger Inc. (Pasadena, California).

RECEIVED October 28, 1994

Chapter 26

Theoretical Study of the Nitriding Process on Cr(100), Fe(100), and Ni(100) Surfaces

Hansong Cheng, David B. Reiser, Paul M. Mathias, Kenneth Baumert, and Sheldon W. Dean, Jr.

Air Products and Chemicals, Inc., 7201 Hamilton Boulevard, Allentown, PA 18195–1501

We present a theoretical study of nitriding, an important corrosion process in chemical plants. Nitriding at metal surfaces occurs due to ammonia chemisorption and subsequent decomposition into nitrogen and hydrogen adatoms. The theoretical approach, which combines density functional theory and the tight-binding extended Hückel method, systematically investigates the nitriding mechanism at Cr(100), Fe(100) and Ni(100) surfaces. The present study includes evaluation of adsorption geometries, estimation of binding energies, and comparison of nitriding properties for different metal surfaces. The calculated surface band structures are in qualitative agreement with UPS photoelectron spectroscopic measurements. Population analysis at each step of the chemisorption/decomposition process yields significant insight into the electron transfer mechanism in the nitriding process. The theoretical results provide an explanation of the experimental observation that Ni is a better nitriding-resisting material than Fe metal and the latter is superior to Cr. Overall, the present study suggests that theoretical chemistry is a useful tool in the analysis of corrosion problems.

Background

The prevailing method for the analysis of corrosion problems involves a combination of experimentation and phenomenological models.(1) This method is limited because the results cannot be used to predict new corrosion-resistant materials, or to analyze new corrosion phenomena. The rapid advances in computer power and widely available software have accelerated the application of computational chemistry to chemical fields covering physical chemistry, organic chemistry and inorganic chemistry. However the application of computational chemistry to corrosion has been slow mainly due to the complicated nature of corrosion processes.(2) This study is an attempt to evaluate computational chemistry as a general tool for use in materials

0097–6156/95/0589–0359$12.00/0
© 1995 American Chemical Society

engineering. We have chosen the nitriding problem for study since it is a well known but little understood problem.

If all the parameters of an alloy and of the interface between that alloy and its environment are known, computational chemistry could, in principle, provide activation energies and corrosion rates for any given liquid or gaseous environment. In the chemical industry, this knowledge could be used to design or select the optimum alloy for any process equipment. Realistically, such an all-encompassing theoretical treatment is still beyond the capabilities of both the algorithms and the computing hardware, but the relentless advances in both hardware and software are making an increasing range of theoretical computations available to the industrial practitioner. However, effective application of computational chemistry requires collaboration among theoreticians, experimentalists and materials engineers, and judicious simplification of the problem. An important goal of this study is to understand the collaborations and simplifications necessary for successful application of computational chemistry to corrosion problems.

Analysis of the Nitriding Process

Nitriding is a chemical corrosion process that usually takes place at high temperature.(1,3-6) Nitriding results from the diffusion of atomic nitrogen into a metal after decomposition of ammonia or molecular nitrogen at the metal surface. Once the nitrogen concentration in the metal exceeds the solubility limit, metal nitrides precipitate. The nitrides make the metal hard and brittle. The hardness increase is beneficial for case hardening of steel, but the brittleness leads to failure of ammonia handling equipment in processes such as nitric acid or amine production. Prevention of nitriding attack is an important consideration in these processes. Practical experience has led to the choice of high nickel alloys to resist nitriding attack in many applications. The purpose of this study was to investigate the mechanism of the reaction that supplies the atomic nitrogen at the metal surface. Understanding this step in the nitriding mechanism at a fundamental level will help explain the practical observations to date and possibly suggest optimum alloy composition for maximizing resistance to nitriding attack.

Nitriding is a complicated gas-solid interaction phenomenon. In general, it involves three major steps: first, ammonia or nitrogen molecules in the gas phase diffuse to and stick on the metal surfaces; next, the adsorbed molecules are decomposed at the surfaces; finally the dissociated species further diffuse into the bulk. Nitriding has become a serious problem in recent years as the average service time of ammonia reactors has been increasing. It presents a great challenge to the materials science community to provide better, less costly alloys to serve as nitriding-resisting materials.

There have been several research efforts to explore the experimental conditions under which the nitriding process on a variety of alloy surfaces takes place.(1) Unfortunately, the vast number of possibilities makes it impractical to experimentally screen all the existing alloys with different compositions to obtain the desired physical/chemical properties. Furthermore, such experiments can be extremely expensive and inefficient. While many experiments have been carefully designed to test the nitriding-resisting properties for specific materials of interest, very little was revealed about the detailed nitriding processes at the metal surfaces. To date, studies on nitriding problem have been limited to a purely empirical description and thus the results cannot be used to predict improved nitriding-resisting materials. As a consequence, the selection of alloy materials for process equipment today is still largely dependent upon practical experience. Theoretical guidelines for the rational selection of optimum materials would be valuable to the chemical industry.

Although the three-step nitriding mechanism has been well known for many years, much work has been primarily concentrated on phenomenological descriptions of the nitriding process. While this approach has been successful to characterize the macroscopic nitriding processes based on thermodynamics and kinetics,(*3-6*) it is unable to provide information about the detailed reaction mechanisms at the microscopic level. Quantum mechanics provides a powerful tool to fulfill this task, and has been widely used for materials development in recent years. The quantum-mechanical approach is not intended to replace the conventional macroscopic methods that have enjoyed a broad usage in the corrosion community; rather, it attempts to complement the conventional approach by yielding information about the microscopic nitriding processes.

While real materials used in an ammonia reactor are alloys of a variety of metals, understanding the ammonia chemisorption and decomposition behavior on the surfaces of various pure metals is an important step towards to understanding the real nitriding process in ammonia plants. Furthermore, it is instructive to compare the nitriding process with the ammonia synthetic process, which has been extensively studied over the past few decades.(*7-19*). In fact, a nitriding process can be viewed as an important part of a reversed ammonia synthesis process catalyzed by a metal surface. Using various surface science and spectroscopic techniques, such as thermal desorption spectroscopy and UPS photoelectron spectroscopy, extensive experiments have been carried out to probe the microscopic processes involved in the catalytic reactions. In parallel, much theoretical work has focused on the nature of chemisorption of NH_3 and N_2 on the metal surfaces; however, very little effort has been devoted to the study of ammonia decomposition processes.(*20-22*) Computationally, it is often exceedingly difficult to calculate the energy structure for systems involving transition metals due to their large number of spin states.

The present work is to investigate the nitriding mechanism at the surface of various pure transition metal by performing quantum-mechanical calculations. An important goal of this study is to explain the difference of reactivity among the pure metals (Fe, Ni and Cr) in response to the nitriding attack, and to test the theoretical techniques against the available experimental results. Ultimately, we would like to perform "computational experiments" to study the nitriding resisting properties for a variety of alloy materials and to provide theoretical guidance for optimal material selections. In this chapter, we present a combined theoretical approach based on the density functional theory (DFT) and the extended Hückel tight binding method (EHTB) to investigate the nitriding process on Cr(100), Fe(100) and Ni(100) surfaces.(*23-25*) Since these three metals are the most important chemical components of stainless steels, it is expected that the investigation on the behavior of ammonia chemisorption/decomposition at their surfaces will yield useful information about the nitriding mechanism in the alloy materials in ammonia reactors.

It should be stressed that it is not our intention to pursue a rigorous quantum-mechanical description of the nitriding system in the present work. Instead, we aim to gain physical insight into the nitriding mechanism at the transition metal surfaces, which will enable us to qualitatively predict the relative order of nitriding-resisting properties for these materials and, ultimately, for the alloys of interests. The DFT method was used to evaluate the chemisorption patterns and geometric parameters; subsequently, using the EHTB theory, the energy band structures of various configurations of chemisorption and decomposition patterns as well as binding energies were calculated. We compare the theoretical results with the photoelectron spectra for the ammonia chemisorption and decomposition species. The predicted relative order of nitriding-resisting properties for different metals will also be compared with material performance observed in practical experience.

Experimental Background

The elements Fe, Cr and Ni are the major components of stainless steels used in ammonia handling equipment. Our practical observation is that under the normal operating conditions, high nickel content materials are most resistant to nitriding, while high chromium content alloys show relatively poor performance for resisting the attack. Unfortunately, alloys with high percentage nickel content are usually much more expensive. In practice, in choosing stainless steel materials for ammonia handling equipment, one needs to consider both the actual service age of the equipment and the nitriding-resisting capability in order to maximize use of the materials at a minimum cost.

Since a nitriding process bears considerable similarities to its reversed process, transition metal catalyzed ammonia synthesis, the results of the detailed studies of experiments at the microscopic level in the latter field should also be useful for study of nitriding. Of all the metal catalysts, the iron surfaces have received the most attention since they are most frequently utilized for ammonia synthesis.(*14-19*) Other metals, such as Ni, Cr and Al, and their alloys with Fe also play an important role in catalyzing the reactions. The ammonia catalytic process at the iron surfaces has been extensively investigated through a series of elegant experimental measurements by Ertl and co-workers.(*7, 16-18*) They showed that the ammonia decomposition at the iron surfaces is a many-step energy down-hill process. These steps include NH_3 adsorption on Fe surfaces and the adsorbate, $NH_3(ad)$, subsequent dissociation into $NH_2(ad)$ and $H(ad)$ adsorbates and eventually into $N(ad)$ and $H(ad)$ adatoms. Based on their experiments and some plausible estimates for the reaction intermediates, Ertl at al. constructed a potential energy diagram for the ammonia decomposition at iron surfaces. Furthermore, they carried out UPS photoelectron spectroscopic measurements to detect the intermediate ammonia fragments. It was shown that ammonia adsorption on a clean iron surface gives rise to two prominent peaks below the Fermi level in the spectrum, which correspond to the $3A_1$ (N lone pair) and $1E$ (N-H bonds) energy levels of gas phase ammonia, respectively. When the system is exposed to ammonia at higher temperature (80°C), these two peaks shift up with relatively weaker intensities, which are associated with the ammonia fragment adsorbates, either $NH_2(ad)$ or $NH(ad)$ or both. At even higher temperature (120°C), only one maximum in the spectrum was observed, which corresponds to the completely decomposed ammonia species: the nitrogen adatoms. The signal of hydrogen adatoms was too weak to be clearly resolved.

In parallel to the work of Ertl and co-workers, experiments were also carried out to investigate the ammonia interactions with Ni and Al surfaces. In particular, Huttinger and Kuppers proposed a many-step decomposition mechanism for NH_3 at Ni(110) surface, similar to the one for NH_3 at iron surfaces. The estimated activation energy for $NH(ad)$ to further dissociate into nitrogen and hydrogen atoms was found to be 47 kcal/mol, which was remarkably high. Furthermore, in addition to the two maxima of ammonia bands ($1E$ and $3A_1$) found in the UPS spectrum for NH_3 on Ni surface, Seabury et al. also resolved the $2A_1$ band of ammonia at about 22 eV below the Fermi level when higher photon frequency was used.(*15*) To our knowledge, very little has been done for ammonia chemisorption and decomposition on chromium surfaces. Nevertheless, in view of the ammonia dissociation mechanism on Fe and Ni surfaces, it is reasonable to assume that the nitriding follows the same decomposition course at Cr(100). The experimental results for Fe and Ni surfaces will be used to benchmark the theoretical calculations and to help improve the quality of extended Hückel parameters to gain better accuracy.

Computational Details and Theoretical Model of a Nitriding System

Computational Methods. The quantum mechanically more rigorous DFT method is known to be capable of providing highly accurate molecular geometries as well as system energetics in many cases. However, it is computationally intensive and thus one can only apply it to relatively small systems. To correctly describe a nitriding system, it is often necessary to model the chemisorption and decomposition of NH_3 on a large portion of a surface for at least two reasons. First, the adsorption of NH_3 and its fragments can take place, in principle, at any site of the surface. The possible adsorption modes include 4-fold hollow site, "bridge" site and "on-top" site, as shown in Fig. 1. Second, upon the decomposition, the fragments of NH_3 may spread out on or defuse into the surface. To describe this phenomenon, it is necessary to include not only a large portion of the surface but also a few top layers of the metal in the modeling. It thus becomes computationally impractical to use the DFT method for such a complicated system. As an alternative, we utilized the computationally more efficient approach, EHTB method, to describe the electronic structures of the adsorption/decomposition system. The extended Hückel method (EH), being computational simple and physically intuitive, is particularly useful for studies of molecules with a large number of atoms and has shown considerable success in study of many chemical and physical processes in condensed phase.(*26-28*) As a very crude approximation, the EH method usually does not yield quantitative results about the system energetics. Instead, it is often used to provide useful physical insight into the reaction mechanism and to qualitatively predicts the trends of a molecular process. Furthermore, it is known that the EH method is not suitable for geometry optimizations of molecular structures due to the lack of repulsive force. Therefore, it is necessary to supply the adsorption/decomposition structures for performing the EH calculations. The combined DFT and EHTB approach allows one to evaluate the adsorption structures for $NH_x(ad)$ (x=0, 1, 2, 3) and H(ad) on a small cluster of the metal surfaces by DFT and subsequently to perform the energy band structure calculations by EHTB. In the present study, the latter calculations were carried out for a sufficiently large surface unit cell with three layer surfaces, on which ammonia is chemisorbed and later decomposed. The required geometric parameters for the EH calculations were supplied by the DFT calculations.

The Biosym DFT code, DMol, was utilized to evaluate the adsorption geometric parameters. We employed the double numerical basis set with inclusion of polarization functions. The Janak-Morruzi-Williams correlation functional was adopted with LYP functional for non-local gradient correction. To enhance the SCF convergence, which appears to be a severe problem for many transition metal calculations, we used the level shifting technique by smearing a small amount of electron density in the energy levels above the Fermi level. Subsequently, we utilized the correlation-exchange potential generated from the previous SCF cycle and gradually reduced the smear value. We repeated this procedure until the smear value becomes negligible and SCF energies converge self-consistently. The specific initial smear values used in our calculations depend on the cluster systems and their sizes. Typically, for a iron cluster with 9 atoms, the smear value can be as high as 0.1 since there are many possible low lying states in the cluster and it becomes very difficult for the SCF to converge. Using the above procedure, the final smear value is reduced to less than 0.005.

The program of the EHTB method was originally obtained from Professor M.-H. Whangbo. The EH parameters for Ni d-orbitals were slightly adjusted to make the d-orbitals more defuse, which is in better agreement with the UPS photoelectron spectrum.(*14*) Except for Ni atom, the parameters used in the EH calculations can be found in Ref. 24. The double zeta values and their coefficients used to describe the 3d orbitals of Ni are listed in Table I.

Table I. Extended Hückel parameters of 3d-orbital of Ni.

orbital	H_{ii} (eV)	ζ_1	ζ_2	c_1	c_2
3d	-9.9	6.50	2.00	0.5683	0.6292

NH_x (x=0-3) (ad) and H(ad) Chemisorption on Metal Clusters and Surfaces. The calculations were performed for rigid metal surfaces before and after the ammonia chemisorption and decomposition, which were directly obtained from the crystal structures. While Ni(100) is a densely packed surface, Cr(100) and Fe(100) exhibit relatively open structures and thus may be chemically more reactive. It has been generally accepted in the literature that the adsorption of NH_3 takes place with nitrogen "end-on" the metal atom and the three hydrogen atoms pointing away from the surfaces.(8,20)   Moreover, Bauschlicher showed that for NH_3 adsorption on Ni(100) surface the rotation around the principal ligand axis requires virtually no energy and such a conclusion was extended by Benndorf et al. to investigate ammonia adsorption and dissociation on a stepped Fe(s)(100) surface in their experiments.(8,20)   In the present work, we assume that the same adsorption pattern and properties are also applicable to Cr(100) and Fe(100) surfaces.   Furthermore, to simplify the calculations, we only consider similar adsorption patterns for the decomposed species, NH_x (x=0,1,2).   It is understood that other adsorption modes, such as the 4-fold hollow sites and the "bridge" sites, can also be accessible by these species, particularly by the smaller ones, as demonstrated by a number of studies.(29-31)   However, it is beyond the scope of present study to include all the possible adsorption modes for these species in the calculations and then to statistically average out the system energies.   Instead, we intend to choose a physically reasonable adsorption pattern to gain insight into the nitriding mechanism.   In addition, the hydrogen adatoms are allowed to be distributed at any site of the surfaces.   Since one of the main goals of the present study is to predict the relative order of nitriding-resisting properties for different metal surfaces, we utilize the same adsorption pattern for all the metal surfaces to facilitate the comparison of band structures as well as binding energies.

To obtain reasonable geometric parameters for the chemisorption/decomposition species, we employed small metal clusters to represent the surfaces and carried out density functional calculations.   Typical structures of the clusters used in the calculations are shown in Fig. 2.   Making use of the assumption that the system energies are insensitive to the rotation about the principal ligand axis of NH_x(ad) (x=0-3), we imposed the point group symmetries in the calculations. The point groups are Cs for NH_3(ad) , C_{2v} for NH_2(ad) and C_{4v} for NH(ad) and N(ad) on the metal surfaces, respectively.   We then optimized the adsorption geometries on rigid metal surfaces.   For small clusters, there appear considerable differences of adsorption binding energies between theory and experiment.   For example, the binding energy of NH_3 on the iron cluster calculated with non-local gradient corrections was found to be about three times higher than the experimental value, which may be largely due to the oversimplified surface model.   Nevertheless, the optimized geometric parameters seem to be in reasonable agreement with the available results in literature and with one's physical expectation.(19, 28)   The calculated adsorption geometric parameters are summarized in Table II.

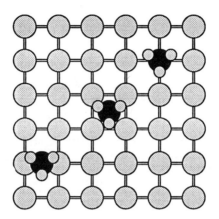

Figure 1. Ammonia chemisorption pattern on metal surfaces.

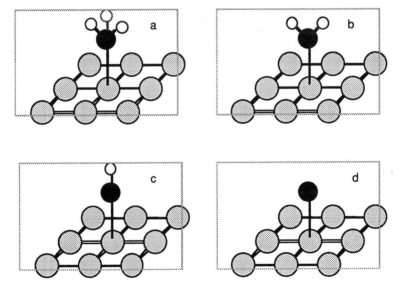

Figure 2. Structures of chemisorption/decomposition species on metal clusters used in density functional calculations. (a). NH_3 in "on-top" mode with C_s symmetry. (b). NH_2 with C_{2v} symmetry . (c). NH with C_{4v} symmetry.. (d) N with C_{4v} symmetry.

Table II. Bond distances (unit: Å).

	N(NH$_3$)	N(NH$_2$)	N(NH)	N	H
Cr	1.99	1.99	1.96	1.85	1.55
Fe	2.00	2.00	1.97	1.77	1.56
Ni	2.01	2.00	1.98	1.80	1.54

The model unit cell structure of ammonia and its composition species on Fe(100) surface used for the EHTB band structure calculations are displayed in Fig. 3. Here we utilized relatively large unit cells, which consist of 2×2 primitive unit cells of the surface, in the computations to facilitate the description of the decomposition at surfaces. In addition, we used three top layers of the metal to represent the surface structures. Our numerical tests with more surface layers indicate that the inclusion of three surface layers into the calculations is not only necessary but also sufficient. Upon the N-H bond dissociation, the adatoms, H(ad) and N(ad), are then distributed in the unit cell at various sites. Figure 4 depicts some of the 4-fold hollow-site adsorption patterns we investigated. Here the hydrogen adatoms and, in the last pattern, the nitrogen adatom are chemisorbed on the 4-fold hollow sites. We found that the band structures do not change significantly as one changes the adsorption sites for H(ad) and N(ad), as will be demonstrated later. The chemisorption and decomposition structures on Ni(100) and Cr(100) surfaces are similar to those shown in Fig. 3. The atom-atom distance of the metal surfaces (a) and the distance between two adjacent surface planes (d) are given in Table III.

Table III. Surface geometry parameters.

Metal	$a(Å)^a$	$d(Å)^b$
Cr(100)	2.88	1.44
Fe(100)	2.87	1.43
Ni(100)	2.49	1.76

[a] a: atom-atom distance
[b] d: distance between two adjacent surface planes

Results and Discussions

Band Structures of NH$_3$ Chemisorption/Decomposition on Cr(100), Fe(100) and Ni(100). For the adsorption pattern shown in Fig. 3, we calculated the energy band structures at each step of the nitriding process. In Fig. 5, we display the band structures of ammonia chemisorption/decomposition on Fe(100) surface. Figure 5(a) depicts the band structure of the clean Fe(100) surface, where the solid line describes the overall density of states and the dashed line represents the contributions from the

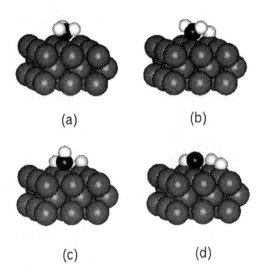

Figure 3. Fe(100) surface unit cell with NH_3 chemisorption/decomposition species in "on-top" modes. (a). NH_3. (b). NH_2 + H. (c). NH + 2H. (d). N + 3H.

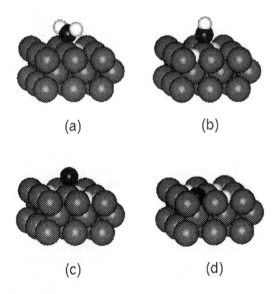

Figure 4. Fe(100) surface unit cell with NH_3 decomposition species with some adatoms in 4-fold hollow modes. (a). NH_2 + H. (b). NH + 2H. (c). N + 3H. (d). N + 3H.

d-orbitals of Fe.   The Fermi level is at E_F = -8.35 eV.   It is seen that the valence bands are overwhelmingly dominated by the d-orbital components of iron.   The 4s- and 4p-bands are pushed above the Fermi level, a typical situation for transition metal band structures.

Figure 5(b) displays the density of states for ammonia chemisorption on the clean Fe(100) surface, where the solid line is for the total density of states while the dotted line projects the contributions to the energy bands from the p-orbitals of nitrogen atoms.   It is seen first that the main features near the Fermi level essentially remain the same.   Once again, they are mainly contributed by the d-orbitals of iron. Underneath the valence bands in the energy spectrum, one observes three maxima. Two of them are primarily contributed by the $3A_1$ (~ -14.5 eV) and $1E$ (~ -16.5 eV) orbitals of ammonia.   The $3A_1$ band is associated with the electron lone-pair of the nitrogen atom and the $1E$ band corresponds to the N-H bond.   This is in good agreement with the experimental observations for ammonia chemisorption on Fe and Ni surfaces.(7,15)   The third maximum observed in the energy band structure (~ -28.7 eV) is associated with the $2A_1$ orbital of ammonia, which qualitatively agrees with the UPS spectrum of ammonia on Ni surface.(7)   The calculated Fermi level is E_F=-8.34 eV, indicating that the Fermi level is essentially not affected by the chemisorption.

We next calculated the density of states for NH_2(ad) and H(ad) on Fe(100).   The results are depicted in Fig. 5(c).   The Fermi level remains the same as in Fig. 5(b). The band structures near the Fermi level again remain essentially unchanged. However, the two maxima of the ammonia $3A_1$ and $1E$ bands observed in Fig. 5(b) now split into three peaks, as indicated by the dotted line in Fig. 5(c).   The sources for the band split come from the following.   First, the maximum for the electron lone-pair of ammonia shifts up (roughly from -14.5 eV to -13.5 eV) with about the same intensity.   However, as one of the three N-H bonds becomes broken, the peak associated with the N-H bonds in Fig. 5(b) splits into two maxima in Fig. 5(c).   One (~-16.7 eV) corresponds to the N-H bonds of NH_2(ad) species and another (~ -14.7eV) reflects the contribution from the broken N-H bond.   The $2A_1$ band shifts up about 0.5 eV.   Furthermore, one observes from Fig. 5(c) that the peak at E=-14.7 eV is twice value of the density of states contributed by the split $1E$ band.   In fact, the additional contribution comes from the 1s orbitals of hydrogen atoms chemisorbed on Fe(100).   Due to the rearrangement of the adsorbates, there appear to be some minor changes in the valence bands even though the basic profile remains similar to the one in Fig. 5(b).

Following the decomposition of NH_2(ad) species into NH(ad) and H(ad), we observe further changes in the energy spectrum, as shown in Fig. 5(d).   As one more hydrogen comes out of the N-H bonds to become an adatom in each unit cell, it adds one more peak adjacent to the one associated with the 1s orbital of hydrogen in Fig. 5(c).   Now there are two electron lone pairs whose energy bands are projected at about -13.5 eV with twice the density of states of the one in Fig. 5(c).   The band associated with the remaining N-H bond shifts up to -15.6 eV, while the 2A1 band shifts up to -27.5 eV.   The Fermi level is at E_F=-8.35 eV.

Finally, when the ammonia is completely decomposed, the energy bands of the $1E$ and $3A_1$ orbitals of ammonia become essentially a single broader maximum located at -13.5 eV, as shown in Fig. 5(e).   In addition, the 1s band of hydrogen adatom represented by the three peaks shoulder by shoulder is further broadened since there are three hydrogen adatoms now in a unit cell.   The $2A_1$ band further shifts up to -26.2 eV and the Fermi level shifts down slightly to E_F=-8.36 eV.

It is seen from Figs. 5(a) to 5(e) that the energy bands contributed by the p-orbitals of the nitrogen atom gradually shift up in the decomposition process, which is in qualitative agreement with experimental results.   The observation that the calculated $2A_1$ and $3A_1$ bands are about 20 eV and 6 eV, respectively, below the

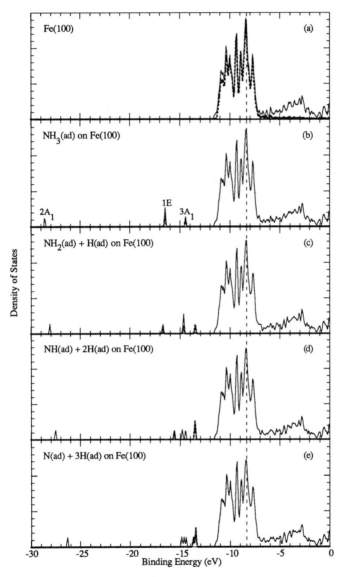

Figure 5. Band structures of NH_3 chemisorption/decomposition species on Fe(100) surface calculated for the adsorption patterns shown in Fig. 3. (a). Fe(100). (b). NH_3 on Fe(100). (c) NH_2 + H on Fe(100). (d). NH + 2H on Fe(100). (e) N + 3H on Fe(100).

Fermi level is remarkably close to the experimental values (~22 eV and 6.5 eV, respectively). Nevertheless, compared with the experimental spectra, it may be observed that the 1E band appears to be too close to the Fermi level (about 8 eV below the E_F while the experimental value is approximately 11 eV). The source of the mismatch may be largely due to the parameterization of the EH theory. Even so, the calculated spectra qualitatively agree with the profiles of UPS photoelectron spectra.

Figure 5 describes the energy band structures only for the chemisorption patterns shown in Fig. 3. It is therefore important to examine the sensitivity of the band structures to the change of adsorption patterns. Our extensive numerical study indicates that the basic features observed in Fig. 5 do not change significantly. Figure 6 shows some typical results for different adsorption patterns depicted in Fig. 4. It is seen that in all cases, the main energy bands described in Fig. 5 are also found in Fig. 6. The bands associated with the decomposition species shift only slightly. The results seem to suggest that photoelectron spectroscopy is incapable of providing information about the chemisorption modes of the nitriding species.

We next examine the band structures of a nitriding system on Cr(100) surface. The calculated results are displayed in Fig. 7. It is seen that the qualitative features of the energy spectra are very similar to those observed in Fig. 5. This is particularly the case for the p-bands of the nitrogen atom. Again, the results presented in Fig. 7 are in qualitative agreement with the experimental spectra of the nitriding system on Fe surfaces. Compared with Fig. 5, one sees that the energy bands in Fig. 7 are relatively broader. This is primarily due to the slightly more loosely packed surface structure of chromium and larger d-orbital overlaps.

Similar energy spectra can be observed in Fig. 8 for the nitriding system at Ni(100) surface. Geometrically, Ni(100) is a much more densely packed surface than both Cr(100) and Fe(100). One would thus expect that the energy bands should be much broader since a densely packed structure usually results in stronger orbital overlap which gives rise to broader energy bands. However, electronically, the d-orbitals of nickel are known to be rather defuse, which leads to smaller orbital overlap. In the EH treatment, the d-orbitals are described with double zeta parameters. In the present work, these parameters were adjusted to qualitatively reproduce the photoelectron spectrum of Ni(111) surface.(14) Compared with the energy band structures of Fe(100) and Cr(100), it is seen that the d-orbital bands of Ni(100) are relatively narrow due to its diffuse d-orbitals.

Chemisorption Binding Energy. At each step of the ammonia decomposition processes on the metal surfaces, one can readily evaluate the total energies of the systems, which are listed in Table III. For the reaction:

$$M + NH_3(g) \rightarrow M + NH_x(ad) + (3-x)H(ad),$$

where $x = 0, \cdots, 3$ and M represents the metal surfaces, the chemisorption binding energy is defined as:

$$\Delta E = E_{M+AD} - E_M - E_{NH_3(g)},$$

where E_{M+AD} is the total energy of the metal surface plus the ammonia decomposition species that stick to the surface, E_{M+AD} the energy of the surface and $E_{NH_3(g)}$ the energy of gas phase ammonia. The calculated binding energies at each step of the chemisorption and decomposition processes on the metal surfaces are depicted in Fig. 9. Here O represents the starting point where the binding energies are zero. The

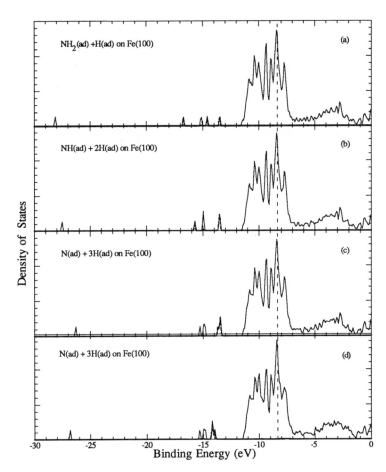

Figure 6. Band structures of NH_3 decomposition species on Fe(100) surface calculated for the adsorption patterns shown in Fig. 4. (a). NH_2 + H on Fe(100). (b). NH + 2H on Fe(100). (c) N + 3H on Fe(100). (d). N + 3H on Fe(100).

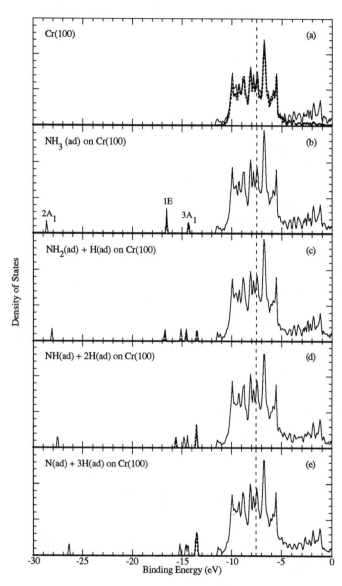

Figure 7. Band structures of NH_3 chemisorption/decomposition species on Cr(100) surface calculated for the adsorption patterns shown in Fig. 3. (a). Cr(100).   (b). NH_3 on Cr(100).   (c) NH_2 + H on Cr(100).   (d). NH + 2H on Cr(100).   (e) N + 3H on Cr(100).

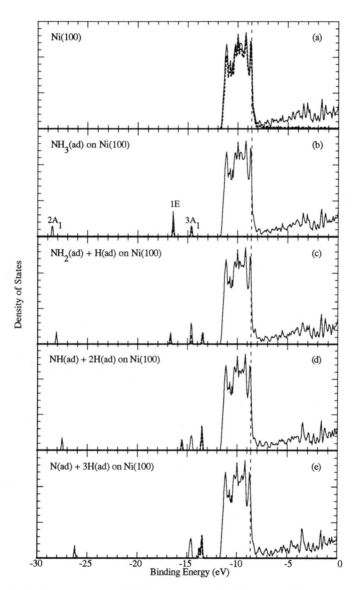

Figure 8. Band structures of NH₃ chemisorption/decomposition species on Ni(100) surface calculated for the adsorption patterns shown in Fig. 3. (a). Ni(100). (b). NH₃ on Ni(100). (c) NH₂ + H on Ni(100). (d). NH + 2H on Ni(100). (e) N + 3H on Ni(100).

first step **I** corresponds to the chemisorption of ammonia on the metal surfaces. The steps from **II** through **IV** are associated with the decomposition. It is seen that the chemisorption/decomposition on the Cr(100) surface is a straight energy down-hill process, indicating nitriding can occur much more easily on chromium. On the other hand, at Ni(100) surface, the ammonia chemisorption is an energetically favorable process. However, it requires a considerable amount of energy to further decompose into the adatoms. Furthermore, the nitriding process at Fe(100) surface appears basically to be energetically accessible except in the last step in which the binding energy is slightly higher than the previous step. Qualitatively, the trend of the nitriding process on Fe(100) predicted from the present calculations agrees reasonably well with the experimental results by Ertl and co-workers.[7] Moreover, we have observed in our ammonia plants that chromium-containing alloys used for materials in our chemical reactors are especially susceptible to the nitriding process, while the addition of nickel improves the resistance to this type of corrosion. The iron metal appears to be superior to chromium in resisting nitriding; however, nickel is the best among the three metals. This observation is in good agreement with the above theoretical results.

The reactivity of the nitriding species on different metals depends on two primary factors. The first one is related to the lattice structures, which will mainly affect the diffusion rate of these species in the bulk. While Cr(100) surface has the largest lattice constants, the Ni(100) surface is most densely packed. Thus the atomic/molecular diffusion is much easier in Cr than in Ni. The second factor is related to the electronic structures of the metals. It is known that the d-orbitals of Ni are more diffuse than those of Cr and Fe. Therefore, their interactions with the nitriding species are the weakest.

Population Analysis. In order to obtain physical insight into the interaction between the metal surfaces and the ammonia decomposition species, we performed Mulliken population analysis at each step of the chemisorption/decomposition processes. The results are displayed in Fig. 10, where the triangles describe the population at Ni(100), spheres for Fe(100) and squares for Cr(100). Here **O** represents either a clean metal surface (Fig. 10(a)) or gas phase ammonia (Fig. 10(b)). **I** to **IV** denote the same steps as in Fig. 9. Figure 10(a) depicts the population of the central metal atom that directly interacts with the nitrogen atom. It is seen that qualitatively, the population of the central atom decreases in the chemisorption and decomposition processes. However, from Fig. 10(b), one observes that upon the ammonia chemisorption on the metal surfaces, the population on the nitrogen atom rapidly decreases. This indicates that the population change on the central metal atom is not due to the electron transfer from the metal to ammonia. In contrast, it is the electron lone pair in ammonia that is attracted to the metal surfaces, which can be readily delocalized into the other surface atoms through the well-overlapped orbitals.

In the ammonia decomposition process on the metal surfaces, the population of the central atom changes slowly, while the population of the nitrogen atom increases very rapidly. This is because the decomposed ammonia species need a specific number of electrons to compensate the broken N-H bonds to stabilize the nitrogen atom. The required electrons can be transferred from the central atom. The empty orbitals of the central atom can then be readily filled out by electrons from other atoms of the metal surfaces. In Fig. 10(c), we show the overlap population between the central atom and the nitrogen atom. It is seen that qualitatively the overlap slowly increases as ammonia becomes decomposed on the surfaces.

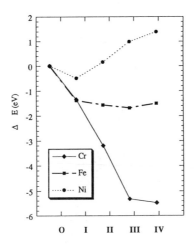

Figure 9. NH₃ chemisorption/decomposition species binding energies on Fe(100), Cr(100) and Ni(100) surfaces.

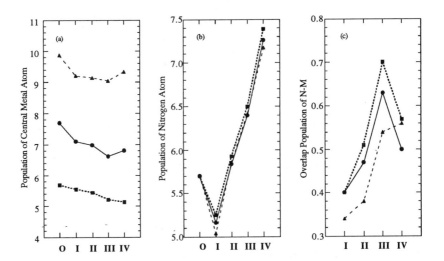

Figure 10. Mulliken electron population analysis. (a). Population of the metal atom that directly interacts with the N atom in the nitriding process. (b).. Population change of the N atom in the nitriding process. (c). Overlap population between the metal atom and the N atom.

Conclusions

In this chapter, we have presented a combined approach based on the DFT method and the EHTB theory to investigate the nitriding process on Cr(100), Fe(100) and Ni(100) surfaces. The nitriding process is broken into a series of chemisorption and decomposition steps, at each of which we evaluated the chemisorption geometries by using the DFT method. The calculated geometric parameters are then used for the EH calculations to obtain the surface energy band structures and the chemisorption binding energies. The EH method allows computations for extended structures with a relatively large unit cell, which is necessary for detailed modeling of chemisorption and decomposition systems. We showed that the calculated energy band structures are in good agreement with the UPS photoelectron spectra. The calculated chemisorption binding energies for nitriding on Fe(100) surface also qualitatively agree with the experimentally estimated results. The population analysis at each step of the chemisorption/decomposition process on the metal surfaces yields useful information about the electron transfer mechanism in the nitriding process. The theoretical results suggest that Ni is a better nitriding-resisting material than Fe metal and the latter is superior to Cr. This conclusion is consistent with our experimental observations.

The present study deals only with nitriding at pure transition metal surfaces. The insight gained from this work should be useful for assisting selection of nitriding-resisting materials used for ammonia handling equipment. In practice, a real nitriding process takes place at alloy surfaces. A more realistic physical model for those surfaces would be an alloy surface model. In principle, the present approach can be readily extended to this type of systems to help selection of nitriding-resisting materials. This work is currently being undertaken. Furthermore, in the present study, we employed a rigid surface model to simplify the calculations. For a strong chemisorption system, the lattice relaxation may be a significant process. In addition, to correctly describe the decomposition process, it is in principle necessary to sample a sufficiently large number of adsorption configurations on the surfaces. However, in the present work, we found no significant changes in the band structures as well as the binding energies for the cases we studied.

Acknowledgment

We thank Professor M.-H. Whangbo of North Carolina State University for kindly providing us the tight binding extended Hückel program. H.C gratefully acknowledges Dr. J. Ren and Professor W. Hehre for helpful discussions and Dr. R. Pearstein for assistance with the KaleidaGraph program.

References

1. Sathe, S.Y.; O'Connor, T. M., *AIChE Ammonia Symposium,* **1987,** Minneapolis.
2. Sekigawa, K. *Corrosion Engineering,* **1993,** *41,* 59.
3. Marsch, H. D. *Plant/Operations Progress* **1982,** *1,* 3.
4. Strafford, K. N. *Corrosion Sci.* **1979,** *49,* 49.
5. Verma, K. M.; Ghosh, H. and Rai, J. S. *Br. Corros. J.* **1978,** *13,* 172.
6. Frisk, K. *Metall. Trans. A* **1992,** *23A,* 1271.
7. Ertl, G. *Catalysis: Science and Technology* 4, Eds. J. R. Anderson and M. oudart, Spring, Berlin, **1983;** *Catal. Rev. - Sci. Eng.* **1980,** *21,* 201.
8. Benndorf, C.; Madey. T. E.; Johnson, A. L.; *Surface Sci.* **1987,** *187,* 434.
9. Grunze, M. *The Chemical Physics of Solid Surfaces and Heterogeneous Catalysis* 4, Eds. D. A. King and D. P. Woodruff, Elsevier, Amsterdam, **1982.**
10. Netzer, F. P.; Madey, T. E. *Surface Sci.* **1982,** *119,* 422.

11. Benndorf, C; Madney, T. E. *Surface Sci.* **1983,** *135,* 164; *Chem. Phys. Lett.* **1983,** *101,* 59.
12. Madey, T. E.; Benndorf, C. *Surface Sci.* **1985,** *152/153,* 587.
13. Ceyer, S.T.; Yates, J. T., Jr. *Surface Sci.* **1985,** *155,* 584.
14. Hüttinger, M.; Küppers, J. *Surface Sci.* **1983,** *130,* L277.
15. Seabury, C. W.; Rhodin, T. N.; Purtell, R. J.; Merrill, R. P. *Surface Sci.* **1980,** *93,* 117.
16. Grunze, M.; Bozso, F.; Ertl, G.; Weiss, M. *Appl. Surface. Sci.* **1978,** *1,* 241.
17. Drechsler, M.; Hoinkes, H.; Kaarmann, H.; Wilsch, H.; Ertl, G.; Weiss, G. *Appl. Surface Sci.* **1979,** *3,* 217.
18. Weiss, M.; Ertl, G.; Nitschke, F. *Appl. Surface Sci.* **1979,** *0,* 614.
19. Erley, W.; IBach, H. *Surface Sci.* **1982,** *119,* L357.
20. Bauschlicher, C. W., Jr., *J. Chem. Phys.* **1985,** *83,* 3129; **1985,** *83,* 2619.
21. Bagus, P. S.; Hermann, K.; Bauschlicher, C. W., Jr., *J. Chem. Phys.* **1984,** *81,* 1966; **1984,** *80,* 4378.
22. Hermann, K.; Bagus, P. S.; Bauschlicher, C. W., Jr., *Phys. Rev. B* **1985,** *31,* 6371; **1983,** *28,* 560.
23. Hohenberg, P.; Kohn, W. *Phys. Rev. B* **1964,** 136.
24. Kohn, W.; Sham, L. J. *Phys. Rev. A,* **1965,** 1133.
25. Saillard, J.-Y.; Hoffmann, R. *J. Am. Chem. Soc.* **1984,** *106,* 2006.
26. Canadell, E.; Whangbo, M.-H. *Chem. Rev.* **1991,** *91,* 965.
27. Ward, T.R.; Alemany, P.; Hoffmann, R. *J. Phys. Chem.* **1993,** *97,* 7691.
28. Cheng, H.; Kipp, D. O. unpublished.
29. Fournier, R.; Russo, N.; Salahub, D. R.; Toscano, M. in: *Cluster Models for Surface and Bulk Phenomena, Eds. G. Pacchioni et al.,* Plenum Press, New York **1992.**
30. Oed, W.; Lindner, H.; Starke, U.; Heinz, K.; Muller, K.; Saldin, D. K.; de Andres, P.; Pendry, J. B. *Surf. Sci.* **1990,** *225,* 242.
31. Muller, J.E.; Wuttig, M.; Ibach, H. *Phys. Rev. Lett.* **1986,** *56,* 1583.

RECEIVED October 20, 1994

Chapter 27

Computational Analysis of Azine-*N*-oxides as Energetic Materials

James P. Ritchie

Los Alamos National Laboratory, Mail Stop B214, Los Alamos, NM 87544

A BKW equation of state in a 1-dimensional hydrodynamic simulation of the cylinder test can be used to estimate the performance of explosives. Using this approach, the novel explosive 1,4-diamino-2,3,5,6-tetrazine-2,5-dioxide (TZX) was analyzed. Despite a high detonation velocity and a predicted CJ pressure comparable to that of RDX, TZX performs relatively poorly in the cylinder test. Theoretical and computational analysis shows this to be the result of a low heat of detonation. A conceptual strategy is proposed to remedy this problem. In order to predict the required heats of formation, new *ab initio* group equivalents were developed. Crystal structure calculations are also described that show hydrogen-bonding is important in determining the density of TZX and related compounds.

ZND theory identifies the Chapman-Jouget (CJ) state as uniquely characteristic of an explosive (1). Consequently, the detonation velocity and pressure associated with this state are frequently cited as figures of merit for an explosive. This state point is found as the point of tangency of the Rayleigh line, the slope of which is related to the detonation velocity, and the Hugoniot of the detonation products, as shown in Figure 1. (A more complete description may be found in Reference 1.) Thus, in this simple one-dimensional model, the explosive is shocked up the solid Hugoniot and decomposes as it decays down the Rayleigh line to the CJ state. The explosive products then expand from this state along an isentrope.

With this theory, it is possbile to use BKW (2) or many other chemical equations of state to locate the CJ state and compute a variety of explosive product properties, including their shock hugoniots and expansion isentropes. The detonation velocity, D_{CJ}, is especially important as it is the steady propagation rate of the detonation and can be accurately measured. Thus, a knowledge of a candidate explosive's composition, bulk density, and heat of formation is

0097–6156/95/0589–0378$12.00/0
© 1995 American Chemical Society

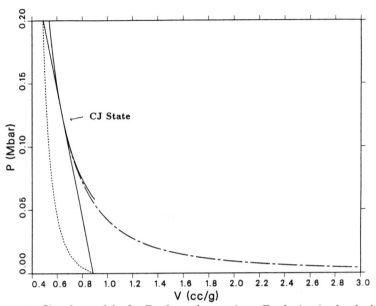

Figure 1. Simple model of 1-D plane detonation. Explosive is shocked up the solid Hugoniot (– – –) then decomposes as it decays down the Rayleigh line (straight ———) to the CJ state, where the decomposition is complete. The CJ state can be found as the point of tangency between the Rayleigh line and the detonation product Hugoniot (curved ———). The explosive expands along the isentrope (——— – ———) through the CJ state.

sufficient to allow the use of BKW for its characterization. This emphasizes the need for accurate models of heats of formation and crystal density.

Although the ZND model and the thermodynamic quantities it employs are important, as are their accurate calculation, the performance of an explosive is usually measured experimentally by its ability to push metal. Thus, an assessment of the performance of an explosive requires further consideration. The cylinder test is one experimental measure of the ability of an explosive to push metal (3). Typically a foot long one-inch i.d. copper cylinder with a .1 inch thick wall is loaded with HE. The explosive is initiated at one end and a detonation propagates along the cylindrical axis causing the walls to expand radially. The wall velocity is measured at several distances of expansion. It has become standard practice to refer to the wall velocities at 5 and 19 mm expansion as a measure of the performance of an explosive. These values have been tabulated for common explosives(4).

Two-dimensional hydrodynamic simulations of the cylinder test have been performed using chemical equations of state. The simulation proceeds by finite-difference solution of partial differential equations derived from consideration of the conservation of mass, momentum, and energy and yields the time evolution of an initial configuration. Using this approach, Souers and Kury find (5) that, when appropriately scaled, BKWR (6) gives the best fit of the experimental data compared with JCZ3 (7) and CHEQ (8). Kerley and Christian-Frear (9) have used the PANDA model (10) to realistically simulate cylinder tests.

Doherty, Short, and Kamlet (11) have related cylinder wall velocities to simple chemical and physical properties; namely, the heat of detonation, number of gas moles created, the sample density, and mole fraction of water. Although these relations are simple, they lack an theoretical basis and, in fact, are empirical fits of existing data.

A one-dimensional hydrodynamic simulation of the cylinder test can be quickly and easily performed to obtain wall velocities at various expansions. In order to make the simulation one-dimensional, the initial state of the explosive is taken to be completely burned at its initial density. In Figure 1, this corresponds to a vertical Rayleigh line and infinite detonation velocity. This procedure neglects the portion of the expansion isentrope above the initial density of the explosive. In addition to providing quantities directly comparable with experiment, the simulation also samples a large proportion of the expansion isentrope, effectively measuring the work performed by the explosive over many volume expansions. This approach also gives consideration to the initial density of the explosive, the number and kinds of gas molecules produced, and the energy of decomposition. All these factors influence the description of the isentrope and hence the performance of an explosive. These hydrodynamic simulations provide a different measure of an explosive's performance as opposed to consideration of the CJ state, which is only a single point on the expansion isentrope, but is weighted more heavily in small volume expansions and in other geometries. Because the hydrodynamic simulation is one-dimensional, the absolute values of the wall velocity are not expected to be realistic. Nonetheless, a correlation of calculated and observed wall velocities at 19 mm expansion proves to be a useful measure of explosive performance.

An interesting case history in the development of explosives is provided by 1,4-diamino-2,3,5,6-tetrazine-2,5-dioxide (TZX), which was first synthesized by M.D. Coburn at LANL. It has been discussed previously (12). Its structure is shown in Figure 2, along with those of some common HE's. Consideration of the CJ state alone for TZX, however, proves to be very misleading in terms of its performance in the cylinder test. In this report, the use of one-dimensional hydrodynamic simulations using a BKW equation of state is described. The performance of TZX is accurately described using this approach. A conceptual strategy is outlined to improve the performance of some nitrogen containing heterocycles. In order to facilitate assessments of such explosives, methods for obtaining their heats of formation and crystal density are examined and extended.

Methods and Results

BKW and Hydrodynamic Simulations: Some observed and calculated properties of TZX and RDX are shown in Table I. Comparison of these quantities shows that TZX has a higher predicted CJ pressure and detonation velocity

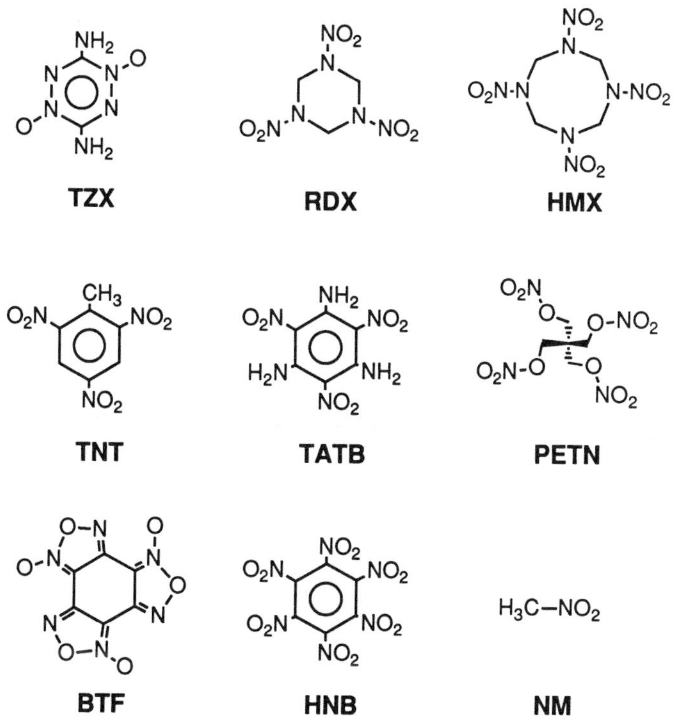

Figure 2. Structures of explosives referred to in this work.

than RDX, which is an explosive of considerable practical importance. In addition, TZX has a higher density and heat of formation than RDX. Another characteristic of an explosive is its heat of detonation. An idealized value of this quantity for C,H,N,O containing explosives can be obtained by assuming a standard set of detonation products (4). First, it is assumed that as much water (as a gas, $H_f = -57.5$ kcal/mol) as possible is formed; the remaining oxygen is then used to create carbon dioxide ($H_f = -94.05$ kcal/mol). The remaining atoms are combined to yield the elements in their standard states. Calculation of this quantity for HMX, RDX, and TZX yields: 1.48, 1.48, and 1.07 kcal/g, respectively. Thus, TZX compares unfavorably with other high-performance HE's on this basis.

The one-diemnsional hydrodynamic computer code SIN (2) was used in the manner described in the introduction to perform simulations of the cylinder test. The explosives shown in Table II were studied. Sample BKW and SIN input decks are shown in appendixes A and B for reference. In the BKW calculations, the RDX parameter set and identical molecular co-volumes were used throughout. SIN produced plots of the cylinder wall velocity versus time. Wall velocities at 19 mm expansion were determined from these plots and are shown for the explosives examined in Table II under $V_{19}^{calc.}$.

NOTE: Appendix A appears on page 392; Appendix B appears on page 393.

Table I. Observed and calculated quantities for TZX and RDX

	TZX	RDX
Observed Density (g/cc)	1.86	1.80
Observed H_f (kcal/mol)(c)	39.0	14.7
D_{CJ} (mm/μs)		
Obs.		8.75
BKW Calc.	8.85	8.75
BKW Calculated P_{CJ} (kbar)	354	347
BKW Calculated T_{CJ} (K)	1702	2587

Experimentally measured D_{CJ} and P_{CJ} have been compared before with those computed using BKW and are not repeated in Table II (2). The calculated values are listed there for reference. For TNT, the calculations were also performed using the TNT parameter set in the BKW run. As shown in Table II, this had a noticeable effect on the CJ state, but a much lesser one on $V_{19}^{calc.}$. HNB also deserves comment because its calculated CJ parameters are in remarkably poor agreement with those observed: D_{CJ} =9.34 mm/μs and P_{CJ} ≈400 kbar (13).

Comparison of the calculated and observed wall velocities shows the calculated values to be systematically too high. When the observed and calculated values of V_{19} are plotted, a good linear correlation between them is found, as shown in Figure 3. V_{19} is poorly predicted for TATB. In this instance, however, diameter effects are likely to be important. Thus the large deviation from that predicted has its basis in a real physical effect. Consequently, a correlation for ideally detonating explosives should neglect this point, until data at larger cylinder diameters becomes available. This was done in determining the correlation equation shown along the bottom of the plot. BTF has the next largest error in predicted and obsered values. This was the only explosive examined in establishing the correlation that produces no water on detonation. The error in the predicted V_{19} is less than 3%.

After the correlation had been established, HNB was considered. The predicted wall velocity from the correlation equation is 1.91 mm/μs; the observed value is 1.95 mm/μs, a difference of less than 3%.

The cylinder test for X-0535, which contains predominately TZX, shows that its performance is much less than that of RDX. This large difference cannot be accounted for by the fact that the comparison is performed for pure RDX versus a formulation containing TZX. Further information about the poor performance of TZX was obtained by performing calculations for two hypothetical forms of TZX, one in which the density was increased to 1.95 g/cc and one in which the heat of formation was raised to 102. kcal/mol, which results in a heat

Table II. Particulars of 1D 1 in. cylinder test calculations performed with BKW EOS.

HE	Molar Composition C	H	N	O	ρ (g/cc)	E_0^a (kcal/fw)	P_{CJ} (kbar)	D_{CJ} (mm/μs)	γ_{CJ}	$V_{19}^{calc.}$ (mm/μs)	$V_{19}^{obs.}$ (mm/μs)	Ref.
PBX-9501	1.470	2.855	2.602	2.689	1.834	10.8	358	8.85	3.0	1.93	1.78	b
Comp-B	2.092	2.619	2.129	2.651	1.717	9.6	283	8.06	2.9	1.76	1.63	c
HMX	4	8	8	8	1.891	36.3	389	9.11	3.0	1.99	1.86	b
RDX	3	6	6	6	1.800	33.5	347	8.75	3.0	1.93	1.78	c
X-0535^d	1.431	2.721	3.956	1.319	1.787	25.7	308	8.42	3.1	1.67	1.52	e
TATB	6	6	6	6	1.860	-13.9	311	8.28	3.1	1.70	1.45	b
TNT^f	7	5	3	6	1.630	3.8	212	7.18	3.0	1.54	1.40	b
NM	1	3	1	2	1.140	-20.8	130	6.43	2.6	1.35	1.22	c
PETN	5	8	4	12	1.765	-103.5	315	8.39	2.9	1.88	1.77	c
PETN	5	8	4	12	1.266	-103.5	162	6.79	2.6	1.49	1.38	g
BTF	6	0	6	6	1.859	158.0	325	8.16	2.8	1.92	1.83	c
HNB^h	6	0	6	12	1.965	39.7	362	8.51	2.8	2.04	1.95	i

a. $E_0 = H_f^{298} + \sum_i^{elements} k_i(H_i^{298} - H_f^0)$; where for formula $C_a H_b N_c O_d$, k=a for C and k=b/2 for H, k=c/2 for N, k=d/2 for O. $(H_f^{298} - H_f^0)$ is closely approximated as .25 kcal/mol of C and 2 kcal/mol of H_2, N_2, and O_2. So, for C,H,N,O explosives, E_0(kcal/fw) = H_f^{298}(kcal/fw) + .25a + b + c + d.

b. *LASL Explosive Property Data*, Gibbs, T.R.; Popolato, A., Eds., University of California Press:Berkeley, 1980.

c. *LLNL Explosives Handbook*, Dobratz, B.M., Ed.; UCRL-52997, 1981.

d. 95/5 weight % TZX/FPC (polymeric binder). Composition also includes .0559 moles Cl and .0838 moles F.

e. Dallman, J.C.; Explosive Technology Group, LANL, personal communication.

f. TNT parameters gave P_{CJ}=205, D_{CJ}=6.95, γ_{CJ} = 2.8, and $V_{19}^{calc.}$ = 1.56.

g. LLNL Shot K260-261.

h. Simpson, R.L., LLNL, provided density and heat of formation.

i. Ref. 12, p.949.

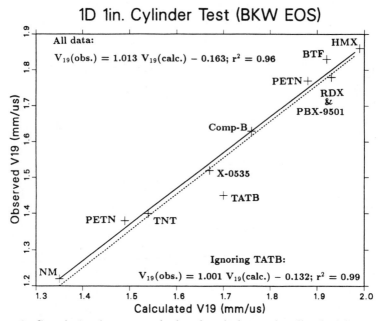

Figure 3. Correlation between calculated and observed wall velocities.

of detonation of 1.5 kcal/g. The artificial values used in these calculations are comparable to those of RDX and HMX. All other quantities were fixed at values corresponding to those of the real material. The BKW and SIN calculations were repeated and the results are summarized in Table III. The predicted V_{19}'s for both of these hypothetical forms show improvements over that found for the real material, but the improvement in performance is much larger when the heat of formation is increased. Thus, the poor performance of TZX is clearly due to a low heat of detonation, as the earlier argument above suggested.

Table III. Results of BKW Calculations for a Hypothetical TZX-like Explosive.

	High Density "TZX"	High Energy "TZX"
Assumed Density (g/cc)	1.95	1.86
Assumed H_f (kcal/mol)	39	102
Calculated P_{CJ} (kbar)	402	394
Calculated D_{CJ} (mm/μs)	9.23	9.12
Predicted[a] V_{19} (mm/μs)	1.71	1.87

[a] Predicted value is found using calculated value in the equation shown in Figure 3. Predicted $V_{19} = 1.62$ mm/μs for pure TZX.

One means of improving the heat of detonation of TZX or other nitrogen-containing heterocycles is to introduce additional oxygen. This allows excess carbon to be oxidized to form carbon dioxide. TZX of course represents the embodiment of this stragegy in that 1-4-diamino-2,3,5,6-tetrazine(DAT) is not expected to be an explosive. Thus simply converting aromatic nitrogen atoms to the corresponding N-oxides is anticipated to raise the heat of detonation of a molecule, if it originally produced an excess of carbon. In this regard, molecules **I** (14) and **II** (15), below, are interesting to consider. These molecules contain adjacent N-oxides. They also have been synthesized. The crowding of the adjacent N-oxides is of interest because of the additional energy it is likely to impart to the molecule. An evaluation of this factor and an estimate of these and related molecules as energetic materials requires heats of formation and crystal densities. In order to obtain these quantities for the molecules under consideration, additional calculations were performed as described next.

I   **II**

Ab Initio **Group Equivalents for Heats of Formation:** Wiberg, Nakaji, and Breneman (16) derived *ab initio* group equivalents to estimate the gaseous heats of formation of azines. Equation 1 is used for this purpose, where E_{tot} is the total energy from the *ab initio* calculations and N_i is the number of groups corresponding to the energy equivalent E_i; E_{tot} and E_i are in a.u., while H_f is obtained in kcal/mol.

$$H_f{}^{298} = 627.51 \times (E_{tot} + \sum_i (N_i \times E_i)) \tag{1}$$

We adopted the $C_{ar} - H$ and N_{ar} equivalents previously determined at the MP2 and MP3 level.(15) In addition, we performed calculations using the MP4(SDTQ) method and developed group equivalents at this level. The $C_{ar} - H$ and N_{ar} MP4 equivalents were determined using only the observed heats of formation of benzene and pyridine. Thus the equivalents at this level are not as well calibrated as those at MP2 and MP3, which were determined using a larger experimental database. $N_{ar} - O$ and $C_{ar} - NH_2$ equivalents are required to calculate a H_f for TZX and other azine-N-oxides. Measured H_f's of aniline and pyridine-N-oxide are available for this purpose. Unfortunately, this uniquely determines the equivalents, which is undesirable. The experimental values appear to be accurate, but it is likely that the resulting errors in the calculated H_f's will be larger than those found in the earlier work.

Total calculated energies for several calibration compounds, azines, and azine-N-oxides are shown in Table IV. In all instances, molecular geometries were optimized at the RHF/6-31G* level. With the exception of aniline, all the molecules were constrained to be planar. Aniline is known to possess a pyramidal amino group, although the planar-pyramidal energy difference is small, as shown in Table IV. So, it seems likely that the optimal geometry of the other amino-group containing molecules may also be non-planar. The group equivalent, however, was derived using the pyramidal geometry. This may cause the H_f of such molecules to be somewhat overestimated.

Table IV. Calculated energies obtained using the indicated theoretical method. Energies were obtained using the RHF/6-31G* optimized planar geometries and 6-31G* basis set throughout, except for the aniline (py.) calculation, in which the amino group was pyramidal.

	RHF	MP2	MP3	MP4(SDTQ)
Benzene	−230.703137	−231.456505	−231.485222	−231.529688
Aniline (py.)	−285.730822	−286.648505	−286.680310	−286.731673
Aniline (pl.)	−285.728227	−286.645308	−286.677003	−286.728158
Pyridine	−246.695820	−247.480360	−247.502442	−247.549709
Pyridine-N-oxide	−321.473768	−322.452900	−322.466279	−322.530149
Pyrazine	−262.683005	−263.500213	−263.515707	−263.565549
Pyrimidine	−262.693488	−263.506478	−263.522476	−263.572534
Pyridazine	−262.650029	−263.469882	−263.485466	−263.536251
s-Triazine	−278.695843	−279.535017	−279.545486	−279.598109
Melamine	−443.860580	−445.174943	−445.197915	−445.267108
s-Tetrazine	−294.592926	−295.477477	−295.480813	−295.537874
1,4-Diamino-2,3,5,6-tetrazine	−404.671096	−405.889518	−405.895167	−405.966694
1,4-Diamino-2,3,5,6-tetrazine-2,5-dioxide	−554.242424	−555.861070	−555.840745	−555.950532
Pyrazine-di-N-oxide	−412.234811	−413.443425	−413.439431	−413.524027
Pyrimidine-di-N-oxide	−412.230887	−413.438271	−413.435156	−413.520431
Pyridazine-di-N-oxide	−412.203017	−413.423889	−413.413820	−413.504075

The resulting group equivalents and calculated heats of formation are shown in Tables V and VI, respectively. Because the MP2 and MP3 calculations are better calibrated and include some accounting for electron correlation, these results are preferred. As shown in Table VI, the difference between these two methods is typically small, but reaches a maximum of 8.3 kcal/mol for TZX; other differences are typically half this size or less. The MP4 heats of formation show several differences from those at MP2 and MP3. In general, the differences, where they exist, are about 2-3 kcal/mol, which is not surprising because these equivalents are less well calibrated than the others. This suggests that the perturbation expansion is converging reasonably and that the accounting for electron correlation at MP2 and MP3 is adequate for our purposes. This behavior is in marked contrast to the results obtained at the RHF level, which show several differences, some as large as 20 kcal/mol from those at MP2 and MP3.

Table V. Group increments (a.u.) for obtaining heats of formation using the indicated theoretical method with the 6-31G* basis set.

	RHF	MP2	MP3	MP4
C_{ar}-H	38.45546^a	38.58138^a	38.58611^a	38.59354
N_{ar}	54.46976^a	54.62523^a	54.62378^a	54.63555
C_{ar}-NH_2	93.48667	93.77475	93.78291	93.79712
$N_{ar} \rightarrow O$	129.22993	129.57947	129.56920	129.59592

[a] Reference 14.

There is only one new directly pertinent theory-experiment comparison shown in Table VI. For melamine, the MP2 and MP3 H_f's are somewhat higher than that observed. This difference could be due to the use of a planar geometry in the calculations, but the difference is not large and is a useful check of the $C_{ar} - NH_2$ equivalent, which was determined with only one experimental datum. The H_f's of diaminotetrazine and TZX have been measured at LANL. The values obtained, however, are for a polycrystalline sample, not a gaseous molecule, as the calculations give. The difference between these numbers is the heat of sublimation. For these two molecules, the difference between the calculated and observed H_f values gives numbers that are sensible for this quantity for molecules of this size.

The PM3 semi-empirical molecular orbital method (17) was also used to calculate H_f's. Comparison of the results obtained at this level with those at MP2 and MP3 show several large differences. One especially pertinent example is the relative H_f's of diaminotetrazine and TZX. According to the *ab initio* calculations, oxidation of the former to give the latter lowers the H_f by more than 30 kcal/mol. PM3 on the other hand, shows a small increase. This difference results in a significantly higher heat of detonation for TZX and an overestimate of its detonation performance.

Predicted H_f's of the di-N-oxides of pyrazine, pyrimidine, and pyridazine show an increase in the heat of formation with decreasing distance between the N-oxides. The additional energy available in pyridazine-di-N-oxide is approximately 12.5 ± 3 kcal/mol. This increase is less than that found in pyridazine itself of about 23 kcal/mol. So the adjacent lone-pairs are more destabilizing than adjacent N-oxides. Also, the destabilization found from the *ab initio* calculations is much less than that found with the PM3 method.

Crystal Structures: Electrostatic interactions are important in molecular crystal structures (18). Atomic charges determined by fitting the electrostatic

Table VI. Calculated and observed standard gaseous H_f's (kcal/mol).

	RHF	MP2	MP3	MP4	PM3	Obs.[a]
Benzene	18.6[b]	19.9[b]	19.7[b]	19.8[c]	23.5	19.8 ± .2
Aniline[c] (py.)	20.8	20.8	20.8	20.8	21.4	20.8 ± .6
Pyridine	32.3[b]	32.5[b]	32.6[b]	33.6[c]	30.4	33.6 ± .2
Pyridine-N-oxide[c]	21.0	21.0	21.0	21.0	27.3	21.0[d]
Pyrazine	49.2[b]	47.5[b]	47.9[b]	50.0	39.4	46.9 ± .3
Pyrimidine	42.6[b]	43.6[b]	43.6[b]	45.6	38.0	46.8 ± .3
Pyridazine	69.9[b]	66.6[b]	66.9[b]	68.4	56.0	66.5 ± .2
s-Triazine	50.1[b]	53.2[b]	52.8[b]	55.9	46.3	54.2 ± .2
Melamine	5.5	15.7	13.9	19.4	28.8	12.4 ± 1.
s-Tetrazine	123.6[b]	116.8[b]	117.0[b]	120.1	94.7	(117.0[b])
1,4-Diamino-2,3,5,6-tetrazine	113.8	101.0	104.0	106.5	87.4	(71.2)[e]
1,4-Diamino-2,3,5,6-tetrazine-2,5-dioxide	81.8	61.4	69.7	66.9	93.4	(39.2)[e]
Pyrazine-di-N-oxide	29.4	25.7	27.2	26.3	39.5	
Pyrimidine-di-N-oxide	31.9	29.0	29.9	28.6	30.3	
Pyridazine-di-N-oxide	49.4	38.0	43.3	38.9	70.1	

[a] Unless noted otherwise, values from J.B. Pedley, R.D. Naylor, S.P. Kirby *Thermochemical Data of Organic Compounds*, 2nd Ed., Chapman:New York, 1986.
[b] Reference 14.
[c] Uniquely determined by the available data.
[d] L. Shaofeng, G. Pilcher, *J. Chem. Thermody.*, **1988**, *20*, 463.
[e] Value is for the polycrystalline material. Determination by M. Stinecipher, LANL.

potential are frequently used for calculating this quantity. The CHELPG procedure was used in this investigation (19). Alternatively, Hirshfeld (20) atom-centered multipole expansions (ACME's) can be used that are obtained from direct integration of the electron density (21). Figure 4 shows a comparison of the electrostatic potential surrounding TZX obtained using atomic charges from the CHELPG procedure and ACME's truncated at quadrupoles (l=2) and octapoles (l=3).

The ability of the charge models to reproduce the potential is measured by the root-mean square (RMS) and relative root-mean square (RRMS) errors. These quantities were evaluated on a rectilinear mesh that enveloped the molecules. The mesh points were spaced .25 Å apart and the closest approach of the boundary of the mesh to an atom of the molecule was 5 Å. At each point of the mesh, the potential was evaluated using either the model charge distributions or rigorous formula as implemented in GAUSSIAN92 (22). The errors in the fit were binned according to the value of the potential. Three

TZX PDC's   TZX L = 2   TZX L = 3

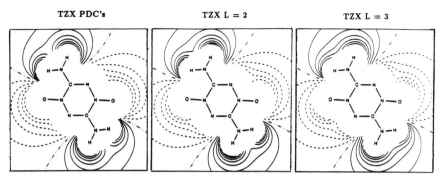

Figure 4. Electrostatic potential of TZX from potential derived charges (PDC) and ACME's truncated at quadrupoles (l=2) and octopoles (l=3). Results from RHF/3-21G//RHF/3-21G calculations are shown. Positive values (line), negative values (dash), and zero (chain–dot) are shown at 10 kcal/mol/e intervals.

bins were used corresponding to values of the potential less than -4.18 kj/mol, between -4.18 and 4.18 kj/mol, and larger than 4.18 kj/mol. This procedure was followed to allow the fits of positive and negative regions to be examined separately.

The results for diaminotetrazine and TZX are shown in Table VII. It is found that ACME's at successively higher levels reduce the RMS and RRMS errors. In the cases examined, the use of quadrupoles results in lower RMS and RRMS errors than obtained with PDC's. Further improvements are obtained upon the inclusion of octapole terms. Only relatively small differences in the RMS and RRMS errors are found in positive and negative regions. The use of the multipoles provides the greatest improvement for diaminotetrazine. It has many lone-pairs of electrons and consequently an atomic charge model does not reproduce the surrounding potential very well.

The effect of improvements in the fit of the electrostatic potential in the calculation of the corresponding crystal structure of these molecules was studied next. To accomplish this, we have modified D.E. Williams PCK/83 program to calculate multipole-multipole interactions including terms up to quadrupoles (23). The program uses Equation 44 of Nijboer and DeWette (24) to compute the electrostatic lattice sums; numerical tests using only real-space sums have supported the use of these formulas, but a formal derivation is lacking. Several sets of calculations were performed, which are summarized in Table VIII. Use of Williams original potential function parameters (18) supplemented by either PDC's or ACME's led only to small, but noticeable, differences in the calculated crystal structures. In addition, the calculated energies, which are poor estimates of the heat of sublimation, were also noticeably different, but the difference was not large. More serious from the standpoint of HE performance, however, was the difference between the observed and calculated densities. This quantity was significantly underestimated. Hydrogen-bonding was believed to be important in these crystal structures. This leads to unusually short hydrogen-heavy atom

Table VII. Error analysis of electrostatic potential fit with different models. PDC refers to the use of potential derived charges. ME n indicates the multipole expansion was carried out to n=0, charges; 1, dipoles; 2, quadrupoles; 3, octapoles. RMS (kj/mol) and RRMS computed as described in the text. Range refers to potential values (kj/mol) for which the error analyses were performed.

Diaminotetrazine

Range: −185.0	→		−4.18,	−4.18	→	4.18,	4.18	→	161.0	
	Npts	RMS	RRMS	Npts		RMS	Npts	RMS	RRMS	
PDC	33730	7.80e+00	3.54e−01	21772	2.14e+00	43230	4.86e+00	2.69e−01		
ME 0	33730	1.85e+01	8.24e−01	21772	6.21e+00	43230	1.41e+01	7.27e−01		
ME 1	33730	4.55e+00	2.51e−01	21772	2.02e+00	43230	3.98e+00	2.39e−01		
ME 2	33730	2.23e+00	1.33e−01	21772	1.17e+00	43230	2.38e+00	1.50e−01		
ME 3	33730	8.76e−01	6.83e−02	21772	6.75e−01	43230	1.34e+00	8.43e−02		

TZX

Range: −194.0	→		−4.18,	−4.18	→	4.18,	4.18	→	296.0	
	Npts	RMS	RRMS	Npts		RMS	Npts	RMS	RRMS	
PDC	43912	3.35e+00	1.41e−01	15384	1.40e+00	54956	2.54e+00	1.13e−01		
ME 0	43912	1.34e+01	4.90e−01	15384	4.67e+00	54956	1.23e+01	4.61e−01		
ME 1	43912	4.47e+00	1.60e−01	15384	2.10e+00	54956	3.54e+00	1.71e−01		
ME 2	43912	1.24e+00	6.64e−02	15384	7.60e−01	54956	2.28e+00	8.82e−02		
ME 3	43912	1.06e+00	7.66e−02	15384	6.31e−01	54956	1.43e+00	6.55e−02		

distances. To simulate this effect, the preexponential parameter of hydrogen atoms covalently bonded to nitrogen or oxygen were reduced by 30%. When the calculations were repeated with this parameter set, the crystal densities were reproduced much more accurately, as shown in Table VIII. These calculations show that molecular modeling can be used to predict the crystal densities of azine-N-oxides. More experimental work is required to produce greater variations in molecular and crystal structures as a further test of the method.

Conclusions

A model for predicting the performance of explosives in the cylinder test has been presented. When accurate heats of formation and crystal densities are available, the largest errors in the predicted wall velocity at 19 mm is about 3%. TATB is worse, but it is predicted that larger diameter cylinders will reduce the error. TZX is treated accurately in the model. Models for obtaining heats of formation and crystal densities have also been investigated. These can be used to estimate these quantities more accurately than before for proposed azine-N-oxides.

Table VIII. Energy optimized lattice parameters (Å and degrees) and energy contributions (kj/mol) of molecular crystal structures for differing parameter sets. Percent errors in parenthesis. PDC and ACME's refer to the use of the indicated charge model with William's original parameters. P(N,O) and A(N,O) refer to the use of the analogous charge model with a reduced pre-exponential factor for hydrogen on nitrogen or oxygen. E(e) is the electrostatic (PDC or ACME), E(d) the dispersion (R^{-6}), and E(r) the repulsion (Aexp(-CR_{ij})) components of the total energy (E(t)). ρ is the rotation angle (degrees) between optimized and original asymmetric unit. T(Å) is the translation of the optimized asymetric unit relative to that observed. D is crystal density (g/cc).

1,4-Diamino-2,3,5,6-tetrazine-2,5-dioxide
Space Group P2$_1$/a (#14), cell choice #3, Z=2.[a]

	Obs.[a]	PDC	P(N,O)	ACME's	A(N,O)
a	6.77	6.72 (-0.7)	6.63 (-2.1)	6.81 (0.6)	6.73 (-0.6)
b	7.35	7.55 (2.7)	7.47 (1.6)	7.55 (2.7)	7.46 (1.5)
c	5.25	5.38 (2.5)	5.25 (-0.0)	5.38 (2.5)	5.26 (0.2)
β	100.5	97.8 (-2.7)	98.2 (-2.3)	98.9 (-1.6)	99.3 (-1.1)
D	1.86	1.77 (-4.8)	1.86 (0.0)	1.75 (-6.0)	1.84 (-1.5)
ρ		5.3	4.7	4.8	4.0
E(e)		-88.78	-97.60	-85.52	-93.92
E(d)		-118.83	-134.93	-115.20	-130.87
E(r)		75.93	87.46	72.28	83.07
E(t)		-131.68	-145.07	-128.44	-141.72

1,4-Diamino-2,3,5,6-tetrazine
Space Group Amam (#63). Z=4.[b]

	Obs.[b]	PDC	P(N,O)	ACME's	A(N,O)
a	6.26	6.17 (-1.4)	6.14 (-1.9)	6.24 (-0.3)	6.20 (-1.0)
b	7.84	8.89 (13.4)	8.60 (9.7)	8.45 (7.8)	8.10 (3.3)
c	9.42	8.94 (-5.1)	8.77 (-6.9)	9.46 (0.4)	9.35 (-0.7)
D	1.61	1.52 (-5.6)	1.61 (0.0)	1.49 (-7.4)	1.59 (-1.5)
T		0.03	0.04	0.28	0.40
E(e)		-49.61	-56.88	-62.30	-78.73
E(d)		-90.29	-104.32	-88.03	-105.16
E(r)		53.02	63.14	57.98	75.91
E(t)		-86.88	-98.06	-92.35	-107.98

[a] Ryan, R. R., et al., Los Alamos National Laboratory, private communication.
[b] Krieger, C.; Fischer, H.; and Neugebauer, F. A., *Acta Cryst.* **1987**, *C43*, 1320.

Appendix A. Sample BKW input deck for X-0535.

```
  0     1     1     1     1     0     0     0     0     0
X-0535  TZX/FPC-461  95/5
  6    18    19
0.50000000000e+00 0.16000000000e+00 0.40000000000e+03 0.10909778444e+02
c      h      n      o      cl     f
0.14310000000e+01 0.27210000000e+01 0.39560000000e+01 0.13190000000e+01
0.55875376900e-01 0.83813065351e-01
0.17870000000e+01 0.10000000000e+03 0.25700000000e+05
0.30000000000e+04 0.30000000000e+00
h2o    co2    n2     h2     o2     co     nh3    h      no     oh     ch4
hcl    ccl4   cl2    hf     cf4    f2     cof2   sol c
0.13185912061e+01 0.10000000000e-01 0.19778868091e+01 0.41906532675e-01
0.10000000000e-01 0.10000000000e-01 0.10000000000e-01 0.10000000000e-01
0.10000000000e-01 0.10000000000e-01 0.10000000000e-01 0.10000000000e-01
0.10000000000e-01 0.10000000000e-01 0.10000000000e-01 0.10000000000e-01
0.16906532675e-01 0.10000000000e-01 0.14003419599e+01
0.42588420868e+02 0.14808050357e-01-0.26391810479e-05 0.19204529633e-09
0.00000000000e+00 0.13428283691e+04-0.57107000000e+05 0.25000000000e+03
0.47481121063e+02 0.19544629380e-01-0.37212960251e-05 0.27703001004e-09
0.00000000000e+00 0.74628094482e+03-0.93968000000e+05 0.60000000000e+03
0.43923400879e+02 0.12225010432e-01-0.23790050818e-05 0.17983220468e-09
0.00000000000e+00 0.11391613770e+04 0.00000000000e+00 0.38000000000e+03
0.29703470230e+02 0.11438289657e-01-0.22012220597e-05 0.16777609857e-09
0.00000000000e+00 0.11758961182e+04 0.00000000000e+00 0.80000000000e+02
0.47030899048e+02 0.12871470302e-01-0.25002170787e-05 0.19015700126e-09
0.00000000000e+00 0.10353764648e+04 0.00000000000e+00 0.35000000000e+03
0.45330818176e+02 0.12381610461e-01-0.24164030492e-05 0.18281809400e-09
0.00000000000e+00 0.11215882568e+04-0.27201000000e+05 0.39000000000e+03
0.42018161774e+02 0.19116619602e-01-0.31643301099e-05 0.21978009690e-09
0.00000000000e+00 0.12069611816e+04-0.93680000000e+04 0.47600000000e+03
0.26391099930e+02 0.81213722005e-02-0.16907399640e-05 0.13168230395e-09
0.00000000000e+00 0.79463159180e+03 0.51619000000e+05 0.76000000000e+02
0.48414981842e+02 0.12693859637e-01-0.24946000394e-05 0.18932129475e-09
0.00000000000e+00 0.12092497559e+04 0.21477000000e+05 0.38600000000e+03
0.42417919159e+02 0.11568469927e-01-0.22266590349e-05 0.16891550658e-09
0.00000000000e+00 0.11835175781e+04 0.35600000000e+04 0.41300000000e+03
0.38756858826e+02 0.23640129715e-01-0.37079569211e-05 0.24707139112e-09
0.00000000000e+00 0.10424278564e+04-0.16000000000e+05 0.52800000000e+03
0.42936641693e+02 0.11874400079e-01-0.22864060156e-05 0.17278070641e-09
0.00000000000e+00 0.11737486572e+04-0.22000000000e+05 0.63700000000e+03
0.67947479248e+02 0.39482139051e-01-0.79941555668e-05 0.61362437354e-09
0.00000000000e+00 0.82403247070e+03-0.25500000000e+05 0.20000000000e+04
0.51320999146e+02 0.14395629987e-01-0.29638829346e-05 0.23498719637e-09
0.00000000000e+00 0.99022290039e+03 0.00000000000e+00 0.95600000000e+03
0.40000709534e+02 0.11445820332e-01-0.22104300115e-05 0.16817050530e-09
0.00000000000e+00 0.11790649414e+04-0.64200000000e+05 0.38900000000e+03
0.55680858612e+02 0.35636339337e-01-0.68981721597e-05 0.51695342451e-09
0.00000000000e+00 0.20372969055e+03-0.21800000000e+06 0.13300000000e+04
0.46372489929e+02 0.13958250172e-01-0.28047579690e-05 0.21583379528e-09
0.00000000000e+00 0.95607006836e+03 0.00000000000e+00 0.38700000000e+03
0.56752330780e+02 0.26767069474e-01-0.51359011195e-05 0.38333131114e-09
0.00000000000e+00 0.56443182373e+03-0.15000000000e+06 0.13300000000e+04
-0.24615189433e+00 0.71798549034e-02-0.12975500567e-05 0.93499950837e-10
0.00000000000e+00-0.25820437622e+03 0.00000000000e+00 0.00000000000e+00

0.44444444776e+00 0.83093583584e+00-0.13938181400e+01 0.67256969213e+00
-0.11353725940e+00 0.64915586263e-02-0.22670534253e+00 0.12051656842e+00
0.83159998059e-01-0.17558999360e+00 0.15531000495e+00 0.12010000229e+02
0.00000000000e+00 0.20000000000e+01 0.00000000000e+00 0.10000000000e+01
0.00000000000e+00 0.00000000000e+00 0.10000000000e+01 0.00000000000e+00
0.00000000000e+00 0.20000000000e+01 0.00000000000e+00 0.00000000000e+00
0.00000000000e+00 0.00000000000e+00 0.20000000000e+01 0.00000000000e+00
0.00000000000e+00 0.00000000000e+00 0.25000000000e+05 0.20000000000e+01
0.00000000000e+00 0.00000000000e+00 0.00000000000e+00 0.00000000000e+00
0.00000000000e+00 0.00000000000e+00 0.00000000000e+00 0.20000000000e+01
0.00000000000e+00 0.00000000000e+00 0.10000000000e+01 0.00000000000e+00
```

Appendix B. Sample SIN input deck for RDX.

```
    3  127  124    1    1    0    0    1    1    0    0    0
Cylinder Test
0.20000000000E+01 0.50000000000E+04 0.20000000000E+03 99999.0000000E+00
RDX
  100    1    1    0
0.12700000000E-01 0.25400000000E-02 0.20000000000E+01 0.18000000000E+01
0.10000000000E-05 0.30000000000E+03 0.00000000000E+00 0.00000000000E+00
0.24230000000E+00 0.18830000000E+01 0.10000000000E-01 0.00000000000E+00
0.00000000000E+00-0.90418722204E+01-0.71318525243E+02-0.12520497936E+03
-0.92042417760E+02-0.22189382573E+02 0.67500000000E+00 0.40000000000E+00
0.54229934924E+00 0.50000000000E-04 0.00000000000E+00 0.00000000000E+00
0.30000000000E+03 0.10000000000E-05 0.00000000000E+00 0.00000000000E+00
0.00000000000E+00 0.00000000000E+00 0.00000000000E+00
0.00000000000E+00 0.00000000000E+00 0.00000000000E+00 4.15944771549E-01
-0.34939402446E+01-0.25521450345E+01 0.25510714285E+00 0.11234997249E-01
-0.96121668096E-02-0.16050449713E+01 0.51908426867E+00 0.65757361251E-01
0.40261694558E-02 0.94796916331E-04 0.74073902496E+01-0.49641482820E+00
0.39721126724E-01 0.30458400580E-01-0.10680694321E-01 0.50000000000E+00
0.10000000000E+00
copper
   25    0    0    0
0.10160000000E-01 0.20320000000E-02 0.20000000000E+01 0.89029999999E+01
0.10000000000E-05 0.30000000000E+03 0.00000000000E+00 0.00000000000E+00
0.39580000000E+00 0.14970000000E+01 0.00000000000E+00 0.00000000000E+00
0.00000000000E+00-0.31983419917E+04-0.55743953279E+04-0.36248840041E+04
-0.10433967951E+04-0.11206726387E+03 0.20000000000E+01 0.93000000000E-01
0.11232168932E+00 0.17670000000E-04 0.00000000000E+00 0.00000000000E+00
0.30000000000E+03 0.10000000000E-05 0.00450000000E+00 0.46600000000E+00
0.05000000000E+00 0.00000000000E+00 0.00000000000E+00
0.10000000000E+01 0.00000000000E+00 0.00000000000E+00 0.00000000000E+00
air
0002
+5.0000      +000+0.02      +000+0.000001   +000+0.000001    +001
+1.0         -006+3.0       +002+0.0        +000+0.0         +000
+0.254       +000+1.5       +000+0.0001     +000+0.0         +000
+0.0         +000+8.17773414663+002+6.10099511390+002+9.09042692454+001
-2.20373162776+001-4.87356368431+000+0.0        +000+0.0276       +000
+5.2882072   -002+1.1666      -005
+3.0         +002+1.0         -006+0.0        +0.0
+0.0          +1.0           -006
+1.0
```

Acknowledgment

Charles Mader provided helpful comments and other assistance in the use of BKW and SIN. Ann Copenhaver and Edward Kober contributed to the development of PCK/ME. Funding from DoE/DoD, Office of Munitions, MOU and LANL CX program, P. Howe, manager, are gratefully acknowledged.

Literature Cited

1. Fickett, W.; Davis, W.C. *Detonation*; UC Press; Berkeley,CA, 1979.
2. Mader, C.L. *Numerical Modeling of Detonation*; UC Press; Berkeley, CA, 1979.

3. a)Kury, J.W.; Hornig, H.C.; Lee, E.L.; McDonnel, J.L.; Ornellas, D.L.; Finger, M.; Strange, F.M.; Wilkins, M.L. In *Proceedings of the Fourth Symposium (International) on Detonation*, White Oak, MD, 1965, p. 3-12. b) Lee, E.; Breithaupt, D.; McMillan, C.; Parker, N.; Kury, J; Tarver, C.; Quirk, W.; Walton, J. *Proceedings of the Eighth Symposium (International) on Detonation*, Albuquerque, NM, p. 613-624.

4. *LLNL Explosives Handbook*; B.M. Dobratz, Ed.; UCRL-52997.

5. Souers, P.C.; Kury, J.W. *Propellants, Explosives, Pyrotechnics*, **1993**, *18*, 175-183.

6. Finger, M.; Lee, E.; Helm, F.H.; Hayes, B.; Hornig, H.; McGuire, R.; Kahara, M.; Guidry, M. *Proceedings of the Sixth Symposium (International) on Detonation*, Coronado, CA, 1976, p.710-722.

7. Cowperthwaite, M.; Zwisler, W.H. Ref. 6, p. 162-172.

8. a)Ree, F.H. *J. Chem. Phys.*, **1984**, *81*, 1251-63. b) Ree, F.H. *ibid.* **1986**, *84*, 5845-56.

9. Kerley, G.I.; Christian-Frear, T.L. SAND93-2131, Sandia National Laboratory, Albuquerque, NM 87185, 1993.

10. Kerley, G.I. *Proceedings of the Eighth Symposium (International) on Detonation*, Albuquerque, NM, 1985, p. 540-547.

11. Doherty, R.M.; Short, J.M.; Kamlet, M.J. *Combustion and Flame*, **1989**, *76*, 297-306.

12. Stine, J.R. In *Structure and Properties of Energetic Materials*, Liebenberg, D.H.; Armstrong, R.W.; Gilman, J.J., Eds.; MRS Symposium Proceedings 296; Boston, MA, 1992; p.3-12.

13. McGuire, R.R.; Ornellas, D.L.; Helm,F.H.; Coon, C.L.; Finger, M. *Proceedings of the Seventh Symposium (International) on Detonation*, Annapolis, Md., June, 1981, p.940-951.

14. Suzuki, I.; Nakadate, M.; Sueyoshi, W. *Tetrahedron Lett.*, **1968**, 1855-8.

15. Ullman, E.F.; Singh, P. *J. Am. Chem. Soc.*, **1972**, *94*, 5077-78.

16. Wiberg, K.B.; Nakaji, D.; Breneman, C.M. *J. Am. Chem. Soc.*, **1989**, *111*, 4182-90.

17. Stewart, J.J.P. *J. Comp. Chem.*, **1989**, *10*, 221-264.

18. a)Williams, D.E.; Cox, S.R. *Acta Cryst.*, **1984**, *B40*, 404-417. b) Cox, S.R.; Hsu, L.-Y.; Williams, *Acta Cryst.*, **1981**, *A37*, 293-301.

19. Breneman, C.M.; Wiberg, K.B. *J. Comp. Chem.*, **1990**, *11*, 361-373.

20. Hirshfeld, F.L. *Theor. Chim. Acta*, **1977**, *44*, 129-138.

21. Ritchie, J.P. *J. Am. Chem. Soc.*, **1985**, *107*, 1829-37.

22. GAUSSIAN92, Rev. A, Frisch, M.J.; Trucks, G.W.; Head-Gordon, M.; Gill, P.M.W.; Wong, M.W.; Foresman, J.B.; Johnson, B.G.; Schlegel, H.B.; Robb, M.A.; Replogle, E.S.; Gomperts, R.; Andres, J.L.; Raghavachari, K.; Binkley, J.S.; Gonzales, C.; Martin, R.L.; Fox, D.J.; DeFrees, D.J.; Baker,J.; Stewart, J.J.P.; Pople, J.A., Gaussian, Inc., Pittsburgh, PA, 1992. The authors thank Dr. R. Martin for his assistance in implementing the program at Los Alamos.

23. Ritchie, J.P.; Kober, E.M.; Copenhaver, A.S. *Proceedings of the Tenth Symposium (International) on Detonation*, Boston, MA, July, 1993, in print.
24. Nijboer, B.R.A.; DeWette, F.W. *Physica*, **1957**, *23*, 309-321.

RECEIVED October 20, 1994

Chapter 28

Genetic Algorithmic Approach for Computer-Aided Molecular Design

Venkat Venkatasubramanian, King Chan, and James M. Caruthers

Laboratory for Intelligent Process Systems, School of Chemical Engineering, Purdue University, West Lafayette, IN 47907

The design of new materials possessing desired physical and biological properties is an important endeavor for designers in the chemical, material and pharmaceutical industries. The traditional approach to molecular design involves a laborious and costly hypothesis, synthesis, and evaluation procedure. To alleviate the protracted design cycle, this paper describes an approach using genetic algorithms (GAs) for the molecular design process. Genetic algorithms are general purpose, stochastic, evolutionary search and optimization strategies based on the Darwinian model of natural selection. The mechanics, characteristics, and other issues in using GAs for molecular design are presented with the aid of case studies in polymer and refrigerant design. The merits and potential deficiencies of this approach are also discussed.

The problem of designing new molecules with desired properties is an important and difficult one, encompassing the design of polymers, polymeric composites, blends, paints and varnishes, refrigerants, solvents, drugs, pesticides, and so on. The traditional approach requires the designer to hypothesize a compound, synthesize the material, evaluate to see if it meets the desired design targets, and to reformulate the design if the desired properties are not achieved (Figure 1). This is a laborious and expensive trial-and-error procedure. For example, it may take over one thousand design, synthesis and evaluation cycles before a new drug is designed. Hence, there is considerable incentive in developing computer-assisted approaches towards the automation of molecular design.

Recently, computer-assisted procedures for designing new materials have been introduced to ease the protracted design, synthesis and evaluation cycle. In general, computer-aided molecular design requires the solution of two problems: the *forward* problem, which requires the computation of physical, chemical and biological properties from the molecular structure, and the *inverse* problem, which requires the identification of the appropriate molecular structure given the desired physicochemical properties (Figure 2). While there has been considerable attention and progress made towards the solution of the forward problem (such as group contribution methods, equation of state approaches, quantitative structure-activity relationships (QSAR) and so on), relatively less attention has been paid for the inverse problem *(1-3)*.

The inverse problem has been approached by a variety of methods. The approaches can be divided into five categories: random search *(4)*, exhaustive enumeration *(5-7)*, mathematical formulation *(8)*, knowledge-based *(9,10)*, and graphical reconstruction *(11, 12)* methods. In random search, the design candidates are randomly created from

0097–6156/95/0589–0396$12.00/0
© 1995 American Chemical Society

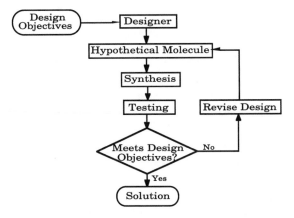

Figure 1. Iterative Molecular Design Process
Reproduced with permission from Ref. [15]. ©1994 Elsevier

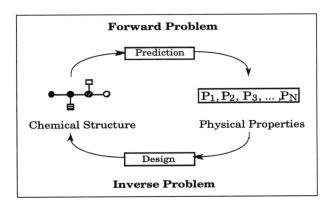

Figure 2. Components of the Molecular Design Problem
Reproduced with permission from Ref. [15]. ©1994 Elsevier

chemical building blocks, called base groups, and then evaluated to see if they satisfy the design constraints. Typically, the solution space or search space associated with any molecular design problem is the set of all possible combinations of the base groups with or without restrictions on the maximum number of base groups (i.e. the "design length") used to create a molecule . Given the combinatorial complexity of the solution space associated with CAMD (7), random search samples a very small region of the total design space in an ad-hoc manner and is thus not a feasible method for large-scale molecular design problems. Another approach is to enumerate all possible designs up to a maximum design length which is obtainable from the base groups. Again the combinatorial explosion, even if one prunes infeasible designs (6,7), makes this approach intractable. Mathematical programming methods consider molecular design as an optimization problem where the objective is to minimize the error between the desired target values and the values attained by the current design. The solutions to mixed integer nonlinear programming (MINLP) formulations are susceptible to local minima traps for problems with nonlinear constraints, which are often found in most structure-property relations. The solutions to these problems are also computationally very expensive, especially for highly nonlinear systems. Knowledge-based methods are artificial intelligence approaches which assume that expert rules exist for manipulating chemical structure to achieve the desired physical properties. However, the nonlinear structure-property relationships cannot be easily quantified, especially when designing for multiple design objectives. Furthermore, extraction of such design expertise from experts on molecular design is not easy, making the knowledge acquisition problem a very difficult one. Lastly, graph reconstruction methods solves the inverse problem of reconstructing the molecular graph(s) of a topological index (13) of quantitative structure-activity relationship (QSAR) equation. For molecular design using this approach, one must express all structure-property relations in terms of topological indices. This may not be appropriate or feasible for all properties. Furthermore, topological indices are not unique and there currently does not exist a general graph reconstruction method for all molecular indices. In addition, since one often deals with a number of design criteria to be satisfied and not just one or two properties, this approach may not be feasible in general. Thus, there exists a critical need to explore alternative strategies for molecular design that can circumvent these problems.

This paper describes a new approach to computer-aided molecular design using *Genetic Algorithms* (GAs). Genetic algorithms are general purpose, stochastic, evolutionary search and optimization strategies based on the Darwinian model of natural selection. The essence of GAs lies in allowing a dynamically evolving population to gradually improve by competing for the best performance. The proficiency with which GAs are known to search noisy, discontinuous, and nonconvex solution spaces motivated us to examine genetic search as a viable approach for molecular design (14, 15). The remainder of the paper is organized in the following manner. First a brief overview of genetic algorithms is given. Next, the adaptation of GAs to molecular design, such as molecular representation, genetic operators, fitness functions, etc. is presented. This is followed by results from the polymer and refrigerant design case studies. Lastly, the summary and conclusion and the relative merits and potential deficiencies of the proposed approach is discussed.

Genetic Algorithms: A Brief Introduction

Genetic search methods have their basis in Darwinian models of natural selection and evolution (16). The general idea behind genetic algorithms is the evolutionary creation of a new population of entities from an earlier generation through certain combinations and changes and by passing on the fittest offsprings to the next generation. Genetic algorithms work by simulating two important characteristics of natural evolution. The first is the "survival of the fittest" idea. By this, we refer to the survival of that part of a population that possesses the necessary characteristics to thrive and reproduce in a given environment. The second is the propagation of the attributes of the mating members by the recombination of genetic material and transmission to the offsprings. For a genetic algorithm, the environment is the set of desired solution characteristics or properties. The population is any set of solution candidates. The representation of the members of a population in the GA is usually in the form of a string, of a finite length, composed of binary bits {0,1} which form the building blocks of a solution. The genetic material of interest in a GA are the building blocks of the population members. Mating typically involves the random combinations of or changes in the building blocks of the parents to produce offsprings. The "fitness" or the ability of a population member to survive in the environment is determined by the closeness of its properties to those desired and is characterized by a non-negative value. The highly fit members are given a greater chance of selection for reproduction. This is called "fitness proportionate selection" and is achieved by the genetic algorithm by weighting the probability of selection of a "parent" candidate based on its fitness value. As a result, individuals who are more fit have a greater chance of producing offsprings and passing on their genetic information onto succeeding generations. The impact of this strategy is that the more promising regions of the solution space are quickly identified and sampling is increasingly focused on these regions at the expense of low performance regions.

An advantage to this search technique is that it is not based on local gradient information and is more resistant to ill-formed solution spaces that are discontinuous or nonconvex. Unlike the traditional optimization methods which move from a single point in the solution space to the next, GAs examine and manipulate a set of potential solutions. The GA uses the information obtained from this set of solution candidates to generate the next set instead of using a single point. Consequently, this significantly reduces the possibility of remaining trapped in local optima in a multimodal search space. In creating offsprings from a given population the sequence of the building blocks of the highly fit members of the current population may be lost. To avoid this, a small percentage of the highly fit members of the current population are retained over to the next generation. This policy of retaining very fit members is called elitism. In general, this approach is ultimately expected to lead to generations that become more and more fit through evolution thus achieving the desired design objective. However, there is no guarantee that this would be achieved, which is the primary drawback of the genetic search.

The process of reproduction involves the modification of the building blocks (genetic material) of the parents to produce offsprings. To this end, the genetic algorithm uses what are called genetic operators. The nature and function of the operators is problem and representation dependent. The transition rule for a GA consists of the plan of reproduction and the genetic operators. In its simplest form the transition rule of a GA consists of fitness proportionate reproduction and two genetic operators: crossover and mutation (*17*). The basic genetic operators as well as the ones developed additionally for molecular design are described in the next section.

There is much theoretical and empirical evidence showing that GAs are effective in navigating through large and complex search spaces (*16,18*). However, there exists no theoretical proof guaranteeing convergence to global optimum. Additional discussions on GA fundamentals and applications can be found in Rawlins (*19*), Davis (*20*), Michalewicz (*21*), Koza (*22*), and Androulakis and Venkatasubramanian (*23*). A summary of GA applications in chemistry can be found in Hibbert (*24*) and Lucasius et al. (*25*).

Genetic Algorithmic Approach to CAMD

In our approach to CAMD, the forward problem, is solved by using the standard group contribution methods. This is discussed in the case study section. Genetic algorithms were used for the inverse problem and is presented in this section. The standard genetic algorithm framework needs to be adapted for the molecular design problem. The modifications introduced were in the representation of molecules, creation of new operators to facilitate the rich chemistry of molecular interactions and rearrangements, and fitness functions to handle property constraints. Figure 3 shows the proposed overall genetic algorithm approach for CAMD. The GA loop can be described as steady-state reproduction with elitism. The former maintains a constant-sized population and the later saves the best designs from the previous generation,. The GA is initiated randomly with feasible designs (i.e., fitness > 0) to provide starting points. To maintain solution diversity, the elitist policy saves only chemically distinct designs from the previous generation. The GA requires values for a number of tunable parameters, such as population size, genetic operator probability rates, and fitness function constants. The GA parameters are provided in the case studies.

Molecular Structure Representation In the present system, an alphabet of chemical building blocks called base groups {e.g., >C<, -H, -O-, >C=O, etc.} was developed as it is conceptually easier to use. The base groups can be elemental (consisting of single atoms or molecules), substructural (consisting of small combinations of molecules), or larger molecular units (such as a methyl group, phenyl group etc.). A design is then a connected list of symbols from the base group alphabet. The base groups are selected such that they have the capacity to produce, in general, a rich variety of molecular combinations that would be of potential interest. The selection of base groups is also influenced by the choice of the predictive method employed for property estimation. For example, in order to apply the group contribution methods for property prediction the selected groups must have all the appropriate group contribution parameters available. Complex molecules will require chemical knowledge-specific rules to construct and alter compounds to result in feasible candidates. These molecular structures are then represented as nested lists in Lisp.

In the nested list representation, the first list represents all the mainchain groups and the sequence in which they appear. The subsequent lists correspond, in order, to the sidechain groups attached to each of the mainchain groups present in the first list. For example, CF_3CHCl_2 in Figure 4, the first list of two C's represents the two carbon mainchain groups. The subsequent list contain sidechain substituents for the corresponding mainchain units.

Adapting Genetic Operators In addition to the standard genetic operators of mutation and 1-point crossover, we have developed new operators to facilitate the rich interactions among molecular groups. They are described below.

    One-Point Crossover Modification from the binary one-point crossover allows operations between variable length strings. This is necessary as molecule size is an important design variable. As a result, variable length symbol-based crossover involves information exchange between two parent strings at randomly generated positions on each molecule. The position are within the interval of 1 and length-1 of the respective strings. Figure 5 illustrates the one-point crossover operator. In the figure, the offsprings are created by crossing over the segments of the parents as shown by the dotted lines.

    Two-Point Crossover A shortcoming of one-point crossover is that it cannot combine certain combinations of features encoded on the string. For instance, important groups located at the endpoints of a string will always be separated by a one-point crossover operation. A solution to this problem is to use two-point crossover. This operator is like one-point crossover, except that two cut points rather than one are selected at random, and the groups are swapped between the two points. Permissible cut points are located from 2 to length-1. An illustrative example of this operation is shown in Figure 6.

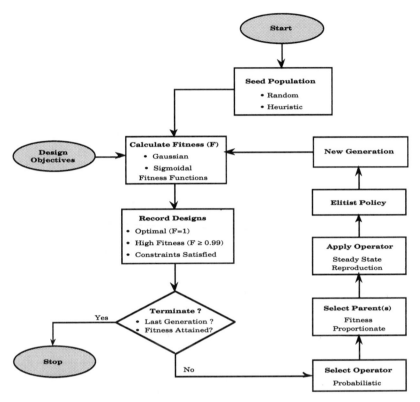

Figure 3. Genetic Algorithm Flowchart for CAMD
Reproduced with permission from Ref. [15]. ©1994 Elsevier

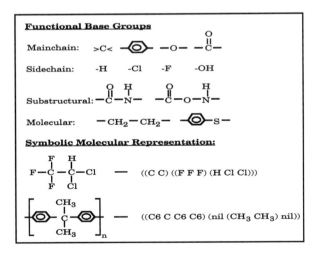

Figure 4. Molecular Structure Representation
Reproduced with permission from Ref. [15]. ©1994 Elsevier

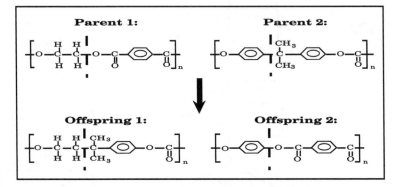

Figure 5. Illustrative One-Point Crossover Operation
Reproduced with permission from Ref. [15]. ©1994 Elsevier

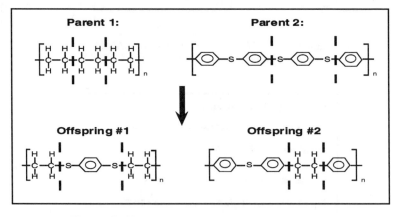

Figure 6. Illustrative Two-Point Crossover Operation

Blending Operator The blending operator produces one offspring from the end-to-end connection of two parents. The motivation for this operator is to overcome the shortcoming of the crossover operator which is to propagate only a random (good or bad) characteristic of the parents (in terms of the sequence of building blocks transmitted) to the offspring. This essentially combines the attributes from both parents. This works like the insertion operator, but is more guided in the sense that it adds a highly fit member to another highly fit member and is not totally random like the insertion operator. The molecular length is increased by applying the blending operator. An example of the blending operation is shown in Figure 7.

Mainchain Mutation and Sidechain Mutation Mainchain and sidechain mutations are analogous to bit mutations. These operations are applied to the molecule as each groups along the main or side chains may be replaced by a randomly selected group. Figure 8 shows examples of mainchain and sidechain mutation operators.

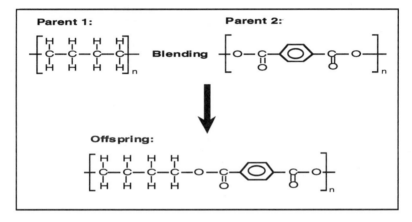

Figure 7. An Example of a Blending Operator
Reproduced with permission from Ref. [15]. ©1994 Elsevier

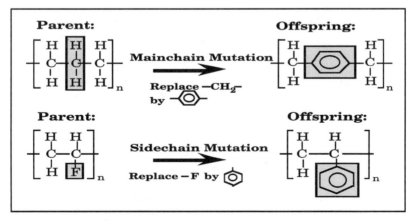

Figure 8. Mainchain and Sidechain Mutation Operators
Reproduced with permission from Ref. [15]. ©1994 Elsevier

Mainchain Insertion and Deletion Operators The insertion operator randomly inserts a group at a single mainchain location. Similarly, the deletion operator randomly removes a group along the mainchain.

Hop Operator The hop operator moves ('hops') a randomly selected group on the mainchain to a randomly selected location on the mainchain. This facilitates small rearrangements in the ordering of the groups which may increase the fitness. Figure 9 illustrates the mainchain insertion, deletion, and hop operators.

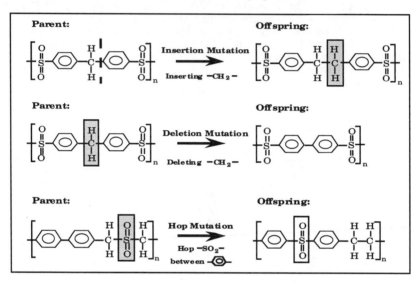

Figure 9. Insertion, Deletion, and Hop Operators

Reproductive Plan To evaluate how well a candidate molecule satisfies the desired target properties, one needs a fitness function that returns a single numerical "fitness" or "measure of merit". The genetic framework searches for design candidates by evaluating how close (or far) a given candidate molecule's macroscopic properties are from the desired target. Thus, a measure of fitness one developes should take as its input the desired target properties, the candidate's properties, the desired level of tolerance in meeting the target specifications, and some tuning parameters. And the output from this fitness fuction should range smoothly from 0 to 1 such that candidates whose properties are far from the target have a fitness close to 0 (and are thus penalized) and those candidates closer to the target have a fitness closer to 1 (and are thus encouraged). The tuning parameter is needed to adjust how strongly one would like the function to penalize (or encourage) deviations from the target. In addition, one would like this function to be as simple as possible. A Gaussian-like function, satisfies these requirements and was used in our approach and is given below:

$$F = \exp\left(-\alpha\left[\sum_{i=1}^{n} \frac{(P_i - \overline{P_i})^2}{|P_{imax} - P_{imin}|^2}\right]\right) \tag{1}$$

where P_i is the ith property value, $\overline{P_i}$ is the average of the maximum and minimum acceptable property values, $P_{i,max}$ and $P_{i,min}$, respectively, which are used to normalize the property values. The index i ranges over all the property constraints that are applied. Thus, the fitness function, F, is a measure of the distance of the property values of the current candidate molecule from the desired values. The function F ranges from 0 to 1, 1 being the target molecule's fitness. The parameter α is the fitness decay rate that determines how the fitness values fall off as the candidates move away from the center of the property constraints. Larger values of α will make the fitness fall off more rapidly as the candidate's properties deviate from the target and thus will enforce tighter constraints. A smaller value of α will be more tolerant of such deviations.

Molecular Design Case Studies

Two molecular design case studies using genetic algorithms are presented in this paper. These problems have been investigated before by Joback and Stephanopoulos (6) using their heuristic-guided enumeration approach. We investigated these case studies using our genetic algorithm approach for comparison. The first case study reported here is the same as the one discussed in Venkatasubramanian et. al (15), but this paper includes some additional results on the near optimal designs. The second study is the problem of designing refrigerant molecules subject to target property constraints.

Case Study 1 The aim of this case study was to investigate the ability of the GA based CAMD framework to solve a practical polymer design problem, compare with the previous attempt of Joback and Stephanopoulos and benchmark the efficiency of the genetic search algorithm against random search. This study also explores the ability of the GA based approach to produce near optimal design alternatives. The problem in this case study is to design polymeric materials for semiconductor encapsulates so that they may be functional under various environmental conditions. In the investigations by Joback and Stephanopoulos, polymers were designed subject to constraints that had either a lower bound or an upper bound. The properties considered were glass transition temperature, volume resistivity, thermal conductivity, and permeability to oxygen. The polymer building block groups employed by Joback are listed in Table I. The constraints on the properties are as given in Table II. Joback reported that there were about 18,000 feasible molecules which satisfy the constraints.

Table I. Base Groups for Polymer Design Case Study I
Reproduced with permission from Ref. 15. ∘1994 Elsevier

$-CH_2-$	$-CH(CH_3)-$	$-CH(C_6H_5)-$
$-C(CH_3)_2-$	$-C(CH_3)(C_6H_5)$	$-\langle O \rangle-$
$-\langle O \rangle-\langle O \rangle-$	$-\langle O \rangle- CH_2-$	$-CH_2-\langle O \rangle- CH_2-$
$-\langle O \rangle- CH_2 -\langle O \rangle-$	$-O-$	$-O-C-$ $\overset{\parallel}{O}$
$-O-C-O-$ $\overset{\parallel}{O}$	$-C-NH-$ $\overset{\parallel}{O}$	$O-C-NH-$ $\overset{\parallel}{O}$
$-CHF-$	$-CF_2-$	$-CHCl-$
$-CCl_2-$	$-CFCl-$	

Since the constraints were not bounded at both the upper and lower ends, this is a relatively easier design problem and the genetic search was able to discover the desired polymers quite easily and quickly as reported elsewhere(15). One would expect the design problem to become more difficult if the constraints were to be tightened by providing both a lower bound and an upper bound for each constraint. The results of this more difficult test are discussed in this study.

Table II. Physical Property Constraints for the Polymer Design Case Study 1

Glass Transition Temperature	Tg > 400K
Volume Resistivity	R > 1x1016 ohm-cm
Thermal Conductivity	L > 1.6x10-7 W / (m - K)
Permeability to Oxygen	P(O2) < 1.0 cc-mil/100 in^2/day/atm

In this investigation, Joback's design study was revisited with tighter constraints. Six polymers of different length and complexity from Joback's list were chosen as the design targets by submitting their properties as the target values. This time, however, the constraints were tightened by specifying that the successful candidate molecule should have its property values within ±0.5% of the target properties. If the genetic search were to be successful, it should discover all the six molecules, one at a time, by combining the groups shown in Table I.

Table III. Parameters for the Design Case Study 1

Genetic Algorithm Parameters	
Population size	100
Number of generations	500
Maximum Design Length	1-7
Fitness Gain (α)	0.001
Total Runs per Case Study	10, 25
Elitist Policy Keep Best	10%
Genetic Operator Probabilities	
Crossover	0.20
Mainchain Mutation	0.20
Sidechain Mutation	0.20
Insertion	0.00
Deletion	0.10
Blending	0.10
Hop	0.20

The steady state reproduction with elitism model (Figure 3) was employed. Polymer design lengths varied from 1 to 7 to coincide with Joback's study. For statistical significance, independent GA runs were repeated ten times with runs terminating after 500 generations. In each run, the starting population consisted of random polymers made of groups from Table I. The parameters for the design case studies are listed in Table III above.

Results and Discussion for Case Study 1 The results are shown in Table IV. As is evident from the success rate, the genetic search did quite well in general. All six molecules were discovered by the genetic search, though some more often and quickly than others. For example, molecule #5 was found fairly quickly ($\approx$ 40 generations on an average) and also was found in every run. On the other hand, some other polymer like #1 took much longer ($\approx$ 113 generations on an average) and also with a smaller success rate (found only in 6 out of the 10 runs). Typically, a molecule that is longer and has more diversity of groups in its structure takes a longer time to discover. Other factors such as how close the alternative solutions are to the target molecule also matter as the genetic search could get stuck in such a local optima.

Table IV. Results for Polymer Design Case Study 1

Target Polymer	Target Length	Avg. Gen. No. when 1st soln. was found	Avg. No. of solns. at the end of the GA search	Random Search Success Rate	GA Search Success Rate
1. (-CHCH3-) (-CF2-) (-CONH-) (-◎-◎-)	4	113.0	13.3	60%	60%
2. (-CHCl-) (-CONH-) 2(-CH2-◎-CH2-) (-◎-◎-)	5	68.8	12.0	20%	50%
3. (-CH2-) (-CHF-) 3(-◎-CH2-◎-)	5	149.7	12.2	0%	90%
4. (-O-) (-CHF-) 2 (-◎-◎-) (-◎-CH2-◎-)	5	110.0	12.3	30%	100%
5. 2(-CCH3-◎) 2(-CH-◎) 2(-OCONH-)	6	40.5	12.6	0%	100%
6. (-CH2-) (-CHF-) (-CCH3-◎-) (-◎-◎-) (-OCONH-)	5	163.0	12.7	20%	60%

Near Optimal Solutions Whenever the genetic search failed to discover the target polymer, it always came up with several solution structures that had very high fitness values (typically 0.95 or higher). Many of these alternatives were structurally quite similar to the target. As an example, Table V shows such near-optimal design candidates that were generated for target polymers #1, #2 and #6 for which the GA had low success rates.

Table V. Near-Optimal Design Candidates from Case Study 1

Target Polymers and Next Best Polymer Solutions Total Generations = 500	Relative % Error	Fitness
Target Polymer: # 1 (-CHCH3 -CF2-CONH-◎-◎-)	(0 0 0 0)	1.0
3 Next Best Solutions: (-CCH3-◎-)(CH-◎-)(-CFCl-)(-OCONH)	(3.33,0.064, 2.44,-2.76)	0.9939
(-CCl2-) (-◎-)(-OCONH-)(-◎-◎-)(-CHCl-) (-CHCH3-)	(3.99,0.694,-1.73,-3.66)	0.9919
(-CFCl-)(-CCl2-)(CH -◎-)(-COO-)(CH-◎-) (-OCONH)(-◎-◎-)	(-3.15, 6.21, 3.90,-1.59)	0.9837
Target Polymer: # 2 (-CHCl-) (-CONH-) 2(-CH2-◎-CH2-) (-◎-◎-)	(0 0 0 0)	1.0
3 Next Best Solutions: (-◎-)(-OCONH-)(-CFCl-)(-◎-◎-)(-CF2)(CH-◎-)	(2.07, 1.71, -2.74,-1.11)	0.9960
(-◎-CH2-) (-◎-CH2-◎-)(-CFCl-) 2(-◎-◎-) (-COO-)	(5.55, -0.77, 1.80,-0.32)	0.9914
(-CONH-)(-◎-CH2-◎-)(-◎-CH2-)(-CFCl-) (-◎-◎-)(-CF2)(COO)(CONH-)	(-2.55,-5.04,-1.92,-0.32)	0.9912
Target Polymer: #6 (-CH2-) (-CHF-) (-CCH3-◎) (-◎-◎-) (-OCONH-)	(0 0 0 0)	1.0
3 Next Best Solutions: (-◎-◎-)(-◎-)(-◎-◎-)(CH-◎-)(-◎-)(-O-) (-OCONH)	(-1.07,-0.80,4.16,-5.90)	0.9867
(-◎-◎-)(-◎-)(-◎-◎-)(-CCH3 -◎-)(-CFCl-) (-CONH-)	(1.46, 4.70, 2.91, 4.97)	0.9858
(-CHF-)2(-◎-◎-)(CH-◎-)(-◎-)(-CONH-) (-CHCl-)	(-3.08, 3.98, 3.49, 5.25)	0.9839

All design alternative have high fitness values exceeding 0.98. The relative percent error for each property ranged from about -6% to 6%. Given that the group contribution methods are not very accurate and may have errors of the order of ±5%, it is important to develop a group of alternate molecules that are nearly as good as the target. In addition, these alternative solutions may be more acceptable from the point of view of synthesis and cost effectiveness. It is important to note that this ability of the genetic design framework to provide a collection of possible alternatives that are of high fitness is a very useful design feature.

Comparison with Random Search For the purposes of comparison, a random search technique was implemented for the case study with doubly bounded design constraints. Analogous to the genetic algorithm case study, 100 polymers per generation were generated for 500 generations for a total of 50,000 design candidates in a single run. This was repeated for 10 runs for each target molecule. The random search procedure consisted of randomly selecting a length of 1 through 7 groups and then randomly filling in the mainchain elements with groups from Table I. On an average, one can expect 7,143 candidate solutions to be generated for each length (50,000·1/7). The total number of design candidates associated with exhaustive enumeration, is given by, $C(m+L-1,L) = (m+L-1)!/L!(m-1)!$, where m is the number of mainchain groups and L is the chain length (15). Consequently, of the six target polymers consisting of 4, 5, and 6 base groups, there are, 8855, 42504, and 177100 total designs for each design length, respectively. Thus, for targets polymers of length 4, 5, and 6 the success rates are 81% (7,143/8,855), 17% (7,143/42,504), and 4% (7,143/177,100), respectively. The success rate, however, drops off to nearly zero for designs which employ many base groups and for longer molecular designs. For example, increasing the maximum length of the design target to 8 decreases the probability of success to 0.28% (50,000 · (1/8) / 2.22x10^6).

The results from the random search are given in Table IV. As expected, the random search performed well on target polymer #1 (60% success rate) which is shorter with a length of 4 base groups and compares favorably with the genetic search. However, the average success rate for the target polymers of length 5 base groups (#2,#3,#4 and #6) is lower at 30%. For polymers #3 and #5, which are of length 6, the random search performed poorly (0% success rate). In addition, random search for the 6 target designs is only able to locate on the average, 2.5 solutions with a fitness greater than or equal to 0.9. This is in contrast to the genetic search which is able to routinely locate dozens of near optimal solutions.

Case Study 2 : Refrigerant Design

Joback and Stephanopoulos (6) also reported a replacement refrigerant molecular design case study subjected to target property constraints using their heuristic-enumeration methodology. The present case study solves the same problem with the GA approach. The target property constraints as specified by Joback and Stephanopoulos are given below and summarized in Table VI. The constraints on the property values are imposed to ensure efficient operating conditions for the refrigeration process (6).

Table VI. Target Property Constraints for the Refrigerant Design Problem

Property	Temperature	Constraint
1. $P_{v\text{-low}}$ (Lowest Vapor Pressure)	272.05K	≥ 1.4 bar
2. $P_{v\text{-high}}$ (Highest Vapor Pressure)	316.45K	≤ 14 bar
3. ΔH_v (Heat of Vaporization)	272.05K	≥ 18.4 kJ/g-mol
4. C_p (Heat Capacity)	294.25K	≤ 32.2 cal/g-mol

The current work also examines designing refrigerants which offer high thermal efficiency as well as satisfying property constraints. An ideal vapor-compression refrigeration system was used since it takes into account the thermodynamic properties of the refrigerant. The thermal efficiency of the refrigerant was calculated based on its "Coefficient Of Performance" (COP) in an ideal vapor compression cycle and the Carnot's cycle which is the performance limit of any refrigeration system. The basic refrigeration cycle and the calculation of thermal efficiency are as given in Figures 10a and 10b. For the properties themselves, group contribution property estimation procedures of Joback and Reid (3) and Joback (7) were employed. The fundamental units or base groups for constructing a refrigerant molecule is a variant and subset of those used in Joback's case study. The current GA does not consider ring compounds as many are toxic and thus environmentally unfavorable. The refrigerant design mainchain and sidechain groups are illustrated in Table VII. Furthermore, the base groups are redefined such that all group combinations result in feasible compounds. This removes the enumeration step to determine chemically feasible designs as is required by Joback's method.

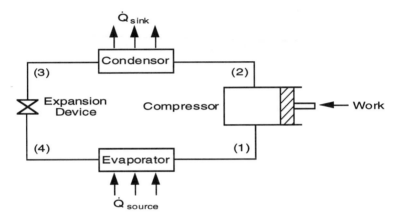

Figure 10 (a). A Typical Refrigeration Cycle

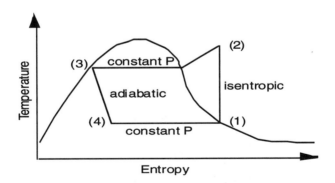

$h_2 - h_1$ = Enthalpy difference across compressor
$h_1 - h_4$ = Enthalpy difference across evaporator
$COP = (h_1 - h_4) / (h_2 - h_1)$
Feff (thermal efficiency) = $COP_{ideal} / COP_{carnot}$

Figure 10 (b). Thermodynamics of an Ideal Vapor Compression Cycle

The base groups are divided into 3 parts: i) mainchain, ii) sidechain, and iii) endgroups for cases when Lengths =1 and ≥2. This division is necessary since some mainchaingroups can act only as endgroups or as non-endgroups. Moreover, the number of sidechains for a particular mainchain depend on the chain length as well as group location. This imposes certain restrictions on the genetic operators that were used for the polymer design case study in terms of creating feasible offsprings. These operators are discussed in the next section.

Table VII. Refrigerant Design Base Groups

• 10 Mainchain Groups, 13 Sidechain Groups

Mainchain Groups	Sidechain no. for non-Endgroup	Sidechain no. for Endgroup, Length=1	Sidechain no. for Endgroup, Length ≥2
>C<	2	4	3
>C=C<	2	4	3
>C=C=C<	2	4	3
−C≡C−	0	2	1
>N-	1	3	1
-O-	0	−	−
-S-	0	−	−
$-\overset{\overset{\text{O}}{\|}}{\text{C}}-$	0	−	−
$>C=\overset{\overset{\text{O}}{\|}}{\text{C}}$	−	2	1
$-\overset{\overset{\text{O}}{\|}}{\text{C}}-O-$	0	2	1

Sidechain Groups				
-H	-CH₃	-F	-Cl	-F
-Br	-I	-OH	-CN	-SH
-NH₂	-NO₂	-COOH		

Modified Genetic Operators The previous case study used a set of base groups so chosen that any combination of the base groups would be feasible and the sidechain requirements for the all base groups on the main chain were the same. This eliminated the possibility of the creation of infeasible molecules during the genetic search. In this case study, however, the base group set consists of certain members with restrictions on the positions in which they can appear in a molecule as shown in Table VII. This means that chemistry imposes restrictions on the number of sidechains that a base group can have depending upon its position in the mainchain. Thus infeasible designs can be created by the random combination of the base groups. To overcome this problem, some of the genetic operators were modified to include "repair" operators which maintain chemical feasibility by correcting the infeasible designs during the genetic search. Another approach to this problem is to use chemical knowledge-based operators that use appropriate molecular construction and alteration rules that do not create infeasible molecules. This technique is more efficient when complex designs with multiple branch points are considered. In this case study, however, the simpler repair operators were employed. The following operators were modified :

• *Blending:* To ensure chemical feasibility, randomly remove one sidechain from each refrigerant endgroup to be blended.
• *Mainchain Mutation:* This operator is as defined previously but replaces endgroups and non-endgroups correspondingly.
• *Insertion:* If insertion is at an endgroup position this operator randomly removes the required number of sidechains from former endgroup.
• *Deletion:* If the group being deleted is an endgroup and the adjacent group is a suitable endgroup this operator adds the required number of sidechains to the adjacent

group. When the adjacent group is not a suitable endgroup the deletion operation (as was defined for Case Study 1) is performed again.

Fitness Function for Refrigerant Design The fitness of a design candidate is expressed as the average of the individual constrained fitness values. The property fitness function brackets off a lower and upper bound in which no constraints are violated and the fitness $=1$. The properties which fall outside these bounds have their fitness reduced by a "sigmoidal" shaped fitness function. The fitness functions employed for the refrigerant design study are given below:

$$F_i = \begin{array}{l} 1, \text{ if } P_{r,min} \leq P_i \leq P_{r,max} \\ Sig_i, \text{ if } P_{r,min} > P_i \text{ or } P_i < P_{r,max} \end{array} \tag{2}$$

$$Sig_i = \exp\left(-\alpha_i\left[\frac{(P_i-P_{i,min})^2}{(P_{r,max}-P_{r,min})^2}\right]\right) \tag{3}$$

$$F_{prop} = \frac{1}{n}\sum_{i=1}^{n} F_i \tag{4}$$

were $P_{r,min}$ and $P_{r,max}$ are the absolute minimum and maximum property values to design for. Their difference also normalizes the property values to remove possible bias towards any single property. $P_{i,min}$ is the minimum property value which yields a fitness of 1. The function F ranges from 0 to 1, 1 when P_i is within $P_{r,min}$ and $P_{r,max}$.

The fitness decay rate is controlled by α_i. For example, α_i for a C_p value greater than 32.3 is 100 since it gives a rapid decay in fitness for constraint violations. On the other hand, α_i for C_p values less than $P_{cp,min}$ is 10 which allows a more gradual decay of fitness. The parameters for equations 3 and 4 are listed in Table VIII.

The thermal efficiency fitness is defined as the ratio of the coefficient of performance (COP) of the ideal vapor-compression cycle vs. the COP of the Carnot cycle (F_{eff} =COPvap-comp/COPCarnot). For thermal efficiency calculations, the refrigeration system must maintain an environment at 260K for available cooling water at 295K. The refrigeration coils and condenser are of sufficient size that a 5K approach can be realized.

Table VIII. Fitness Function Design Constants

Property	$P_{i,min}$	$P_{r,min}$	$P_{r,max}$	$a_{i,rmin}$	$a_{i,rmax}$
1. $P_{v\text{-low}}$	1.4	0	10.0	10	100
2. $P_{v\text{-high}}$	5.0	0	20.0	100	10
3. ΔH_v	18.4	0	60.0	10	100
4. C_p	17.0	0	40.0	100	10

More detailed thermodynamic theory on the refrigeration process is described in Cerepnalkovski (*26*). The overall fitness when considering both property and thermal efficiency constraints is given by the linear combination of F_{prop} and F_{eff}.

$$F_{total} = \beta_1 F_{prop} + \beta_2 F_{eff} \tag{5}$$

where, β_1 and β_2 are the weight factors for F_{prop} and F_{eff}, respectively. The sum of β_1 and β_2 is unity so that F_{total} ranges from 0 to 1. Equal weight is given to F_{prop} and F_{eff} as both β_1 and β_2 are equal to 0.50.

The genetic search is required to design for refrigerants subjected to property and thermal efficiency constraints. The population size is 100, the elitist policy keeps 10%

of the previous best population members. The initial population consist of refrigerants with random mainchain and sidechain groups and random design lengths of 1 to 7. For statistical significance, results were compiled after 25 GA runs of 500 generations each. The genetic operator probability values are as follows: i) crossover = 0.30, ii) mainchain mutation = 0.20, iii) sidechain mutation = 0.20, iv) insertion = 0.10, v) deletion = 0.10, vi) blend = 0.10, and vii) hop = 0.00. The hop operator was not required as the order of the base groups does not affect the predicted physical properties for this case study.

Results and Discussion for Case Study 2

The GA approach found 19 molecules that satisfied all property constraints. 15 of 19 design candidates had efficiency fitness values of 0.6 to 0.7. The other four were between 0.7 and 0.8. Table IX provides an illustrative example of designs which satisfy all property constraints and have high thermal efficiencies.

Table IX. Refrigerant Design Results for Property and Thermal Efficiency Constraints

Refrigerant Molecule	Pv-low	Pv-high	ΔH_v	C_p	F_{eff}
1.H_3C-CH=C=O	1.51	6.67	27.77	28.97	.820
2.CH_3-C=O-CH_3	1.73	7.34	27.14	30.17	.818
3.H_3C-CF=C=O	1.57	6.97	26.96	31.39	.789
4.FH_2CCH=C=O	1.58	7.12	26.88	31.83	.774
5. HC≡CCH_3	1.69	6.35	21.44	24.18	.690
6. CH_3Cl	1.61	6.25	21.58	20.10	.663
7. H_3C=C=CH_3	1.55	5.97	20.85	25.03	.657
8. CH_2FCl	1.67	6.66	20.71	23.00	.630
9. H_3C-CH_3	2.74	9.55	18.67	22.58	.622
10. H_3C-N-FCH_3	1.83	7.33	20.42	30.15	.610

Table X. Illustrative Near-Optimal Refrigerants for Property and Thermal Efficiency Constraints

Refrigerant Molecule	Pvp (272.05)	Pvp (316.45)	ΔH_v (272.05)	C_p (294.25)	Property/ Thermal Eff. Fitness
1. H_2C=$CHCH_3$	1.33	5.24	21.24	25.47	.999 / .652
2. $CF_2(CH_3)_2$	1.40	5.51	18.92	32.53	.993 / .616
3. CH_3FHC-CH_3	1.23	5.01	20.80	30.65	.993 / .616
4. H_3C-O-CH_3	1.20	5.04	22.23	26.85	.990 / .657
5. H_3C-CH_2-CH_3	1.17	4.77	22.07	28.30	.987 / .666
6. $FClN$-CH_3	1.03	4.66	23.41	27.77	.966 / .660
7. C=C=O-NF_2	2.62	11.56	24.96	33.88	.960 / .713
8. HC≡C-C=O-C≡CH	.59	3.02	33.14	33.72	.817 / .914
9. H_3C-O-C=O-CH_3	.68	3.59	31.20	34.72	.803 / .823
10. H_3C-$(C$=$O)_2$-CH_3	.987	5.29	36.19	38.10	.714 / .963

Refrigerants molecules # 6,8 and 9 in Table IX are used in current practice. The GA was able to identify the correct refrigerant molecule length as no designs exceeded three base groups even though longer designs were allowed. Refrigerant designs with efficiency fitness values over 0.70 had the -C=O- group in common. Since the program does not consider cost, stability, toxicity or flammability constraints, etc., the proposed designs must be screened further for these properties. However, such additional constraints can be easily integrated into the genetic search. Furthermore, the GA also found numerous near-optimal solutions with high property and efficiency fitness values (71 refrigerants with $F_{tot} \geq 0.80$). An illustrative set is shown in Table X. The values shown in italics and underlined in Table X are the ones that are out of their specified bounds indicated in Table VI.

Summary and Conclusions

The complexity of the molecular design problem requires the use of computer-assisted design procedures. However, the current CAMD approaches such as exhaustive enumeration and heuristics-guided enumeration suffer from the problem of combinatorial complexity of dealing with millions of possible candidates for large, complex molecules. In the knowledge-based approach, acquisition of appropriate heuristic knowledge is a serious problem. While the MINLP formalism has a certain appeal due to its rigorous mathematical foundation, it suffers from local minima traps.

In contrast, the genetic algorithm framework offers a number of advantages: it is a multiple point search technique that examines a set of solutions and not just one solution-- this and the stochastic nature of the algorithm helps the search to escape local minima traps; it is not derivative-based, thus avoiding the difficulties faced by math programming techniques in this respect; it is relatively easy to express the rich and complex chemistry of molecules; one can integrate whatever heuristic knowledge one might have about the problem quite easily into the genetic framework to speed up the design process-- for instance, instead of starting the initial population at random, the designer can start with structures that are good guesses based on his or her experience; it provides a set of alternate solutions and not just one. This would be particularly useful for the design of complex molecules where the forward problem results may not be completely reliable.

This paper described the various issues involved in the application of the genetic algorithm approach CAMD with the aid of two case studies from polymer and refrigerant design. The genetic design approach was found to be quite successful and it fared much better than the random search. In addition, the genetic search was able to identify several near optimal solutions.

On the down side, the proposed approach suffers from two main drawbacks. One is the heuristic nature of the search and that there is no guarantee of finding the best solution. But then, this criticism would be applicable to the other heuristic approaches mentioned above as well. Even in the MINLP approach, there is no guarantee of a global optimum solution. The other drawback is that the selection of the parameter values would require some experimentation.

While the results from this study are quite promising, considerable further research needs to be carried out to address issues such as parameter selection, sensitivity, algorithm convergence, parameter scheduling, incorporation of higher-level knowledge and related constraints, representing more complex entities like composites and so on. Despite its main drawback as a heuristic procedure that has no convergence guarantees, it appears that this approach opens up a new avenue of possibilities in rational approaches to molecular design that could have a significant impact on a wide variety of molecular design problems.

Literature Cited

1. Reid, R.C.; Prausnitz, J.M.; Poling B.E. *The Properties of Gases and Liquids*; McGraw-Hill: New York, 1987.
2. Lyman, W.J.; Reehl, W.F.; Rosenblatt D.H. *Handbook of Chemical Property Estimation Methods;* McGraw-Hill: New York, NY, 1982.
3. Joback, K.G.; Reid R.C. *Chem. Eng. Comm.* **1987**, 57, 233-243.
4. Derringer, G. C.; Markham R. L. *J. Appl. Polymer Sci.* **1985**, 30, 4609-4617.
5. Brignole, E.A.; Bottlini, S.; Gani, R. *Fluid Phase Equilibria.* **1986**, 29, 125.
6. Joback, K.G.; Stephanopoulos G. In *Proc. FOCAPD '89;* Snowmass,CO,1989; pp. 363-387.
7. Joback, K.G. Ph.D. Dissertation, Volumes 1 and 2, MIT, 1989.
8. Macchietto, S.; Odele, O.; Omatsone,O. *Trans. IChemE.* **1990**, 68, Part A, 429-433
9. Nagasaka, K.; Wada, H; Yoshimitsu, H; Yasuda, H.; Yamanouchi, T. *AIChE Annual Meeting.* Chicago, IL, November, 1990; 39e.
10. Gani, R.; Nielsen, B; Fredenslund Aa. *AIChE J.* **1991**, 37,9 1318-1332.
11. Gordeeva, E.V.; Molchanova, M.S.; Zefirov, N.S. *Tetrahedron Comput. Methodol.*, **1990**, 3,6B, 389-415.
12. Kier, L.B.; Lowell, H.H.; Frazer, J.F. *J. Chem. Inf. Comput. Sci.* **1993**, 33,1, 143-147.
13. Kier, L.B.; Hall, L.H. *Molecular Connectivity in Chemistry and Drug Research;* Academic Press: New York, 1976.
14. Venkatasubramanian, V.; Chan, K.; and Caruthers, J.M. *AIChE Annual Meeting*, Miami, FL, November, 1992; 140d.
15. Venkatasubramanian, V.; Chan, K.; Caruthers J.M. *Computers Chem. Engng*, **1994**,*18*, 4,833-844
16. Holland, J.H. *Adaptation in Natural and Artificial Systems*; The University of Michigan Press: Ann Arbor, 1975.
17. Baker, J.E. In *Proceedings of the Second International Conference on Genetic Algorithms;* Grefenstette, J.J., Ed.; Lawrence Erlbaum Associates, Hillsdale, NJ , 1987; pp 14-21.
18. Goldberg, D.E. *Genetic Algorithms in Search, Optimization, and Machine Learning*; Addison-Wesley: Reading, Mass., 1989.
19. Rawlins, G.J.E., Ed.; *Foundations of Genetic Algorithms*; Kaufmann Publishers: San Mateo, Calif., 1991.
20. Davis, D., Ed.; *Handbook of Genetic Algorithms*; Van Nostrand Reinhold: New York, 1991.
21. Michalewicz, Z. *Genetic Algorithms + Data Structures = Evolution Programs;* Springer-Verlag: Berlin, 1992.
22. Koza, J.R. *Genetic Programming: On the Programming of Computer by Means of Natural Selection;* MIT Press: Cambridge, Mass, 1992
23. Androulakis, I.P.; Venkatasubramanian, V. *Computers chem. Engng.* **1991**, 15,4, 217-228 .
24. Hibbert, D.B. Genetic Algorithm in Chemistry. *Chemometrics and Intelligent Laboratory Systems* **1993**, 19, 277-293.
25. Lucasius, C.B.; Kateman, G. *Chemometrics and Intelligent Laboratory Systems* **1993**, 19, 1-33.
26. Cerepnalkovski, I., *Modern Refrigerating Machines,* Elsevier: Amsterdam, 1991

RECEIVED January 13, 1995

Author Index

415

Affiliation Index

Subject Index

Production: Susan Antigone
Indexing: Deborah H. Steiner
Acquisition: Rhonda Bitterli

Printed and bound by Maple Press, York, PA